Human Apolipoprotein Mutants III

Diagnosis and Treatment

NATO ASI Series

Advanced Science Institutes Series

A series presenting the results of activities sponsored by the NATO Science Committee, which aims at the dissemination of advanced scientific and technological knowledge, with a view to strengthening links between scientific communities.

The Series is published by an international board of publishers in conjunction with the NATO Scientific Affairs Division

A Life Sciences **B Physics**	Plenum Publishing Corporation London and New York
C Mathematical and Physical Sciences **D Behavioural and Social Sciences** **E Applied Sciences**	Kluwer Academic Publishers Dordrecht, Boston and London
F Computer and Systems Sciences **G Ecological Sciences** **H Cell Biology** **I Global Environmental Change**	Springer-Verlag Berlin Heidelberg New York London Paris Tokyo Hong Kong Barcelona Budapest

NATO-PCO DATABASE

The electronic index to the NATO ASI Series provides full bibliographical references (with keywords and/or abstracts) to more than 30000 contributions from international scientists published in all sections of the NATO ASI Series. Access to the NATO-PCO DATABASE compiled by the NATO Publication Coordination Office is possible in two ways:

- via online FILE 128 (NATO-PCO DATABASE) hosted by ESRIN, Via Galileo Galilei, I-00044 Frascati, Italy.
- via CD-ROM "NATO Science & Technology Disk" with user-friendly retrieval software in English, French and German (© WTV GmbH and DATAWARE Technologies Inc. 1992).

The CD-ROM can be ordered through any member of the Board of Publishers or through NATO-PCO, Overijse, Belgium.

Series H: Cell Biology, Vol. 73

Human Apolipoprotein Mutants III

Diagnosis and Treatment

Edited by

Cesare R. Sirtori
Guido Franceschini

Center E. Grossi Paoletti
Institute of Pharmacological Sciences
University of Milano
Via Balzaretti 9
20133 Milano, Italy

Bryan H. Brewer Jr.

Molecular Disease Branch
National Institutes of Health
NHLBI
Bethesda, MD 20205, U.S.A.

Springer-Verlag
Berlin Heidelberg New York London Paris Tokyo
Hong Kong Barcelona Budapest
Published in cooperation with NATO Scientific Affairs Division

Proceedings of the NATO Advanced Research Workshop on Human Apolipoprotein Mutants III. Apolipoproteins in the Diagnosis and Treatment of Disease, held at Limone sul Garda, Italy, April 5–8, 1992

Additional material to this book can be downloaded from http://extra.springer.com.

ISBN-13:978-3-642-84636-6 e-ISBN-13:978-3-642-84634-2
DOI: 10.1007/978-3-642-84634-2

Library of Congress Cataloging-in-Publication Data.

Human apolipoprotein mutants III: in diagnosis and treatment / edited by Cesare R. Sirtori, Guido Franceschini, Bryan H. Brewer Jr. p. cm. - (NATO series; Series H. Cell biology; vol. 73)
"Proceedings of the NATO Advanced Workshop on Human Apolipoprotein Mutants III. Apolipoproteins in diagnosis and Treatment of Disease, held at Limone sul Garda, Italy, April 5-8, 1992" - T.p. verso.
"Published in cooperation with NATO Scientific Affairs division."
Includes bibliographical references and index.
ISBN-13:978-3-642-84636-6
1. Apolipoproteins-Congresses. 2. Recombinant blood proteins-Congresses. I. Sirtori, Cesare R. II. North Atlantic Treaty Organization. Scientific Affairs division. III. NATO Advanced Research Workshop on Human Apolipoprotein Mutants III: Apolipoproteins in Diagnosis and Treatment of Disease (1992: Limone sul Garda, Italy) IV. Title: Human apolipoprotein mutants 3. V. Title: Human apolipoprotein mutants three. VI. Series.
(DNLM: 1. Apolipoproteins-diagnostic use-congresses.
2. Apolipoproteins-therapeutic use-congresses.
3. Apolipoproteins-genetics-congresses. QU 55 H918 1992) QP99.3.A65H85 1993
612' .015754-dc20 DNLM/DLC for Library of Congress 93-20507 CIP

Softcover reprint of the hardcover 1st edition 1993

Typesetting: Camera ready by authors
31/3145 - 5 4 3 2 1 0 - Printed on acid-free paper

Presentation

Limone sul Garda gave hospitality from the 5th to the 9th of April 1992, under the sponsorship of NATO-ARW and of the National Research Council of Italy, to an International scientific congress on the mutants of apolipoproteins. This time, the practical use of apolipoproteins in the diagnosis and treatment of disease was discussed. Once more the interest in the clinic of the apo AI-Milano, discovered in the blood of some of our citizens by a scientific team headed by Proffs Cesare Sirtori, Valter Gualandri and Guido Franceschini, was underlined by all participants.

As in the previous scientific congresses held in 1984 and 1988, always under the NATO shield, Limone sul Garda, through the City Admnistration, local operators and all citizens, gave factual demonstration of a special attention to the major results of scientifc research, and particularly to studies carried out in various parts of the world on topics, in some way related to the Limone protein.

It is everybody's wish that Medicine will be able soon to take further advantage from this discovery for the benefit of all of us.

From Limone a very hearty gratitude is owed to prof. Cesare Sirtori and to all of his co-workers, who very brilliantly succeeded in organizing the meeting and carrying it through to the best of success, as also witnessed by this Book of Proceedings.

Guido Tosi
Mayor of Limone sul Garda

FOREWORD

The apolipoprotein season is reaching vintage time. Hopes to develop apolipoproteins for diagnostic or, even better, therapeutic purposes are growing. Areas of use may range from arterial disease to AIDS, fertilization programs, neurological or inflammatory disorders, and the like.

However, these young members of the large family of circulating proteins still deserve careful study, both in terms of structural and functional properties and of their pathological changes. In addition, related molecules, in particular belonging to the family of enzymes, such as lipases and acyltransferases, offer exciting insight into the mechanism/s of regulation of lipid transport and exchange.

The Third Symposium on Apolipoprotein Mutants was plagued, for the first time after the earlier very enjoyable previous gatherings, by foul wheater. Rather unique for Limone sul Garda, that, however, still offered its incomparable hospitality to all those participating.

The ARW started, this time, from the clinical observations, slowly moving toward structural studies, to the new potential therapeutic/diagnostic applications for apolipoproteins and finally ending with an overview on lipoprotein transformation processes, expression systems and natural mutants.

Today, the biotechnological production of apolipoproteins offers a formidable approach to the direct evaluation of the physiological properties of these molecules. Membrane interaction studies, the still whimsical receptor for HDL, the in vitro vs in vivo behaviour of apolipoproteins when faced with lipids and, of course, an enticing potential to develop therapeutically active molecules, offer a wide breadth to the field. Limone sul Garda may still play the role of the nurturing cot for this explosive field.

Cesare R. Sirtori
Guido Franceschini

CONTENTS

ENZYMES AND APOLIPOPROTEINS IN THE DIAGNOSIS OF DISEASE

STRUCTURAL AND FUNCTIONAL RELATIONSHIPS IN APOLIPOPROTEINS

ANIMAL MODELS IN THE APOLIPOPROTEIN MUTANT RESEARCH

RECOMBINANT APOLIPOPROTEINS AS DRUGS

MOLECULAR ABNORMALITIES OF APOLIPOPROTEINS

Lipoprotein lipase: Gene variants and coronary atherosclerosis

Raj K. Mattu and David J. Galton

Department of Human Metabolism and Genetics,
Medical Professorial Unit,
The Medical College of St. Bartholomew's Hospital,
Clinical Sciences Building,
Charterhouse Square,
London. EC1M 6BQ

INTRODUCTION

Initial reports, such as the Framingham data, questioned the role of hypertriglyceridaemia in coronary artery disease (CAD), because of the inconsistent association between elevated plasma triglycerides and CAD using multivariate analysis (1,2). More recently, there has been growing support for the view that the use of multivariate analysis in these circumstances is inappropriate due to the metabolic interconversions between triglyceride and high density lipoprotein (HDL) (3). However, it is now widely accepted that hypertriglyceridaemia, particularly in the presence of low HDL, is a potent risk factor for CAD.

Lipoprotein Lipase (LPL) is the rate limiting enzyme for the hydrolysis of triglyceride-rich lipoproteins in the circulation. LPL activity is known to affect plasma lipid levels. For example; *in vitro* post-heparin LPL activity has been shown to correlate with the degree of post-prandial lipaemia (4), rate of clearance of very low density lipoproteins (VLDL) (5) and HDL_2 levels (6). LPL deficient states give rise to Type I Hyperlipidaemia. This rare, autosomal recessive disorder has been used for studying the LPL gene, LPL and metabolic sequelae of genetic alterations. By influencing plasma triglyceride and HDL levels, LPL activity may affect susceptibility to CAD. Hence the LPL gene warrants investigation for an aetiological role in CAD.

The LPL gene is located on chromosome 8p22 (7). It spans approximately

NATO ASI Series, Vol. H 73
Human Apolipoprotein Mutants III
Edited by C. R. Sirtori, G. Franceschini, B.H. Brewer Jr.

30kb and comprises 10 exons. Exon 1 contains the signal peptide, whilst exon 10 is untranslated. Exons 3, 4, and 5 probably code for the catalytic site (8,9). The gene encodes for a mature protein possessing 448 amino acids and a molecular weight of 65,000 Daltons (10).

Genetic variants that increase susceptibility to CAD, may do so through an 'intermediate phenotype' such as a dyslipidaemia. It is unlikely that CAD in the general population is a consequence of marked 'intermediate phenotypic' effects arising from underlying genetic variability, but more likely to result from a small number of polymorphic genes each having minor 'intermediate phenotypic' effects. For example, the mutation in the low density lipoprotein (LDL) receptor gene (11) has a major impact in affected individuals but, owing to it's low frequency, it does not contribute a major influence upon CAD in the population. Therefore, improvement of our understanding into the role of the LPL gene in CAD in the population, is likely to require elucidation of those genetic defects giving rise to minor 'intermediate' functional abnormalities.

In this paper we discuss data from our and other laboratories, to demonstrate some current approaches to study the phenotypic effects arising from genetic variants of the LPL gene and consider this gene as a candidate for susceptibility to CAD.

METHODS AND RESULTS

Fig. 1.

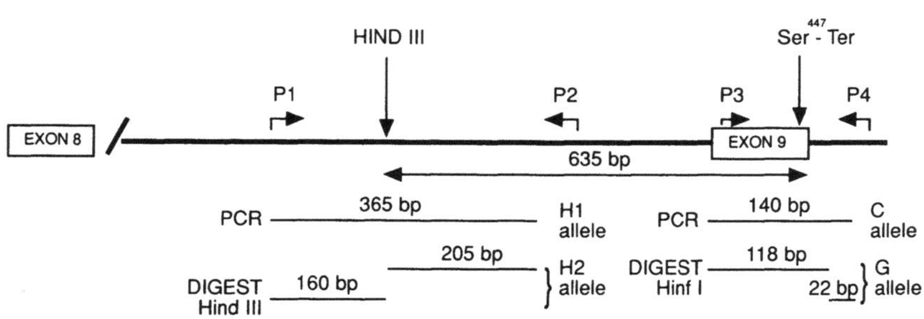

1. <u>Restriction fragment length polymorphism (RFLP) analyses</u>

We have analysed two polymorphisms identified by the restriction enzymes *Pvu-II* (intron 6) and *Hind-III* (intron 8). Although both show positive associations, the more striking observations involve the latter.

Fig. 2.

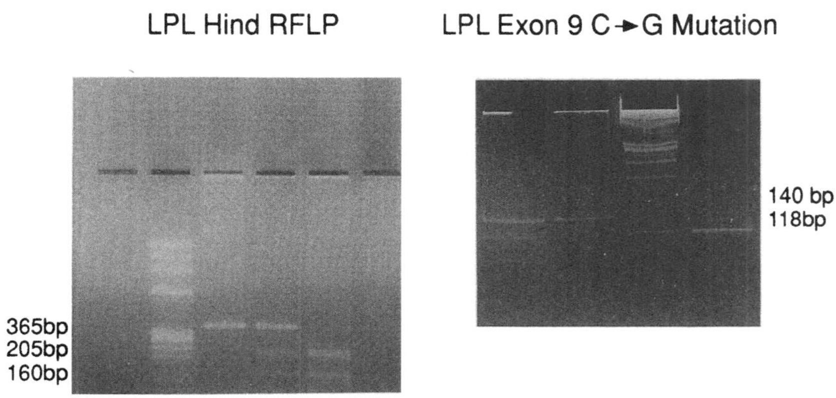

The *Hind-III* RFLP was analysed after polymerase chain reaction (PCR) amplification had generated a 365bp product. In the presence of the common allele (*H2*), *Hind-III* digestion gave rise to two fragments of 160bp and 205bp (Fig. 1). Following digestion, electrophoresis in a 2% agarose gel enabled subjects to be genotyped. Homozygote *H1H1*'s possess a single band of 365bp; homozygote *H2H2*'s two bands of 160bp & 205bp and heterozygotes three bands of 160bp, 205bp and 365bp (Fig. 2).

In a case control study, involving hypertriglyceridaemic Caucasian subjects and matched controls, we found the frequency of the *H2H2* genotype significantly higher than *H1H1* in the hypertriglyceridaemic group, $p<0.01$. The *H2* allele was also significantly more frequent in the disease subjects, $p<0.01$ (Fig. 3) (12).

Fig. 3.

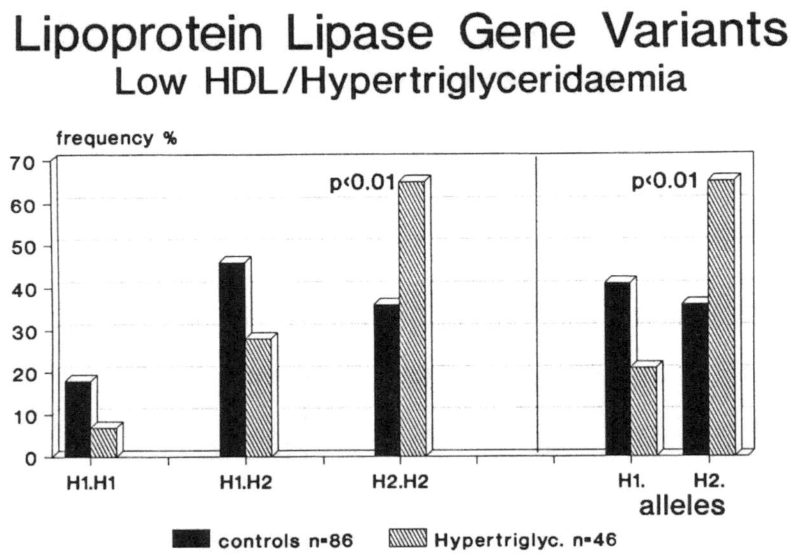

In a further case-control study, we found that the *H2* allele was significantly more frequent in patients with premature CAD, $p<0.001$. Similarly, the genotype *H2H2* when compared to *H1H1* was significantly more frequent in CAD subjects, $p<0.01$ (Fig. 4) (13).

Heinzmann *et al.* analysed four RFLP's in the LPL gene, using southern hybridisation, identified by the restriction enzymes *Pvu II* (intron 6), *Hind III* (intron 8), *Xba I* (intron 9) and *Bam HI* (3' to exon 10). They investigated Mediterraneans, U.S blacks and white Americans. The latter involved patients and their spouses from the John's Hopkins University CAD study. They found that the *Xba I* polymorphism was only present in the black population and the *Hind III* polymorphism was significantly associated with HDL levels in the white population (n=188), p=0.03. The rare *Hind III* allele was associated with significantly higher HDL levels (14).

Fig. 4.

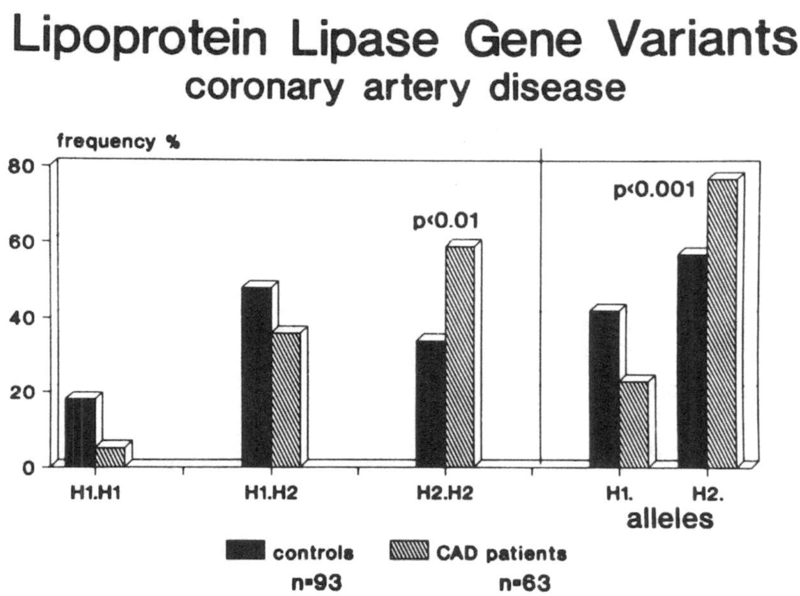

2. <u>The *Ser*^{447}*-Ter* mutation in exon 9</u>

Whilst testing their ability to identify nucleotide substitutions using single-strand conformation polymorphism (SSCP) analysis, Hata *et al.* discovered a nonsense mutation at position 1595 in exon 9 of the LPL gene (15). A *C-G* transversion converts serine 447 to a termination codon, producing a truncated mature enzyme lacking the two carboxyl terminal amino acids (Ser-Gly). This *Ser*^{447}*-Ter* mutation is just 635bp downstream from the *Hind III* polymorphism. They proceeded to use allele specific oligonucleotide (ASO) hybridisation to ascertain the frequency of this mutation in subjects with hypertriglyceridaemia plus low HDL-cholesterol and random unrelated population controls. They found that the mutation was significantly more frequent in the population controls, $p=0.037$.

Our laboratory embarked upon analysis of this mutation in a case-control study to investigate primary hypertriglyceridaemia. We wanted to use a reliable yet quicker technique. Since the *C-G* transversion does not create or abolish a restriction enzyme site, we could not employ the same technique as for the *Hind III* polymorphism. However, by using a modified 3' amplimer we were able to analyse the exon 9 mutation by PCR amplification followed by restriction digestion. This entailed a modified primer downstream from position 1595, containing a single base mismatch corresponding to position 1598, which generated a PCR product of 140bp. Thereby intentionally introducing a base change 3 nucleotides downstream from the *C-G* transversion in the PCR product. In the presence of the *G* allele this created a restriction site for the enzyme *HinfI*. Genotyping was undertaken following electrophoresis in a 10% non-denaturing polyacrylamide gel, (Fig. 2). The homozygote *CC* has a single band of 140bp, the heterozygote *CG* two bands of 118 & 140bp and the homozygote *GG* a single band of 118bp. The 22bp fragment cleaved, in the presence of the *G* allele, is not visualised since it runs off the gel during electrophoresis.

By using this method we found a significantly higher frequency of the *Ser*^{447}*-Ter* mutation in control subjects compared to patients with primary hypertriglyceridaemia, $p < 0.01$ (16). Analysis of the haplotypes constructed using the *Ser*^{447}*-Ter* mutation plus *Hind III* alleles showed that the *Ser*^{447}*-Ter* was in strong linkage disequilibrium with the *Hind III (H1)* allele, $p < 0.0001$.

3. Functional studies

The *Ser*^{447}*-Ter* mutation was reported by Faustinella *et al.* as functionally insignificant following analysis in a family of Turkish descent, containing two brothers diagnosed as having Type I Hyperlipidaemia (17). They analysed the LPL enzyme activity and immunoreactive mass in COS cells transfected with wild type and mutant cDNA's in expression vectors. The authors reported no differences in the specific activity. However, in their amended table there was a 24% reduction in LPL mass in the COS cell media with the *Ser*^{447}*-Ter* mutation (18). Both patients had also co-inherited a missense mutation involving residue 156 in exon 5 (*Asp*^{156}*-Gly*). The presence of this mutation yielded reduced LPL mass with zero specific activity. The authors concluded that the *Asp*^{156}*-Gly* mutation alone accounted for the familial LPL deficiency in their subjects. Their findings also supported the proposal based on sequence alignment data between LPL and pancreatic lipase plus site directed mutagenesis, that the catalytic triad in human LPL corresponds to *Asp*^{156}*-His*^{241}*-Ser*^{132} (19,20,21,22).

Subsequently, Kobayashi *et al.* investigated a 14 year old girl with Type I Hyperlipidaemia in whom they had previously attributed the cause of her dyslipidaemia to a defect involving the lipid interface recognition site (23). On sequencing exons 1→9 of her LPL gene, the only nucleotide difference found between the proband, her parents and two controls was the *Ser*^{447}-*Ter* mutation. She was heterozygous for the mutation. They transfected COS-1 cells with cDNA containing this *Ser*^{447}-*Ter* mutation and wild type cDNA. Expression in this vector was driven by a SV40 promoter (24). The esterase activity of the LPL produced was analysed using tributyrin; no differences were shown between mutant and wild type LPL. LPL activity was also analysed using triton X-100-triolein and lysophosphatidyl choline as substrates. They found > 50% reduction in hydrolysis of triton X-100-triolein by the mutant LPL, whereas no differences were observed in LPL activity using lysophosphatidyl choline. The mutant LPL in the culture medium of the transfected COS-1 cells exhibited activity properties consistent with the post-heparin LPL of the proband. This led the authors to conclude that the *Ser*^{447}-*Ter* mutation may contribute to the lipid interface recognition defect in their patient.

DISCUSSION

The *H1 (Hind III)* allele at the LPL gene may act as a linkage marker for a mutation influencing susceptibility to hypertriglyceridaemia and CAD. The *Ser*^{447}-*Ter* mutation is in linkage disequilibrium with this allele and could be such a mutation. The distribution of the *H1* allele and *Ser*^{447}-*Ter* mutation suggest that the *Ser*^{447}-*Ter* mutation may be protective against hypertriglyceridaemia and possibly CAD.

The allelic associations of the *Hind III* polymorphism with hypertriglyceridaemia and premature CAD found in our laboratory have not been observed by Heinzmann *et al.*, yet they found significant associations with HDL levels which were not previously shown. The apparent discrepancies may be explicable. The case-control studies from our laboratory had selected subjects satisfying distinct inclusion criteria (for example, fasting plasma triglyceride levels of > 2.3 mmol/l), whereas the study by Heinzmann *et al.* included populations collected randomly from other studies. This fundamental difference in patient selection could account for the discrepant results. The unexpected association in their study of a linkage marker for the LPL gene with HDL levels, without concomitant association with triglyceride levels, also emphasises the complexity of the metabolic relationship between HDL, triglycerides and LPL.

The conclusions of Faustinella *et al.* and Kobayashi *et al.* appear to be contradictory, but their conclusions are based on similar techniques yet different analyses. The latter group determined LPL function using several substrates; only one revealed activity differences between *wild type* and *Ser^{447}-Ter* DNA. This disparity highlights the importance of investigating all phases of the substrate-enzyme interaction when quantifying *in vitro* LPL activity, since the physicochemical nature of the substrate appears to influence it's catabolism.

The LPL C-terminal is highly conserved, with the dipeptide ***Ser-Gly*** present in all six reported mammalian LPL sequences (25-28), yet the function of this region remains undefined. The functional studies to date suggest that this region is unlikely to influence esterase activity *per se* but it possibly influences lipid binding, thereby affecting catalytic activity. Furthermore, there is recent evidence suggesting that the LPL C-terminal region may enhance the binding of chylomicrons to the putative VLDL remnant receptor (29). The *Ser^{447}-Ter* is the first report of a relatively common protein polymorphism of human LPL. Clearly, further studies are needed to determine any effects that the *Ser^{447}-Ter* mutation may have upon LPL secretion, dimer formation, heparin release, apoC-II activation and lipid binding.

We have considered LPL as a 'candidate gene' to reflect some approaches to delineate structure-function relationships and their subsequent involvement with disease. Such an approach, combined with 'reverse genetics', should facilitate and expedite our attempts to unveil and explain the mechanisms by which molecular alterations give rise to dyslipidaemia and atherosclerosis.

ACKNOWLEDGEMENTS

RKM is grateful to the McAlpine Trust for a research grant; DJG expresses gratitude to the Joint Research Board of St. Bartholomew's Hospital for support.

REFERENCES

1. Hulley SB, Rosenman RH, Bawol RD, *et al.* Epidemiology as a guide to clinical decisions; the association between triglycerides and coronary heart disease. *NEJM* 1980;302;1383-9.

2. Castelli WP, Doyle JT, Gordon T *et al.* HDL cholesterol and other lipids in coronary heart disease: The Cooperative Lipoprotein Phenotyping Study. *Circulation* 1977;55;76772.

3. Austin MA. Plasma triglyceride as a risk factor for coronary heart disease: The epidemiologic evidence and beyond. *Am. J. Epidemiol.* 1989;129;249-59.

4. Nikkila E. Familial Lipoprotein Lipase deficiency and related disorders of chylomicron metabolism. *The metabolic basis of inherited disease.* Ed Stanbury JB, Wyngaarden JB, Fredrickson DS, Goldstein JL, Brown MS. 5th Ed 1983;chapter 30;New York, McGraw-Hill.

5. Pykalisto OJ, Smith PH, Brunzell JD. Determinants of adipose tissue lipoprotein lipase. *J. Clin. Invest.* 1975;56;1108-17.

6. Patsch JR, Prasad S, Gotto Jr AM, Patsch W. HDL_2 relationships of the plasma levels of this lipoprotein species to it's composition, to the magnitude of postprandial lipaemia and to the activities of Lipoprotein Lipase and Hepatic Lipase. *J. Clin. Invest.* 1987;80;341-7.

7. Sparkes RS, Zollman S, Klisak I *et al.* Human genes involved in the lipolysis of plasma lipoproteins: Mapping of the loci for Lipoprotein Lipase to 8p22 and Hepatic Lipase to 15q21. *Genomics* 1987;1;138-44.

8. Semenkovich CF, Luo CC, Nakanishi MK *et al.* In vitro expression and site-specific mutagenesis of the cloned human lipoprotein lipase gene. *J. Biol. Chem.* 1990;265;5429-33.

9. Kirchgessner TG, Chaut J-C, Heinzmann C *et al.* Organization of the human lipoprotein lipase gene and evolution of the lipase gene family. *Proc. Natl. Acad. Sci.* 1989;86;9647-51.

10 Wion KL, Kirchgessner TG, Lusis AJ *et al.* Human Lipoprotein Lipase complementary DNA sequence. *Science* 1987;235;1638-41.

11. Brown MS, Goldstein JL. A receptor-mediated pathway for cholesterol homeostasis. *Science* 1986;232;34-47.

12. Chamberlain JC, Thorn JA, Oka K *et al.* DNA polymorphisms at the lipoprotein lipase gene: associations in normal and hypertriglyceridaemic subjects. *Atherosclerosis* 1989;79;85-91.

13. Thorn JA, Chamberlain JC, Alcolado JC *et al.* Lipoprotein lipase and hepatic lipase variants in coronary atherosclerosis. *Atherosclerosis* 1990;85;55-60.

14. Heinzmann C, Kirchgessner T, Kwiterovich PO *et al.* DNA polymorphism haplotypes of the human lipoprotein lipase gene: possible association with high density lipoprotein levels. *Human Genetics* 1991;86;578-584

15. Hata A, Robertson M, Mitsuri E, Lalouel J-M. Direct detection and automated sequencing of individual alleles after electrophoretic strand separation: identification of a common nonsense mutation in exon 9 of the human lipoprotein lipase gene. *Nucl. Acids Res.* 1990;18;5407-11.

16. Stocks J, Thorn JA, Galton DJ. Lipoprotein lipase genotypes for a common premature termination codon mutation detected by PCR-mediated site-directed mutagenesis and restriction digestion. *J. Lipid Res.* 1992;33;853-7.

17. Faustinella F, Chang A, Van Biervliet JP *et al.* Catalytic triad residue mutation ($Asp^{156}\rightarrow Gly$) causing familial lipoprotein lipase deficiency. *J. Biol. Chem.* 1991;266;14418-24.

18. Faustinella F, Chang A, Van Bierliet JP *et al.* Additions and Corrections. *J. Biol. Chem.* 1992;267;7194.

19. Winkler FK, D'Arcy A, Hunziker W. Structure of human pancreatic lipase. *Nature* 1990;343;771-74.

20. Datta S, Luo C-C, Li W-H *et al.* Human hepatic lipase: Cloned cDNA sequence, restriction fragment length polymorphisms, chromosomal localization, and evolutionary relationships with lipoprotein lipase and pancreatic lipase. *J. Biol. Chem.* 1988; 263;1107-10.

21. Faustinella F, Smith LC, Semenkovich CF, Chan L. Structural and functional roles of highly conserved serines in human lipoprotein lipase: Evidence that serine 132 is essential for enzyme catalysis. *J. Biol. Chem.* 1991;266;9481-85.

22. Emmerich J, Beg OU, Peterson J *et al.* Human lipoprotein lipase: Analysis of the catalytic triad by site-directed mutagenesis of Ser^{132}, Asp^{156} and His^{241}. *J. Biol. Chem.* 1992;267;4161-65.

23. Kobayashi J, Shirai K, Saito Y, Yoshida S. Lipoprotein lipase with a defect in lipid interface recognition in a case with type I hyperlipidaemia. *Eur. J. Clin. Invest.* 1989;19;424-32.

24. Kobayashi J, Nishida T, Ameis D *et al.* A heterozygous mutation (the codon for $Ser^{447}\rightarrow$a stop codon) in lipoprotein lipase contributes to a defect in lipid interface recognition in a case with type I hyperlipidaemia. *Biochem. Biophys. Res. Comm.* 1992;182;70-77.

25. Kirchgessner TG, Svenson KL, Lusis AJ, Schotz MC. The sequence of cDNA encoding lipoprotein lipase. A member of a lipase gene family. *J. Biol. Chem.* 1987;262;8463-66.

26. Enerbäck S, Bjursell G. Genomic organisation of the region encoding guinea pig lipoprotein lipase: Evidence for exon fusion and unconventional splicing. *Gene* 1989;84;391-7.

27. Yang CY, Gu ZW, Yang MF *et al.* Structure of bovine milk lipoprotein lipase. *J. Biol. Chem.* 1989;264;16822-7.

28. Beisegel U, Weber W, Bengtsson-Olivecrona G. Lipoprotein lipase enhances the binding of chylomicrons to low density lipoprotein receptor-related protein. *Proc. Natl. Acad. Sci.* 1991;88;8342-6.

APOLIPOPROTEINS IN THE DIAGNOSIS OF DISEASE: CEREBRAL AND PERIPHERAL ATHEROSCLEROSIS

R. Fellin, G.B. Vigna
Institute of Patologia Medica, University of FERRARA,
via Savonarola, 9 44100 Ferrara ITALY

THE "RATIONALE" OF THE APOLIPOPROTEINS

The plasma apolipoprotein concentration furnishes information about the characteristics of the lipoprotein particles transporting them. For example, three main HDL types were characterized on the basis of the constituent apolipoproteins: those containing only apo AI, those containing apo AI and apo AII, and those rich in apo E. The HDL containing only apo AI are preferentially distributed in the HDL2 density range (1.064-1.125g/ml); 4 polypeptides of apo AI go into the constitution of the complex HDL macromolecule. Particles made up of apo AI and apo AII contain 2 polypeptides of each apolipoprotein, and are distributed mainly among the HDL3 (1.125-1.250 g/ml). HDL rich in apo E appear less dense than the HDL2, and even if they are considered important from a metabolic point of view, they are not assayable by common laboratory techniques (Scriver et al 1989, Eisenberg 1984).
A single apo B molecule is contained in each VLDL, IDL or LDL type particle. Hyperapobetalipoproteinemia, a recently identified condition, is associated with the presence of dense, small LDL with a normal lipid contents, the composition of which however is rich percentage-wise in proteins (apo B) (Sniderman et al 1980); in other words, high molar LDL concentrations may escape detection in the assay of plasma cholesterol. The same phenomenon was also observed in the hypertriglyceridemias in general, and in the phenotypical polymorphous condition characterizing the picture of combined familial hyperlipemia (Scriver et al 1989).
The close relationship between the number of apoprotein peptides and lipoprotein particles therefore enables a good estimation of the latter from the former. The apolipoprotein concentration, moreover, allows an evaluation of the subclass distribution of the main circulating lipoproteins. Atherosclerosis may be related to both these conditions. For example, a low HDL2 level seems to be selectively predictive of atherosclerotic lesions, as is a state of hyperapobetaliproteinemia (Sniderman et al 1980). High apo B

NATO ASI Series, Vol. H 73
Human Apolipoprotein Mutants III
Edited by C. R. Sirtori, G. Franceschini, B.H. Brewer Jr.

levels, moreover, may be associated with other conditions, in particular a concomitant rise in apolipoprotein apo (a), which seems to be an important risk factor for early ischemic heart disease (Durrington et al 1988). It is therefore conceivable that evaluation of the apolipoprotein concentration could contribute to a better definition of vascular risk.

THE "LIMITS" IN DETERMINING APOLIPOPROTEINS

Among the several laboratory techniques used to assay apolipoproteins, the following appear important: radial immunodiffusion, rocket electrophoresis, radioimmunoassay, immunoenzymatic (ELISA), immunonephelometric and immunoturbidimetric methods. Unfortunately, many of these procedures are costly and often not very sensitive; at times some require preliminary steps (i.e., wide serum dilutions) that may amplify the analytic error, or are subject to important interferences determined by the characteristics of the material under analysis (i.e., excess of triglycerides, variations in lipoprotein particle size, etc.). Another limit to a correct methodologic approach to this issue, stems from the difficulty in establishing a quality control standard that would make comparisons between different laboratory techniques and workers possible (Cooper et al 1989).

APOLIPOPROTEINS AND PERIPHERAL ATHEROSCLEROSIS (PAD)

A relationship between altered plasma lipoprotein levels and coronary atherosclerosis emerged in several studies; using case-control type investigations, most demonstrated that reduced apo AI and high apo B levels are correlated with the development of vascular injury. Likewise high levels of apo (a) are an important factor of coronary risk. The assay of apo AII and other apolipoproteins instead produced contrasting results (Albers et al 1989). On the whole, it is not yet certain that apo AI and apo AII are better discriminants of ischemic heart disease than HDL cholesterol. A body of concordant data instead indicate that apo B seems to be a better predictive factor of ischemic heart disease itself than total or LDL cholesterol.

Here we will analyze reports evaluating plasma apolipoprotein concentrations in relation to the development of cerebral and peripheral atherosclerosis. As simple and reliable instrumental techniques that could resolve the diagnostic suspect of extracoronary arterial disease are in fact lacking, the number of clinical studies is low. On the other hand, research in coronary atherosclerosis has been favored by electrocardiogram investigation, which enables a rapid although partial evaluation of the coronary circulation. Other reasons for the minor attention of workers to the extracoronary districts are the low immediate mortality due to peripheral obstructive arterial disease, the difficulty in diagnosing the embolic, hemorrhagic or thrombotic cause of a cerebrovascular injury with reasonable certainty (only a thrombus is clearly related to injury of the arterial wall), and, finally, the small number of epidemiologic data relating lipids to the development of atherosclerosis in these districts. The recent introduction of new non-invasive diagnostic methods will further the work of investigators in this field of interest, and allow the verification of analogies and differences between atherosclerosis risk factors in diverse vascular districts.

In Table I, we report the findings of the main studies that have correlated apolipoproteins with PAD.

Although apo AI is frequently correlated negatively with the presence of atherosclerosis at the aorto-femoral level, the findings are not uniform. By means of multivariate analysis, Pilger et al. found that apolipoproteins AI, AII and B are better discriminants than lipids in this disease (1983). However, investigations that strictly relate these apolipoproteins to the severity of the district vascular injury are not available. Indeed, in a recent study, Stringer and Kakkar (1990) found a significant inverse correlation between total cholesterol, LDL cholesterol, apo B, and angiographically evaluated severity of peripheral atherosclerosis.

TABLE I

RELATIONSHIPS BETWEEN PLASMA APOLIPOPROTEINS AND PERIPHERAL ARTERIAL DISEASE.

AUTHOR (YEAR)	ASSOCIATION	WITH	LESION	COMMENTS
	A-I	A-II	B	
Bradby (1978)	-	-	n.s.	
Franceschini (1982)	n.s.	n.s.	n.s.	Increased apo B in VLDLs
Pilger (1983)	-	-	+	
McConathy (1984)	-		n.s.	SAS and SPAD groups; apo C-III +, apo E n.s.
Stringer (1990)	n.s.	n.s.	-	Related to PAD severity; Lp(a) n.s.
Vitale (1990)	(-)		(+/-)	Differences in M and F sex
Sentì (1992)			n.s.	apo E (phenotypes) n.s.

Legend:
- : apolipoprotein levels significantly lower in cases than in controls
+ : apolipoprotein levels significantly higher in cases than in controls
n.s.: apolipoprotein levels in cases not significantly different from controls

Other studies addressed the presence of PAD patient subgroups with peculiar apolipoprotein characteristics. McConathy et al. (1984) evaluated the apolipoprotein profile in women belonging to two distinct groups: one featured localized arterial stenosis (SAS, small aorta syndrome), and the other diffuse segmentary vascular stenosis (SPAD, stenosing peripheral arterial disease). Compared to normal controls, both groups had significantly reduced apo AI and apo CIII levels; apolipoproteins B and E seemed increased but not significantly. In a subsequent study (data not reported in table I) (Mc Conathy et al 1989), the workers compared lipid and apolipoprotein profiles in two groups of male patients with aorto-iliac aneurysmatic dilatation and stenosing disease, respectively; the results indicated lower apo AI levels in the subjects with aneurysmatic dilatation, while apolipoprotein B, CIII and E levels overlapped controls.
A recently published investigation (Sentì et al 1992) could not demonstrate that apo E polymorphism or apoB and apo E concentration, have a major impact in the development of PAD.
Franceschini et al. (1982) found that VLDL in patients with PAD were significantly enriched in apo B, compared to normal controls or non PAD hyperlipemic subjects; moreover, plasma apolipoproteins did not show significant differences compared to the same controls. A subsequent investigation, however, did not completely confirm these results (Carlson et al 1986)).
In subjects with severe, angiographically demonstrated atherosclerosis of the supra-aortic trunks and lower limbs and no manifest signs of coronary disease, we found that apolipoproteins AI and B were only weak indicators of disease (Table II) (Vitale et al 1990). Apo AI was lower in

TABLE II

MEAN VALUES (mg/dl, M±SD) OF APOLIPOPROTEINS A-I, B AND APO A-I/APO B RATIO IN PATIENTS WITH EXTRACORONARY ATHEROSCLEROSIS.

SITE	SEX	No	A-I	B	A-I/B
Lower limbs	M	54	138.7±26.2 *	116.5±31.8 *	1.32±0.55
Supra-aortic trunks	M	21	140.6±16.5 *	122.0±36.1	1.22±0.32
Control subjects	M	39	151±20.6	131.5±32.5	1.22±0.37
Lower limbs	F	18	144.1±22.1	116.2±29.8 *	1.28±0.26
Supra-aortic trunks	F	9	155.7±16.0	99.8±27.9	1.67±0.52
Control subjects	F	25	151.0±16.1	98.5±16.4	1.58±0.39

* $p < 0.05$, patients vs. control subjects

the patients, compared to the controls, but statistical significance ($p < 0.05$ by the t test) was reached only in males. On the contrary, apo B was higher in female patients, compared to controls, but lower among males with PAD.

APOLIPOPROTEINS AND CEREBRAL ATHEROSCLEROSIS

Table III reports the most important studies regarding apolipoproteins and cerebrovascular disease. The clinical manifestation of vascular injury (stroke, TIA, multi-infarctual dementia) was often used to recruit patients, and this places some limits on the interpretation of the results. In other cases, ultrasound examination and angiography more accurately evaluated the type of arterial alteration present. Four of the 7 reported studies demonstrated that apo B and AI were indexes correlated with the presence of cerebral arterial disease. In a recent investigation conducted in China, apo B was not associated with stroke, not even following patient subdivision into subgroups (cerebral, lacunar infarct; cerebral hemorrahge); apo AI instead showed a negative correlation, and Lp(a) a positive correlation (Woo et al 1991). This last parameter was also valued by Költringer and Jürgens (1985). Kostner et al. (1986) further confirmed the importance of apo AI by means of multivariate discriminant analysis that also showed a relationship between low apo AII levels and cerebral vascular disease. Matsuda et al. (1987) studied several plasma apolipoproteins in subjects with cerebral infarct, but only apo B levels were significantly higher than controls. Shimano et al. (1989) evaluated apo AI,

TABLE III

RELATIONSHIPS BETWEEN PLASMA APOLIPOPROTEINS AND CEREBRAL ATHEROSCLEROSIS.

AUTHOR (YEAR)	ASSOCIATION WITH LESION			COMMENTS
	A-I	A-II	B	
Postiglione (1985)	-		+	FH cases
Költringer (1985)				Lp(a) +
Kostner (1986)	-	-	n.s.	
Matsuda (1987)	n.s.	n.s.	+	Apo C-II, C-III, E n.s.
Shimano (1989)	-		+	Apo E n.s. (multi-infarct dementia)
Vitale (1990)	(-)		n.s.	
Woo (1991)	-		n.s.	Chinese patients
Bonithon Kopp (1991)	n.s.		+	Middle aged women

Legend:
- : apolipoprotein levels significantly lower in cases than in controls
+ : apolipoprotein levels significantly higher in cases than in controls
n.s.: apolipoprotein levels in cases not significantly different from controls

B and E levels in subjects with multi-infarct vascular dementia due to cerebral atherosclerosis. Low apo AI and high apo B levels were correlated with disease, while apo E did not differ significantly from control levels, even if the frequency of the E4 isoform, associated with high lipid levels, was elevated. A recent multivariate study conducted in middle aged women showed that apo B levels were significantly correlated with the severity of carotid atherosclerosis evaluated by ultrasound (Bonithon Kopp et al 1991). Data from our investigation (Vitale et al 1990) disclosed that only male subjects with atherosclerosis of the supra-aortic arterial trunks had lower apo AI levels than controls, while apo B showed no significant differences. The lipid parameters in this case were better indexes of vascular damage (in particular, LDL and HDL2 cholesterol).

GENETICS AND APOLIPOPROTEINS IN PERIPHERAL AND CEREBRAL ATHEROSCLEROSIS

DNA variations may be relevant to atherosclerosis risk. Such alterations occasionally may be demonstrated by cutting genomic DNA with restriction enzymes, followed by hybridization with genetic probes, and evaluation of the different fragments obtained (RFLPs). Concerning extracoronary atherosclerosis, very few studies have addressed the relationship between apolipoprotein gene polymorphisms and vascular disease.

Monsalve et al. (1988) evaluated the association between apo B gene polymorphisms in subjects with isolated PAD or associated with arterial involvement in other districts (coronary and carotid). Using XbaI and EcoRI enzymes, a higher frequency of some alleles compared to normal controls (X1 and R2, respectively) was found, with no signficant differences among the various patient subgroups. The results of a similar study in regions close to the locus coding for apo AI (gene cluster AI, CIII, AIV) were inconclusive (Monsalve et al 1991). The same workers then discerned a relationship between gene polymorphisms in the apo B gene or the apo AI, CIII, AIV gene cluster, and the coefficient of intra-individual variation in cholesterol levels; they advanced that elements in the promoter region of some of these genes might be particularly responsive to environmental factors that determine wide fluctations in their nuclear transcription, and consequently in that of the lipoprotein levels. This might be directly relevant to atherosclerosis development (Monsalve et al 1991). Finally, a different type of polymorphism was evidenced in the AI, CIII, AIV gene cluster also by O'Connor et al. (1988), and was associated with both carotid athero-sclerosis and with aorto-femoral atherosclerotic involvement.

CONCLUSIONS

It is now possible to draw some conclusions, even if the lack of prospective studies in this field preclude definitive statements. PAD seems characterized by an abnormal lipoprotein composition (apo B-rich VLDLs), but the relationships with apolipoprotein levels are not clearly defined. On the other hand (with some notable exceptions) the behavior of the plasma apolipoproteins in carotid atherosclerosis and ischemic heart disease is rather similar, especially in males, thus implying that the two conditions might be close on a pathogenetic level.

DNA variations in apolipoprotein genes could be important in extracoronary atherosclerosis development, also if they do not seem to determine the site of disease onset in the arterial system. The effect of these genetic variations could be partially independent of circulating plasma lipid levels.

These investigations are harbingers of further studies that will better define the inter-relationships between genetic or environmental factors, and apolipoprotein levels and characteristics in the genesis of atherosclerosis in the various vascular districts.

REFERENCES

Albers JJ, Brunzell JD, Knopp RH (1989) Apoprotein measurements and their clinical application. Clin Lab Med 9: 137-152

Bonithon Kopp C, Scarabin PY et al. (1991) Risk factors for early carotid atherosclerosis in middle-aged french women. Arterioscler Thromb 11:966-972

Bradby GVH, Valente AJ, Walton KW (1978) Serum high-density lipoproteins in peripheral vascular disease. Lancet ii:1271-1274

Carlson LA, Pauciullo P et al. (1986) Modified composition of VLDL in peripheral vascular disease: evidence for low cholesteryl ester but normal apolipoprotein B content of small VLDL. Monogr Atheroscler 14:31-34

Cooper RC, Henderson LO et al. (1991) Clinical application and standardization of apolipoprotein measurements in the diagnostic workup of lipid disorders. Clin Chem 37:619-620.

Durrington PN, Ishola M et al. (1988) Apolipoproteins (a), AI and B and parental history in men with early onset ischemic heart disease Lancet i:1070-1073
Eisenberg S (1984) High density lipoprotein metabolism. J Lipid Res 25:1017-1058
Franceschini G, Bondioli A et al. (1982) Increased apoprotein B in very low density lipoproteins of patients with peripheral vascular disease. Arteriosclerosis 2:74-80
Költringer P, Jürgens G (1985) A dominant role of lipoprotein (a) in the investigation and evaluation of parameters indicating the the development of cervical atherosclerosis. Atherosclerosis 58:187-198
Kostner GM, Marth E et al.(1986) Laboratory parameters as discriminators for peripheral atherosclerosis and stroke. Monogr Atheroscler 14:119-123
Matsuda M, Miyahra T et al.(1987) Lipoprotein abnormalities in survivors of cerebral infarction with a special reference to apolipoproteins and triglyceride-rich lipoproteins. Atherosclerosis 68:131-136
McConathy WJ, Alaupovic P et al. (1989) Lipids and apolipoprotein profiles in men with aneurismal and stenosing aorto-iliac atherosclerosis. Eur J Vasc Surg 3:511-514
McConathy WJ, Greenhalgh RM et al.(1984) Plasma lipid and apolipoprotein profiles of women with two types of peripheral arterial disease. Atherosclerosis 50:295-306
Monsalve MV, Robinson D et al. (1991) Within-individual variation in serum cholesterol levels: association with DNA polymorphisms at the apolipoprotein B and AI-CIII-AIV loci in patients with peripheral arterial disease. Clin Genet 39, 260-273
Monsalve MV, Young R et al. (1988) DNA polymorphisms of the gene for apolipoprotein B in patients with peripheral arterial disease. Atherosclerosis 70123-129
O'Connor G, Stocks J et al. (1988) A DNA polymorphism of the apolipoprotein C-III gene in extracoronary atherosclerosis. Clin Sci 74:289-292
Pilger E, Pristautz H et al.(1983) Risk factors for peripheral atherosclerosis. Arteriosclerosis 3:57-63
Postiglione A, Rubba P et al. (1985) Carotid atherosclerosis in familial hypercholesterolemia. Stroke 16:658-661
Scriver CR et al (eds) (1989) Lipoprotein and lipid disorders. In: The metabolic basis of inherited disease. Mc Graw Hill, New York:p.1127-1302.
Shimano H, Ishibashi S et al. (1989) Plasma apolipoproteins in patients with multi-infarct dementia. Atherosclerosis 79:257-260
Sniderman AD, Shapiro S et al.(1980) Association of coronary atherosclerosis with hyperapobetalipoproteinaemia. (increased protein, but normal cholesterol levels in human plasma, low density (β)lipoproteins) Proc Natl Acad Sci USA 77:604-608
Stringer MD, Kakkar VV (1990) Markers of disease severity in peripheral atherosclerosis. Eur j Vasc Surg 4:513-518
Vitale E, Zuliani G et al. (1990) Lipoprotein abnormalities in patients with extra-coronary arteriosclerosis. Atherosclerosis 81:95-102
Woo J, Lau E et al. (1991) Hypertension, lipoprotein(a), and apolipoprotein A-I as risk factors for stroke in the chinese. Stroke 22:203-208

FUNCTIONAL PROPERTIES OF DISCOIDAL APOPROTEIN/PHOSPHOLIPID COMPLEXES: CHOLESTEROL EFFLUX FROM MACROPHAGES AND CONVERSION BY LECITHIN-CHOLESTEROL ACYL TRANSFERASE.

M. Rosseneu[1], A. Devreese[1], B. Vanloo[1], J. Baert[2]
Department of clinical chemistry
A.Z. St-Jan
B-8000 Brugge
Belgium

INTRODUCTION.

Cholesterol esters loaded macrophages or "foam cells" have been identified in early atherosclerotic lesions (Ross 1990). and can be generated "in vitro" by incubation of macrophages with chemically modified low density lipoproteins (LDL) (Brown and Goldstein 1983).
High-density lipoproteins, discoidal apoprotein-phospholipid complexes, as well as the apolipoproteins A-I, A-II, C-III and E alone, were reported to efficiently remove cholesterol from laden peritoneal macrophages (Hara and Yokoyama 1991). In vivo, cholesterol would then be esterified by the LCAT enzyme and the discoidal complexes converted into spherical mature HDL (Eisenberg 1984).
In this paper we demonstrate that discoidal apoprotein-lecithin complexes are efficient inductors of cholesterol efflux from cholesterol-laden macrophages. In the presence of the LCAT enzyme in the cell medium, the cholesterol-enriched discoidal complexes are converted into spherical particles with a core of cholesteryl esters.

2: KULAK, Kortrijk

NATO ASI Series, Vol. H 73
Human Apolipoprotein Mutants III
Edited by C. R. Sirtori, G. Franceschini, B.H. Brewer Jr.

EXPERIMENTAL PROCEDURES.

Preparation of the apoprotein-lipid complexes.

Lipoproteins were isolated by preparative ultracentrifugation and LDL was acetylated (Basu et al. 1976). Apo A-I and apo A-II were prepared from HDL by delipidation and ion-exchange chromatography (Van Tornout et al. 1981). Apo A-IV was obtained from chylous ascites fluid (Yang et al. 1989). Complexes were prepared with synthetic dipalmitoylphosphatidylcholine (DPPC) at a 3:1, w/w ratio, by the cholate dialysis procedure (Matz and Jonas 1982). The complexes were isolated by gel chromatography on a Superose 6HR column with an FPLC system (Waters) (Vanloo et al. 1991). After purification, (Jonas 1986), the LCAT enzyme was dialysed against a 10 mM Tris-HCl, 5 mM EDTA, pH 7.6 buffer and stored at -70°C.

Cellular cholesterol efflux.

J774 murine macrophages are grown for 18 h in DMEM (Gibco) with addition of 10 % fetal calf serum (FCS), in a 5% CO_2 atmosphere at 37°C and incubated with acetylated LDL at a concentration of 100 μg apo B/ml for 24h.

The apoprotein-lecithin complexes are then added to the medium at concentrations varying between 10 and 200 μg apoprotein/ml and cells were incubated for 18 h. In some experiments 120 μl LCAT enzyme, dialysed against a 10 mM Tris-HCl buffer pH 8.6, 0.15 M NaCl containing 0.1 mM dithiothreitol (DTT) were added to the medium.

Lipids were extracted from cells and from the culture medium with 5 ml hexane/isopropylalcohol (3:2, v:v). Cholesterol and cholesteryl esters were quantified by HPLC on a reverse-phase column (Vercaemst et al. 1989,a,b).

Cellular proteins were dissolved in 0.1 M NaOH and the proteins were assayed using a kit for protein assay, (Pierce Europe, Nederland) with bovine serum albumin as a standard.

Complex recovery and characterisation.

The apoprotein-phospholipid complexes were separated from the medium by density gradient UCF in a sucrose-NaCl gradient (Rosseneu et al. 1989) and their chemical composition was determined by enzymatic phospholipid and cholesterol assyas and immunological apoprotein measurements (Bury and Rosseneu 1988). The size of the complexes was determined by gradient gel electrophoresis and by electron microscopy after negative staining (Vanloo et al. 1991).

After LCAT conversion of the discoidal into spherical complexes, their properties were compared to those of spherical particles generated "in vitro" (Vanloo et al. 1992).

RESULTS.

Release of intracellular cholesterol by the apoprotein-lipid complexes.

The cholesterol content of the J774 macrophages loaded with acetylated LDL increased from 25 ± 8 up to 114 ± 18 μg/mg cell protein, 47 % of which was esterified (Table 1). After incubation with 100 μg apo A-I-, apo A-II- or apo A-IV-DPPC complex for 24 h, 50, 46 and 39 % of the intracellular cholesteryl esters were hydrolyzed and released into the medium together with 32, 25 and 38 % of the cellular unesterified cholesterol. Cholesterol efflux into the medium occured only as free cholesterol as no cholesteryl esters were detected by HPLC, it increased as a function of the acceptor concentration (Fig. 1). The cholesterol efflux induced by the apo A-I- and apo A-IV-DPPC complexes was significantly higher than that induced by the apo A-II-DPPC complex (Fig. 1).

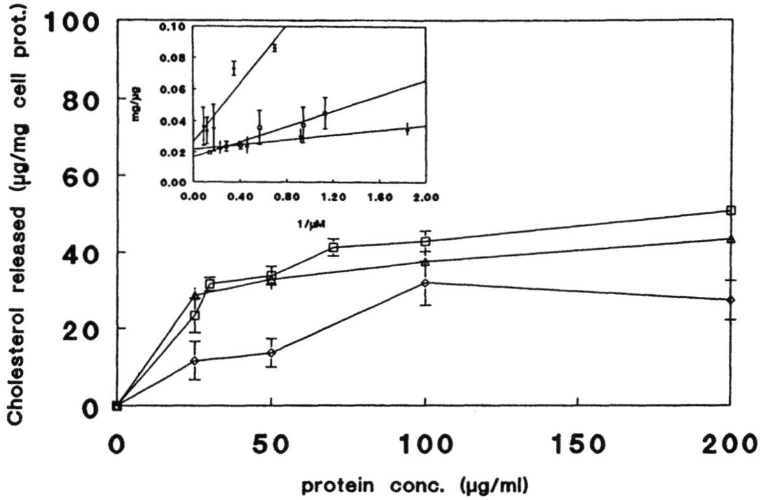

Fig. 1. .Cholesterol released into the medium as a function of the concentration of the apolipoproteins after 24 h incubation time. (□), apo A-I; (◇), apo A-II; (△), apo A-IV. Insert: Lineweaver-Burk plots of the inverse of the cholesterol released into the medium as a function of the inverse of the apolipoprotein concentration. A: (□), apo A-I; (◇), apo A-II; (△), apo A-IV.

As the cholesterol release showed saturation kinetics, the data were analyzed by double reciprocal plots yielding the apparent Km values of 1.4, 3.4, and 0.4 μM for apo A-I, A-II and A-IV respectively (Fig. 1, insert, Table 1). The acceptor efficiency expressed as Vmax/Km is apo A-IV > apo A-I > apo A-II.

Table 1: Cholesterol efflux from J774 macrophages by apoprotein - DPPC complexes.

J774 Macrophages	Acceptor Complex	TC	UC	CE	Vmax ug chol/mg cell protein	Km uM	Vmax/KM
		ug/mg cell protein					
unloaded (n=16)	/	25±8	25±8	/	/	/	/
chol. loaded (n=14)	/	114±18	60±9	54±13	/	/	/
chol. loaded (n=4)	DPPC/A-I	70±11	41±13	27±2	58.9	1.4	36.9
chol. loaded (n=4)	DPPC/A-II	75±31	45±22	29±8	37.4	3.4	9.6
chol. loaded (n=4)	DPPC/A-IV	71±21	37±10	33±12	46.0	0.4	100.8

TC: Total Cholesterol CE: Cholesterol esters
UC: Unesterified Cholesterol n : Number of experiments

Isolation of the apo A-I-DPPC complexes after cellular incubation.

After incubation of the cholesterol-laden J774 macrophages with 100 μg apo A-I-DPPC complex, in the presence or absence of 120 μL LCAT, the complex was recovered by gradient ultracentrifugation at densities around 1.10 g/ml (Fig.2). The composition of the complexes before and after cellular incubation is summarized in Table 2. During incubation with the J774 macrophages, 11 mole cholesterol were incorporated per mole complex. This amount increased up to 29 mole per mole complex, after LCAT was added to the incubation mixture, and 60 % of the cholesterol was esterified and detected as cholesteryl palmitate by HPLC.

By electron microscopy, the complexes enriched with cellular cholesterol appear as "rouleaux" of stacked discs (Fig.2A, insert), about 50 Å thick with a diameter of 122 Å, compared to 109 Å measured by gradient gel electrophoresis (Table 2).

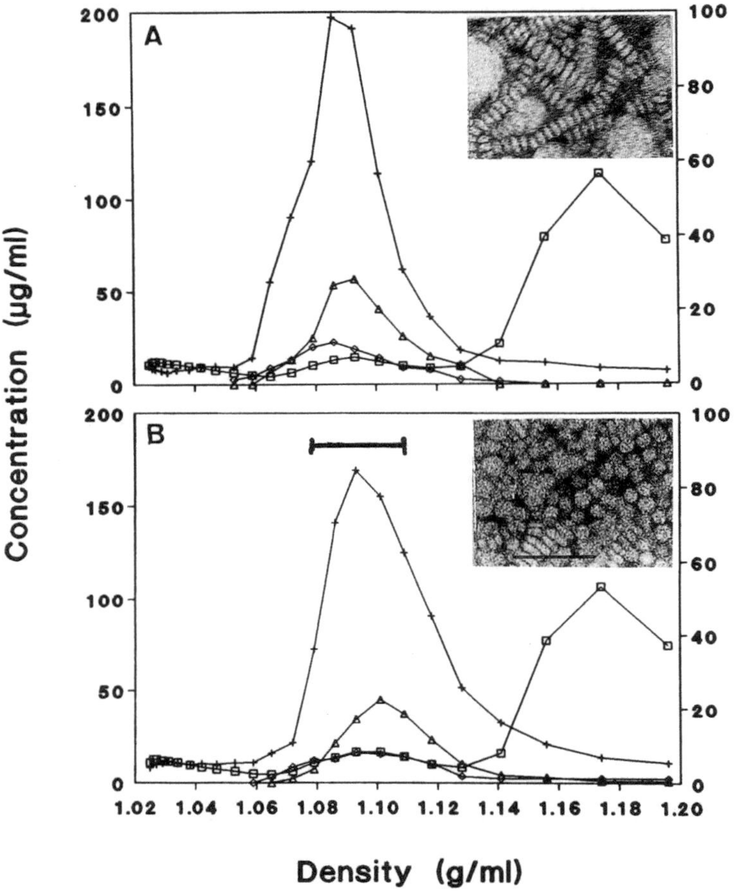

Fig. 2. Isolation of the apo A-I-DPPC complexes (⊢⊣) after cellular incubation, in the absence (A) or presence (B) of LCAT, by density gradient ultracentrifugation. (+), phospholipid; (◇), cholesterol; (△), apo A-I; (□), Trp fluorescence intensity. Insert: Electron micrographs of the apo A-I-DPPC complexes enriched with cellular cholesterol, in the absence (A) or presence (B) of LCAT. The bar marker represents 60 nm.

Table 2: Characterisation of the apo A-I/DPPC complexes before and after incubation with J774 macrophages and LCAT.

Incubation mixture	A-I mol	DPPC	UC	CE	GGE Stokes diam. (A)	EM Diameter (A)
		mol/mol apo A-I				
A-I/DPPC complex	1	114	-	-	108	114 ± 15
A-I/DPPC + J774 macrophages	1	122	11	-	109	122 ± 16
A-I/DPPC + J774 macrophages + LCAT	1	116	12	17	-	112 ± 12
A-I/DPPC/Chol + LDL + LCAT	1	47	3	19	98	110 ± 15

After LCAT incubation for 24 h, spherical HDL-like particles of about 110 A diameter were generated (Fig.2B) enriched in cholesteryl esters representing about 60 % of the total cholesterol content (Table 2). The phospholipid content of the complexes recovered by ultracentrifugation after incubation with the J774 macrophages in the presence of LCAT was higher than in the spherical particles generated "in vitro" (Table 2). During this "in vitro" conversion, cholesterol is transfered from LDL into the complexes while phospholipids are transfered from the spherical particles to LDL (Vanloo et al 1992).

DISCUSSION.

Discoidal complexes generated between DPPC, apo A-I, A-II, A-IV, act as efficient acceptors of cholesterol from cholesterol-laden J774 macrophages, as 45 % of the cellular cholesteryl esters were hydrolyzed and released into the medium, together with 30 % of the cellular free cholesterol. The cellular cholesterol released is incorporated into the discoidal complexes which contain around 10 mole % cholesterol and serve as substrates for the LCAT enzyme, when added to the cellular medium. During this enzymatic process, most of the discoidal complexes are converted into HDL-like spherical structures through the formation of a core of cholesteryl esters.

Cholesterol efflux has been demonstrated for several types of cells including fibroblasts, adipose cells and macrophages, whether or not laden with cholesterol (Mahlberg et al. 1990; Phillips et al. 1987; Nowicka et al. 1990). In this process, lipid-bound apo A-I, A-IV and E behave as the most efficient promotors of cholesterol efflux, compared to apo A-II-containing HDL subfractions (Barkia et al. 1986), and are also the best LCAT activators (Jonas 1986; Vanloo et al. 1992).

The mode of cholesterol insertion in the discoidal lipid-protein complexes is probably dependent upon the structure of the complexes and thus relate to the LCAT activation properties of the apoprotein constituents. The rate of cholesterol exchange between vesicular and micellar complexes is a function of the number of apoprotein segments located along the edge of the particles (Letizia and Phillips 1991). This number is maximal for the apoproteins consisting of helical repeats separated by beta-strands, i.e. apo A-I, A-IV and E (Brasseur et al. 1990, 1992). The apoprotein-phospholipid complexes generated with apo A-I and A-IV should therefore represent optimal acceptors for cholesterol, as observed in this study.

REFERENCES

Barkia A Puchois P Ghalim N Torpier G Barbaras R Ailhaud G and Fruchart JC (1991) Differential role of apolipoprotein A-I-containing particles in cholesterol efflux from adipose cells. Atherosclerosis 87:135-146

Basu SK Goldstein JL Anderson RGW and Brown MS (1976) Degradation of cationized low density lipoprotein and regulation of cholesterol metabolism in homozygous familial hypercholesterolemia fibroblasts. Proc Natl Acad Sci USA 73:3178-3182

Brasseur R De Meutter J Vanloo B Goormaghtigh E Ruysschaert JM and Rosseneu M (1990) Mode of assembly of amphipathic helical segments in model high density lipoproteins. Biochim Biophys Acta 1043:245-252

Brasseur R Lins L Vanloo B Ruysschaert JM and Rosseneu M Molecular modelling of the amphipathic helices of the plasma apolipoproteins. Proteins:.Structure, Function, and Genetics 13:246-257

Brown MS and Goldstein JL (1983) Lipoprotein metabolism in the macrophage- Implication for cholesterol deposition in atherosclerosis. Annu Rev Biochem 52:223-261

Bury J and Rosseneu M (1988) Apolipoprotein quantification by ELISA: Technical aspects and clinical applications. Rev.Immunoassay Techn. 1:1-25.

Eisenberg S (1984) High density lipoprotein metabolism. J. Lipid Res 25:1017-1058

Hara H and Yokoyama S (1991) Interaction of free apolipoproteins with macrophages. Formation of high density lipoprotein-like lipoproteins and reduction of cellular cholesterol. J Biol Chem 266:3080-3086

Jonas A (1986) Synthetic substrates of lecithin cholesterol acyltransferase. J Lipid Res 27:689-698

Letizia JY and Phillips MC (1991) Effects of apolipoproteins on the kinetics of cholesterol exchange. Biochemistry 30:866-873

Mahlberg FH Glick JM Jerome WG and Rothblat GH (1990) Metabolism of cholesteryl ester lipid droplets in a J774 macrophage foam cell model. Biochim Biophys Acta 1045:291-298

Matz CE Jonas A (1982) Micellar complexes of human apolipoprotein A-I with phosphatidylcholines and cholesterol prepared from cholate-lipid dispersions. J Biol Chem 257:4535-4540

Nowicka G Brüning T Böttcher A Kahl G and Schmitz G (1990) Macrophage interaction of HDL subclasses separated by free flow isotachophoresis. J Lipid Res 31:1947-1963

Phillips MC Johnson WJ and Rothblat GH (1987) Mechanisms and consequences of cellular cholesterol exchange and transfer. Biochim Biophys Acta 906:223-276

Ross R (1990) The pathogenesis of atherosclerosis- an update. N Engl J Med 314:488-495

Rosseneu M Van Biervliet JP Bury J and Vinaimont N (1989) Isolation and characterization of lipoprotein profiles in newborns by density-gradient ultracentrifugation. Pediatric Res 17:788-794

Vanloo B Morrison J Fidge N Lorent G Lins L Brasseur R Ruysschaert JM Baert J and Rosseneu M (1991) Characterization of the discoidal complexes formed between apo A-I-CNBr fragments and phosphatidylcholine. J Lipid Res 32:1253-1264
Vanloo B Taveirne J Baert G Lorent G Vercaemst R Lins L Ruysschaert JM and Rosseneu M Activation of Lecithin Cholesterol Acyl Transferase (LCAT) by the CNBr fragments of human apo AI and conversion of the discoidal complexes into spherical particles. Biochim Biophys Acta Accepted
Van Tornout P Caster H Lievens MJ Rosseneu M and Assmann G (1981) In vitro interaction of human HDL with human apolipoprotein A-II. Synthesis of apolipoprotein A-II-rich HDL. Biochim Biophys Acta 663:630-636
Vercaemst R Union A and Rosseneu M (1989a) Separation and quantitation of free cholesterol and cholesteryl esters in a macrophage cell line by high-performance liquid chromatography. J Chromatography 494:43-52
Vercaemst R Rosseneu M De Craene I De Backer G and Kornitzer M (1989b) Quantitation of plasma free cholesterol and cholesteryl esters by high performance liquid chromatography. Study of a normal population. Atherosclerosis 78:245-250
Yang C-y Gu Z-W Chong I Xiong W Rosseneu M Yang H-x Lee B-r Gotto AM and Chan L (1989). The primary structure of human apolipoprotein A-IV. Biochim.Biophys.Acta 1002:231-237.

Lipid Transport by Apolipoprotein AI: Structure and Function

E.L. Gong, A.V. Nichols and A. R. Tall[a]

Lawrence Berkeley Laboratory,
University of California
Berkeley, California 94720
USA

INTRODUCTION

Numerous epidemiologic studies have established a strong inverse relationship between plasma HDL levels and coronary heart disease (CHD) (Miller, 1987). Recent studies suggest that factors influencing levels and metabolism of HDL include both the composition and size of HDL and the fractional catabolic rates of apoAI or apoAII (Brinton et al., 1991). Factors modulating HDL composition and size include lipid transfer proteins (cholesteryl ester transfer protein {CETP} and phospholipid transfer protein {PLTP}) as well as enzymes, such as lecithin:cholesterol acyltransferase {LCAT} and lipases, hepatic triglyceride lipase {HTGL} and lipoprotein lipase {LPL}. Like mature HDL in human plasma, HDL precursors (designated as nascent HDL or nHDL) are also potentially subject to the above modulating agents. However, except for LCAT little is known what effect these factors may have on the structural and metabolic properties of the precursor particles or how such remodeling may influence the channeling of precursors towards specific HDL subpopulations in human plasma. LCAT-mediated transformation of apoAI-containing discoidal analogs of precursor nHDL to HDL-like products is characterized by a strong positive correlation between the size of the precursor particles and the size of their core-containing transformation products (Figure 1) (Nichols, 1990). Thus, modulation-induced change in size of nHDL particles by transfer proteins or enzymes would be expected to correspondingly shift the size of their LCAT-transformation products. In view of the association between HDL level and HDL particle size and composition, reduction in nHDL size may contribute to lower HDL levels and increased risk of CHD. Since nHDL lack core lipids and consist almost exclusively of phospholipid, unesterified cholesterol and apolipoproteins, their remodeling mainly involves phospholipid depletion as facilitated by CETP and PLTP and phospholipid lipolysis as facilitated by HTGL.

[a]Columbia University College of Physicians and Surgeons, 630 168th Street, New York, New York 10032, USA

NATO ASI Series, Vol. H 73
Human Apolipoprotein Mutants III
Edited by C. R. Sirtori, G. Franceschini, B.H. Brewer Jr.

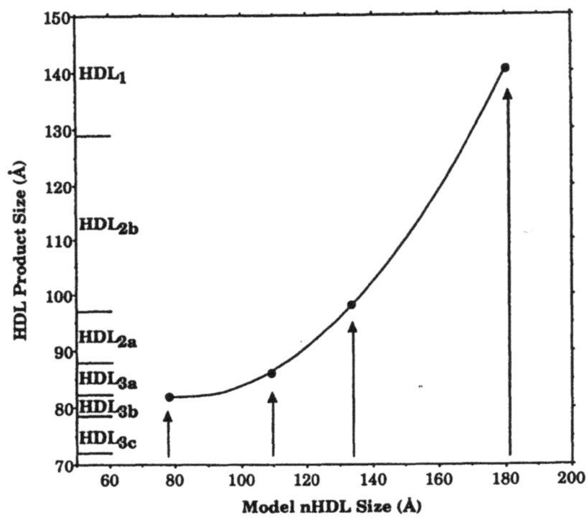

Figure 1. Correlation (r=0.95) of model nHDL discoidal precursor particle size (Å) and size of LCAT-mediated transformation product (Å) following incubation (18hr, 37°C) of nHDL with partially purified LCAT and LDL as a source of cholesterol (Nichols, 1990). Size intervals of HDL subpopulations are indicated for reference.

nHDL from various sources exhibit discrete species of discoidal and small (77Å) particles with unique apolipoprotein stoichiometry (Nichols et al., 1987a). Like plasma HDL, nHDL exhibit apolipoprotein-specific populations including, nHDL(AIw/oAII) {nascent HDL containing apoAI without apoAII} and nHDL(AIwAII) {nascent HDL containing apoAI with apoAII} that are comprised, to a greater or lesser extent, of discoidal and small particles (Nichols et al., 1989). Discoidal nHDL(AIw/oAII) species can be modeled in vitro by reassembly of apoAI with phosphatidylcholine (PC) into complexes with and without unesterified cholesterol (UC) using detergent-dialysis procedures (Matz and Jonas, 1982; Nichols et al., 1983b). The present study is concerned with three classes of complexes identified by their number of apoAI per complex: 2AI-class, 3AI-class and 4AI-class. Subclasses within each class exhibit unique sizes, molar ratios of PC:apoAI and, based on a recently proposed model of discoidal complexes (Jonas et al., 1989), may contain a specific number of helices per apoAI in their structure.The discoidal model hypothesis proposes that the unique size and phospholipid content of a specific subclass is largely determined by its number of apoAI molecules and by the number of helices per apoAI assumed by apoAI upon interaction with the phospholipid bilayer. We have examined the fit of the model to major subclasses we have identified during reassembly of complexes within the 2AI-, 3AI-and 4AI-classes. In addition, we have used the model to predict properties of hypothetical discoidal complexes, having a specific number of apoAI and helices per apoAI, but which appear either to be unstable or to occur as minor or trace components under the reassembly conditions used. Table 1 presents properties of both experimentally-defined and predicted minor subclasses of

Table 1
Phosphatidylcholine-Apolipoprotein AI Discoidal Complex Classes[a]

	4AI-Class			3AI-Class			2AI-Class		
Subclass size (Å)	170±3[b]	(149)[c]	(130)	130±3	(115)	(101)	95±2	86	77
Molar ratio PC/AI	155/1	(124/1)	(87/1)	119/1	(90/1)	(59/1)	83/1	(55/1)	40/1
Helices/AI	(8)	(7)	(6)	(8)	(7)	(6)	(8)	(7)	(6)
Nomenclature:									
Nichols et al.	4AI-170Å	(4AI-149Å)	(4AI-130Å)	3AI-130Å	(3AI-115Å)	(3AI-101Å)	2AI-95Å	(2AI-86Å)	2AI-77Å
Jonas et al. (8)[d]							rHDL3	rHDL2	rHDL1

[a] Experimental procedures: Preparation and characterization of complexes (Matz and Jonas, 1982; Nichols et al., 1983b); gradient gel electrophoresis (GGE) (Nichols et al., 1983a); number of apoAI/complex (Swaney, 1980); PC content/AI in predicted complexes and estimated number of helices/AI in all complexes (Jonas et al., 1989)
[b] ± Standard deviation
[c] Values in parentheses derived from discoidal model (Jonas et al., 1989)
[d] Complexes contain unesterified cholesterol

within each of the three classes. Figure 2 shows the size distributions of the 5 subclasses experimentally produced in different molar ratio mixtures of PC and apoAI. Based on the discoidal complex model, addition or depletion

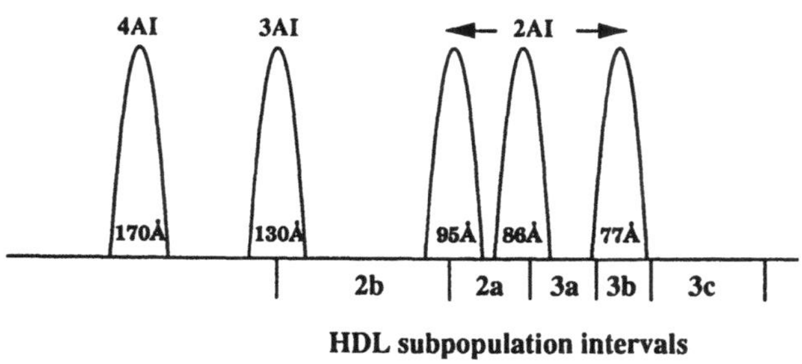

Figure 2. Schematic representation of discoidal complex classes isolated from mixtures of PC:apoAI at different molar ratios. Number of apoAI per particle for each subclass is noted, as well as the size of the complexes relative to HDL subpopulation size.

phospholipid from a subclass, within a specific class of complexes, would be expected to increase or reduce, respectively, the number of helices per apoAI within the structure of the product complexes. At the present time, it is not clear whether phospholipid depletion of 4AI-subclasses or 3AI-subclasses produces intraclass or interclass conversions or both. Interclass conversions may involve particle destabilization leading to production of subclasses with lower number of apoAI per particle.

In the present study, we investigated phospholipid depletion using the largest subclass experimentally identified within each of the classes, specifically the 2AI-95Å, 3AI-130Å and 4AI-170Å subclasses. Depletion was effected by spontaneous and CETP-mediated transfer of phospholipid to human plasma HDL3 (Tall et al., 1987). HDL3 was used as PC acceptor in view of its avidity for vesicle and chylomicron phospholipid. Specific questions addressed in the present study were: [1] does CETP facilitate phospholipid depletion of discoidal complexes, [2] do intraclass conversions occur and are the products consistent with the discoidal complex model, and [3] do interclass conversions occur and what products are formed?

Spontaneous and CETP-Mediated Conversion of 2AI-Class Complexes

According to Table 1, the 2AI-class of PC-apoAI complexes includes 3 subclasses: 2AI-95Å; 2AI-86Å and, 2AI-77Å, with a predicted number of helices per apoAI of 8, 7 and 6, respectively. All of these subclasses are formed in reassembly mixtures with 75:1 PC:apoAI molar ratio. Spontaneous transfer of PC from the 2AI-95Å to HDL3 over a period of 24 hr results in its progressive conversion to a smaller product with size corresponding to the 2AI-77Å subclass (Fig. 3A,B). The size and

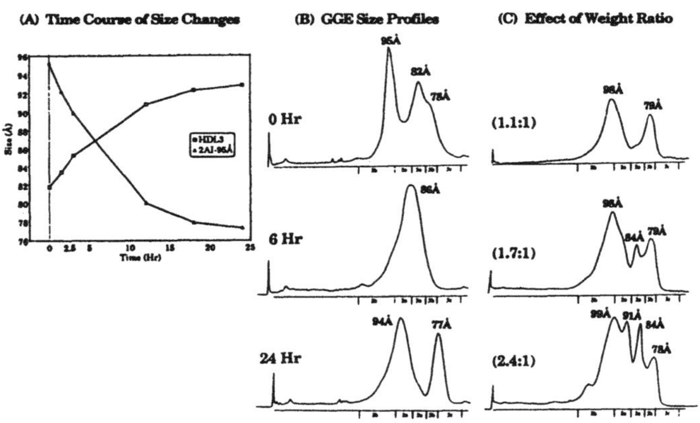

Figure 3. Spontaneous PC transfer between 2AI-95Å subclass and HDL_3. Weight ratio of PC (subclass): protein (HDL_3) in interaction mixture (37°C) was 1.1:1 and was used in all subsequent studies unless otherwise noted. Subclass complexes were prepared as in Table 1; HDL_3 were isolated ultracentrifugally (Anderson et al., 1977). Analyses were performed as in Table 1. (A) Time course of size changes; (B) size profiles at 0, 6 and 24 hr; (C) size profiles at 18 hr when weight ratio of PC (subclass):protein (HDL_3) in mixture was increased.

composition of this product (2AI; 77Å; 33:9:1, PC:UC:AI) are comparable to those of small (77Å) particles identified in native nHDL (Nichols et al., 1987a). The UC observed in the conversion product was picked up from the HDL_3 during PC depletion. Based on area measurements from protein-stained profiles (from gradient gel electrophoresis), there appears to be little dissociation of apoAI during this conversion, indicating that the process is primarily one of PC depletion from the complexes. Reflecting its uptake of PC, HDL_3 size increases and approaches a plateau when conversion of the complexes nears completion. Higher levels of 2AI-95Å in the interaction mixture result in saturation of HDL3 by PC and the appearance of conversion products now including not only the 2AI-77Å but also 2AI-84Å particles (Fig. 3C), both of which are predicted by the discoidal model based on number and conformation of apoAI molecules. These observations suggest that PC transfer can drive sequential conversion of the 2AI-95Å to the other stable subclasses of the 2AI-class (2AI-86Å and 2AI-77Å, with predicted helices/AI of 7 and 6, respectively). The above observations also suggest that small 2 apoAI-particles as observed in native nHDL may arise in part from conversion of discoidal particles driven by PC transfer.

Addition of CETP markedly accelerated PC transfer to HDL3 and conversion of the 2AI-95Å (Fig. 4A,B). Under the same conditions shown in Figure 3, at 3 hr the CETP-mediated decrease in size of the 2AI-95Å was approximately fourfold greater than the decrease from spontaneous transfer. Similar to the conversion observed during spontaneous transfer, CETP-mediated conversion was associated with minimal change in apoAI content of the complexes.

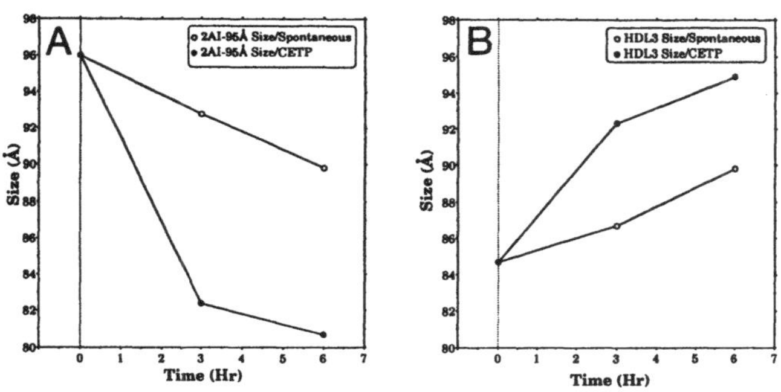

Figure 4. CETP-mediated PC transfer between 2AI-95Å subclass and HDL3. Experimental conditions were as described in Table 1 and Fig. 3. CETP, prepared by Dr. Alan Tall, was recombinant CETP expressed by CHO cell line; partially purified CETP was prepared from human plasma (Ohnishi et al., 1990). Time course of size changes in (A) 2AI-95Å subclass; (B) HDL3.

Spontaneous and CETP-Mediated Conversion of 3AI-Class Complexes

According to Table 1, the 3AI-class of PC-apoAI complexes includes 3 subclasses: 3AI-130Å; 3AI-115Å and, 3AI-101Å, with predicted number of helices of 8, 7 and 6 per apoAI, respectively. The 3AI-130Å is formed in PC-AI reassembly mixtures prepared at a PC:AI molar ratio of 200:1. On the other hand, the 3AI-115Å and 3AI-101Å are not detected under the conditions used. Reasssembly of a stable 3AI-109Å complex requires incorporation of a significant amount of unesterified cholesterol and yields a discoidal complex with molar composition of approx. 78:22:1 (PC:UC:AI)

(Jonas et al., 1989; Nichols et al., 1987a). Thus, except for the 3AI-130Å, the 3AI-subclasses predicted by the discoidal model for PC-apoAI complexes appear unstable under the reassembly conditions used. Smaller 3AI-complexes can be formed under these conditions but their formation requires incorporation of UC.

Spontaneous and CETP-mediated PC transfer from the 3AI-130Å to HDL3 was evaluated in mixtures with the same level of complex PC as in experiments with the 2AI-95Å. Unlike our observations with the 2AI-subclass, the rate of size decrease of the 3AI-subclass was similar both in the presence and absence of CETP (Fig. 5A). In the course of a 6 hr incubation with or without CETP, the size of the 3AI-130Å was reduced to 122Å which may correspond to that of the predicted 3AI-115Å (7 helices/AI) complex. No peaks were observed corresponding to a 3AI-101Å (6 helices/AI) complex, although its peak may have merged with that of the PC-enriched HDL3. Compared to spontaneous transfer, the increase in HDL3 size was markedly greater in the presence of CETP (Fig. 5B), indicating that considerable transfer to the HDL3 was facilitated by CETP. However, this transfer occurred without facilitating size reduction of the 3AI-130Å. Interestingly, the increased transfer to HDL3 was correlated with a dramatic decrease in peak area of the 3AI-130Å (Fig. 6A). Thus, in the presence of CETP, the 3AI-130Å peak area decreased 77% at 6 hr compared to a 20% decrease without CETP. The redistribution of subclass mass appeared mainly as PC enrichment of the HDL3 and appearance of additional peaks (approx. 21% of total profile area) primarily in the size

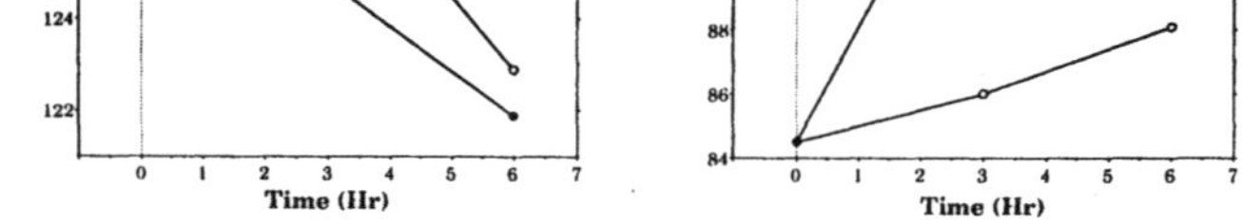

Figure 5. CETP-mediated PC transfer between 3AI-130Å subclass and HDL3. Conditions and analyses as in Table 1 and Fig. 4. Time course of size changes in (A) 3AI-130Å subclass; (B) HDL3.

interval (HDL3b/c) containing the smaller 2AI-subclasses (Fig.6B). CETP-mediated interclass conversion of the 3AI-subclass complexes is indicated, since product particles in the HDL3b/c size interval would be expected to have 2 apoAI per particle. Thus, in the case of the 3AI-130Å, CETP appears to directly facilitate interclass conversion of the subclass with minimal

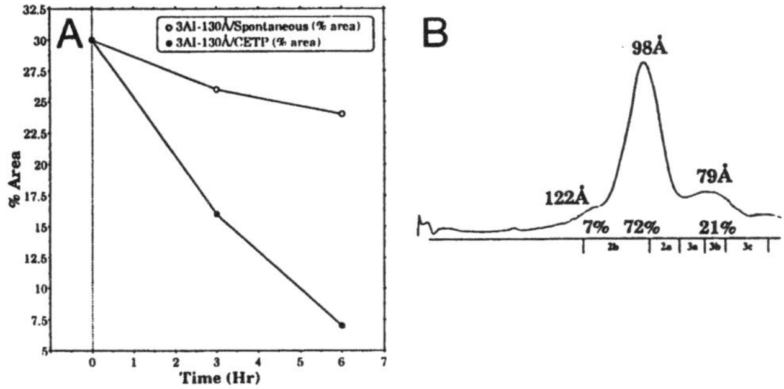

Figure 6. CETP-mediated PC transfer between 3AI-130Å subclass and HDL_3. Conditions and analyses as in Fig. 4. (A) Time course of change in subclass peak area expressed as % of total size profile area; (B) total size profile of interaction mixture at 6 hr, with % distribution of peak areas indicated.

influence on intraclass conversion. This suggests that the transfer mechanism may include CETP-mediated fusion of the 3AI-130Å with HDL_3 resulting in extensive PC transfer and interclass conversion.

Spontaneous and CETP-Mediated Conversion of 4AI-Class Complexes

According to Table 1, the 4AI-class of PC-apoAI complexes includes 3 subclasses: 4AI-170Å; 4AI-149Å; and 4AI-130Å, with a predicted number of helices per apoAI of 8, 7 and 6, respectively. The 4AI-170Å is formed in PC-AI reassembly mixtures prepared at molar ratio of 200:1. The other two 4AI-subclasses are not detected in these mixtures under the reassembly conditions used and their properties in Table 1 were derived using the discoidal model. Possibly, as with 3AI-complexes, incorporation of UC may be required to stabilize smaller 4AI-complexes.

Spontaneous and CETP-mediated transfer of PC from the 4AI-170Å to HDL_3 was investigated in interaction mixtures providing complex PC at the same level as in experiments with the 2AI-95Å. As with the 2AI-subclass, CETP markedly increased the rate of size change in both the 4AI-170Å and the HDL_3 (Fig. 7A,B). CETP-mediated PC transfer promoted intraclass conversion of the 4AI-170Å to the 4AI-147Å whose size corresponds closely to that of the predicted 4AI-149Å (7 helices/AI) (Table 1). During spontaneous transfer, a 149Å conversion product is attained only after 18 hr incubation. At much higher levels of HDL_3 relative to that of the 4AI-170Å in the interaction mixture, a product with size of 135Å is formed after 18 hr of spontaneous transfer. This particle may correspond to the predicted 4AI-130Å (6 helices/AI) shown in Table 1. In the presence of CETP, the size

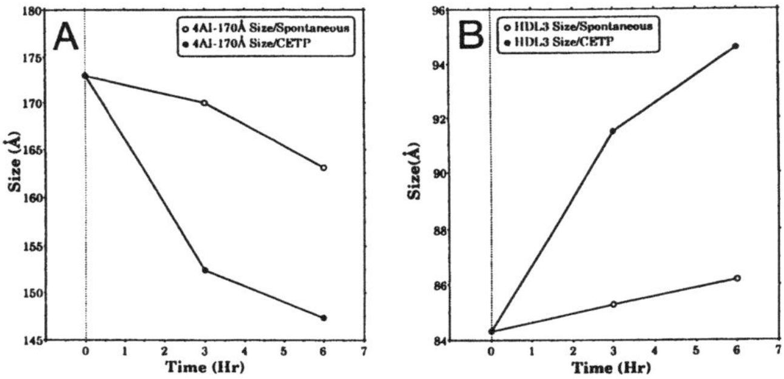

Figure 7. CETP-mediated PC transfer between 4AI-170Å subclass and HDL_3. Conditions and analyses as in Fig. 4. Time course of size changes in (A) 4AI-170Å subclass; (B) HDL_3.

increase in HDL3 approximates that observed with 2AI-95Å indicating comparable PC transfer. At 3 hr, the CETP-mediated decrease in size of the 4AI-170Å, relative to the spontaneous decrease, is approximately five-to-sevenfold greater and at 6 hr approaches a limiting size of 147Å. A major feature of the CETP-mediated intraclass conversion of the 4AI-170Å is a concurrent and marked reduction in its peak area of approx. 60% at 6 hr (Fig. 8A). During spontaneous transfer, the decrease in this subclass' size is considerably less and is associated with a minimal decrease in peak area. In summary, these results suggest that the 4AI-170Å subclass undergoes intraclass conversions which are associated with PC transfer and with changes in AI conformation. However, in addition to promoting intraclass conversion, CETP also promoted extensive interclass redistribution of its protein and phospholipid mass. The size distribution of

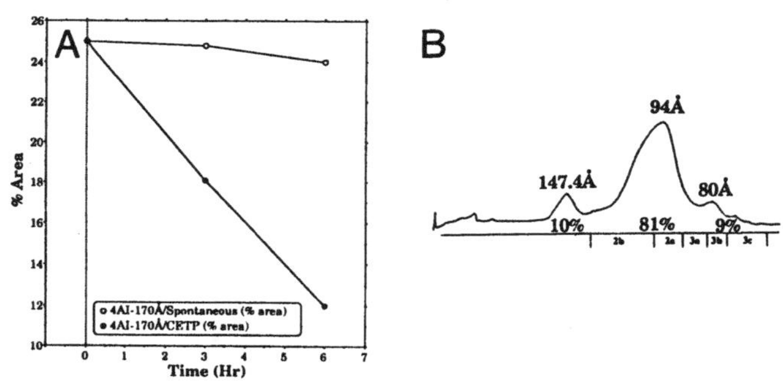

Figure 8. CETP-mediated PC transfer between 4AI-170Å subclass and HDL_3. Conditions and analyses as in Fig. 4. (A) Time course of change in subclass peak area expressed as % of total size profile area; (B) total size profile of interaction mixture at 6 hr, with % distribution of peak areas indicated.

products formed during CETP-mediated conversion of the 4AI-170Å was evaluated from protein-stained gradient gels after electrophoresis of the interaction mixture (Fig. 8B). The reductions in peak area correlate with area increases within the size interval of the PC-enriched HDL_3 and within the size interval of smaller HDL particles, namely the $HDL_{3b/c}$. Since complexes in the $HDL_{3b/c}$ interval are primarily the smaller 2AI-subclasses, these results would be consistent with CETP-induced interclass conversion. The size interval containing the PC-enriched HDL_3 also may include 2AI-subclass products, such as the larger 2AI-95Å.

SUMMARY

The present study utilized model discoidal complexes 2AI-95Å, 3AI-130Å, and 4AI-170Å to evaluate structural remodeling (intraclass and interclass conversion) in response to spontaneous and CETP-mediated PC transfer to HDL_3 (Table 2). Compositional and structural properties of the

Table 2
Intraclass Conversions During PC Depletion

	4AI-Class			3AI-Class			2AI-Class		
Size of conversion products (Å)	170 →	147 →	135	130 →	122 →	?	95 →	84 →	77
Predicted # of helices/AI	8	7	6	8	7	6	8	7	6
Complex sizes (Å) predicted from # of apoAI & helices/AI	168	149	130	130	115	101	91	82	72
Complex sizes (Å) in reassembly mixtures	170	---	---	130	---	---	95	86	77

substrate complexes were consistent with those predicted by the discoidal complex model (Jonas et al., 1989). Properties of product complexes generated by PC depletion were also generally consistent with those predicted by the model. Under the experimental conditions used, CETP markedly facilitated transfer of PC from all of the above complexes to HDL_3. With the 2AI-95Å subclass, CETP facilitated intraclass conversion associated with the PC transfer. With the larger 3AI-130Å and 4AI-170Å subclasses, PC transfer was associated with interclass conversions apparently involving fission with formation of smaller products having a lower number of apoAI per particle. Our studies suggest that CETP-mediated and, most likely, PLTP-mediated remodeling of nHDL can have significant impact on precursor size, number of apoAI and apoAI conformation. By affecting precursor size, remodeling can significantly influence the size of HDL particles (whether HDL_2 or HDL_3) resulting from LCAT-mediated transformation of the remodeled precursor particles.

ACKNOWLEDGEMENTS

This work was supported by National Institutes of Health Grants HL 46281-01 (A.V.N.) and HL 22682 (A.R.T.), and was conducted at the Lawrence Berkeley Laboratory (Department of Energy Contract DE-AC03-76SF00098 to the University of California).

REFERENCES

Anderson DW, Nichols AV, Forte TM, Lindgren FT (1977) Particle distribution of human serum high density lipoproteins. Biochim Biophys Acta; 493: 55-68

Brinton EA, Eisenberg S, Breslow JL (1991) Increased apo A-I and apo A-II fractional catabolic rate in patients with low high density lipoprotein-cholesterol levels with or without hypertriglyceridemia. J Clin Invest; 87: 536-544

Jonas A, Kezdy KE, Wald JH (1989) Defined apolipoprotein A-I conformations in reconstituted high density lipoprotein discs. J Biol Chem; 264: 4818-4824

Matz CE, Jonas A (1982) Micellar complexes of human apolipoprotein A-I with phosphatidylcholines and cholesterol prepared from cholate-lipid dispersions. J Biol Chem; 257: 4535-4540

Miller NE (1987) Associations of high-density lipoprotein subclasses and apolipoproteins with ischemic heart disease and coronary atherosclerosis. Am Heart J; 113: 589-597

Nichols AV, Blanche PJ, Gong EL (1983) Gradient gel electrophoresis of human plasma high density lipoproteins. In: Lewis L, Opplt J, eds. CRC Handbook of Electrophoresis Vol III. Boca Raton: CRC Press; 29-47

Nichols AV, Gong EL, Blanche PJ, Forte TM, Shore VG (1987) Origin and heterogeneity of HDL subspecies. In: Lippel K, ed. Proceedings of the Workshop on Lipoprotein Heterogeneity. NIH Publication No. 87-2646; 331-340

Nichols AV, Gong EL, Blanche PJ, Forte TM (1989) HDL populations: Origins and interconversions. In: Miller NE, ed. Proceedings of the 2nd International Workshop on HDL and Atherosclerosis. Amsterdam: Elsevier; 159-171

Nichols AV, Gong EL, Blanche PJ, Forte TM, Shore VG (1987) Pathways in the formation of human plasma high density lipoprotein subpopulations containing apolipoproteins AI without apolipoprotein AII. J Lipid Res 28: 719-732

Nichols AV, Gong EL, Blanche, Forte TM (1983) Characterization of discoidal complexes of phosphatidylcholine, apolipoprotein A-I and cholesterol by gradient gel electrophoresis. Biochim Biophys Acta 750: 353-364

Nichols AV (1990) Conversions in the origins and metabolism of human plasma HDL. In: Esfahani M, Swaney J, eds. Advances in Cholesterol Research. Caldwell: Telford Press; 315-365

Ohnishi T, Yokoyama S, Yamamoto A (1990) Rapid purification of human plasma transfer proteins. J Lipid Res 31: 397-406

Swaney JB (1980) Properties of lipid-apolipoprotein association products. J Biol Chem 255: 877-881

Tall A, Swenson T, Hesler C, Granot E (1987) Mechansims of facilitated lipid transfer mediated by plasma lipid transfer proteins. In: Gotto AM Jr., ed. Plasma Lipoproteins. Amsterdam: Elsevier; 277-297

CHARACTERIZATION OF LIPOPROTEIN PARTICLES CONTAINING APOLIPOPROTEIN A-IV

G.R. Castro, N. Duverger, P. Duchateau, J.C. Fruchart
SERLIA & INSERM U.325
Institut Pasteur
1, rue du Professeur Calmette
59019 Lille Cédex
France

INTRODUCTION

Human apolipoprotein A-IV (apo A-IV) is a 46,000 KD protein which is synthesized predominantly in the small intestine (Green et al., 1979 and 1980).

Although the precise physiologic function of apo A-IV is not yet known, several lines of evidence suggest that it plays an important role in the transport of cholesterol from peripheral organs to the liver for excretion. Consistent with this hypothesis, apo A-IV has been shown to activate lecithin-cholesterol acyltransferase (Steinmetz et al., 1985 ; Chen et al., 1985), to bind specifically to both hepatocytes (Dvorin et al, 1986 ; Ghiselli et al, 1986) and endothelial cells (Savion et al., 1987 ; Savion et al., 1988), and to promote cholesterol efflux from fibroblasts (Stein et al., 1986) and adipocytes (Steinmetz et al., 1990).

Early reports using traditional methods such as ultracentrifugation or agarose gel filtration (Green et al., 1979 ; Beisiegel et al., 1979 ; Bisgaier et al., 1985) led to the conclusion that very little plasma apo A-IV is lipoprotein-associated. In contrast, Lagrost et al. (1989) using high

NATO ASI Series, Vol. H 73
Human Apolipoprotein Mutants III
Edited by C. R. Sirtori, G. Franceschini, B.H. Brewer Jr.

performance gel filtration techniques recently reported a high proportion of apo A-IV (about 70 %) in the large size high density lipoprotein (HDL) subfraction. However, dissociation of apo A-IV from, or artifactual association with lipoproteins could have been induced by such fractionation techniques of plasma (Weinberg et al., 1992). We have used a mild procedure i.e., immunoaffinity chromatography, to isolate apo A-IV containing particles. This procedure can accurately define the molecular association of the more labile apolipoproteins in native plasma, which are unstable to classical fractionation techniques (Castro et al., 1984).

Apo A-IV containing lipoprotein particles were isolated and characterized from plasma of :

a) 5 normolipidemic subjects with apo A-IV, 1/1 phenotype ;

b) 1 normolipidemic subject with apo A-IV, 2/2 phenotype ;

c) 5 homozygous Tangier patients (provided by Dr. H.B. Brewer).

For each of the characterized fractions, their metabolics activities were measured : lecithin-cholesterol acyltransferase (LCAT) activity, cholesterol ester transfer protein (CETP) activity, and their ability to promote the efflux of cholesterol from cholesterol loaded adipocytes. Results of these studies indicate that these particles may be involved in the reverse cholesterol transport process.

METHODS

<u>Isolation of lipoprotein particles</u>

The plasma samples were sequentially passed through anti-apo A-IV ànd anti apo A-I immunoaffinity columns to isolate the two populations of apo A-IV containing lipoproteins i.e., those particles containing both apo A-IV and A-I [Lp(A-IV w A-I)] and those containing apo A-IV but no apo A-I [Lp(A-IV w/o A-I)]. Both types of particles were finally passed over a heparin-sepharose column to eliminate the E containing particles. Details of the procedure have been published elsewhere (Steinmetz et al., 1990).

After dialysis and concentration, the lipid and apolipoprotein composition of the apo A-IV containing particles was determined. Total cholesterol, free cholesterol, triglycerides and phospholipids were measured by specific test kits (Boehringer Mannheim, Germany). Protein concentration was measured by the method of Lowry (Lowry et al., 1984). Apolipoproteins concentrations were measured using established specific immunoassays (Fruchart et al., 1985). The size of the particles was determined by gel filtration chromatography using a Fast Protein Liquid Chromatography system equipped with a Superose 6B column.

The activity of LCAT was measured by the method of Chen and Albers (1982) and that of CETP by the method of Tollefson et al. (1986). Efflux of cell cholesterol was measured as described previously (Steinmetz et al., 1990).

RESULTS

a) Normolipemics subjects - Phenotype apo A-IV 1/1

The compositions of apo A-IV containing particles from these subjects are shown in Table 1. The particles are relatively poor in lipids. In the Lp(A-IV w/o A-I) particles apo A-IV is the major apolipoprotein whereas in Lp(A-IV w A-I) particles the major apolipoprotein is apo A-I. In addition to the assayed apolipoproteins others proteins apo D, apo H, LCAT and CETP were also detected by specific immunoblotting.

Table 1 : Protein and lipid composition, and apolipoprotein composition of apo A-IV containing particles

		Lp(A-IV w A-I)	Lp(A-IV w/o A-I)
Protein	Weight %	82.8 ± 2.2	82.6 ± 2.3
Total Cholesterol		4.2 ± 0.7	4.6 ± 0.5
Free Cholesterol		1.1 ± 0.5	1.5 ± 0.6
Triglyceride		4.5 ± 1.4	6.6 ± 3.1
Phospholipid		8.4 ± 3.0	5.9 ± 4.0
Apo A-I	Molar %	70.1 ± 7.0	0
Apo A-II		16.9 ± 4.1	4.5 ± 3.4
Apo A-IV		11.8 ± 3.5	94.9 ± 2.1
Apo B		0.2 ± 0.1	0.3 ± 0.1
Apo C-III		0.9 ± 0.8	0.3 ± 0.2
Apo E		0	0

Particles isolated from plasma of 5 normolipidemic subjects phenotype apo AIV 1/1.
Results are given as mean ± SD.

By gel filtration the particles separated into two major subpopulations with Stokes diameters of 9.3 and 10.3 nm. Both populations showed a minor subpopulation of 29.9 nm and the Lp(A-IV w/o A-I) fraction had a further minor subpopulation of 16 nm.

LCAT activity for both particles was similar with a mean value of 0.2 µmol of cholesterol esterified per hour per µg of protein particle.

For CETP activity, the Lp(A-IV w/o A-I) particles showed a higher activity than the Lp(A-IV w A-I), 18,3 and 8,4 % respectively.

Both groups of particles demonstrated ability to remove cholesterol from cholesterol-loaded adipocytes (30-40 % after 3 hours). However, the rate of efflux was not significantly different between the two populations (at 1 and 3 hours incubation).

b) Normolipemic subject - Phenotype apo A-IV 2/2

Genetic polymorphism of apo A-IV has been demonstrated. The gene frequencies of the two most common phenotypes A-IV 1/1 (0.94 - 0.88) and A-IV 2/2 (0.06 - 0.12) have been determined in several populations (Menzel et al., 1982 ; Kamboh et al., 1982). Lohse et al. (1990) have recently reported that the molecular basis for the apo A-IV-2 polymorphism is a G→T mutation in the A-IV-2 allele, which converts the glutamine at position 360 of the mature protein to a histidine. This mutation adds one positive charge unit to the apo A-IV-1 iso-protein. It has been proposed that apo A-IV-2 has more α-helical structure, higher affinity for phospholipid surfaces and an increased (Weinberg et al., 1990) or decreased (Tenkanen et al., 1991) catalytic efficiency of LCAT. Populations studies have shown that the two alleles at the apo A-IV locus influence HDL cholesterol and triglycerides levels (Menzel et al., 1988 ; De Knijff et al., 1988). It was therefore interesting to study the apo A-IV particles in a subject homozygous for the allele A-IV-2.

The particles isolated from the plasma of the apo A-IV 2/2 subject were similar in composition and metabolic activities to the particles from apo A-IV 1/1 subjects. However, the Lp(A-IV w/o A-I) particles of this subject had significantly less lipid bound to the apo A-IV that the phenotype 1/1 particles.

c) Homozygous Tangier patients

In Tangier homozygotes, plasma apo A-I and apo A-II levels are approximately 1% and 7% respectively of the normal values. The resulting low concentration of HDL is also extremely abnormal in composition (Herbert et al., 1983). However the risk of premature coronary artery disease is only modestly increased, suggesting that the pathway of reverse cholesterol transport remains effective. Since we have found the apo A-IV plasma concentration in Tangier patients to be significantly higher than the normal values (0.27 ± 0.12 vs 0.14 ± 0.02 g/l), and because apo A-IV particles in normal subjects are highly efficient in the steps of the reverse transport of cholesterol, we have characterized the potential of particles from plasma of 5 homozygous Tangier patients to promote this process.

Taking into account the extremely low levels of apo A-I of these patients, we omitted the apo A-I immunoaffinity chromatography step.

The Table 2 shows the composition of the Tangier particles, they are very heterogeneous, and contain more triglycerides and less protein than the normal particles. These particles demonstrated lower LCAT activity than the normal particles (0.1 ± 0.05 vs 0.21 ± 0.03 µmol/h/µg) respectively, but they were equally effective in promoting the efflux of cell cholesterol.

Table 2 : Protein and lipid composition, and apolipoprotein composition of apo A-IV containing particles LpA-IV from 5 Tangier patients

Protein	Weight %	59.8 ± 24
Cholesterol		3.6 ± 2
Triglyceride		29.6 ± 20
Phospholipid		6.6 ± 3
Apo A-I	Molar %	8.2 ± 6.7
Apo A-II		16 ± 23
Apo A-IV		60 ± 30
Apo C-III		16 ± 14

These particles were immunologically deplected of apo B and apo E.
Results are given as mean ± SD.

DISCUSSION

We have characterized apo A-IV-containing particles isolated from human plasma by immunoaffinity chromatography. When the normal apo A-IV containing particles were applied to anti apo A-I immunosorbent 30 % of apo A-IV was retained, we called this fraction Lp(A-IV w A-I) even though apo A-I is the major apolipoprotein. The non-retained fraction was called Lp(A-IV w/o A-I). Both types of particles have a low lipid content, which may be due to the reported`weak association of the apolipoprotein A-IV to lipids (Weinberg et al., 1990).

The particles are heterogenous in size and protein composition ; they contain LCAT and CETP and are able to promote the efflux of cholesterol from murine adipocytes preloaded with cholesterol.

The particles isolated from the plasma of a subject with phenotype apo A-IV 2/2 showed no apparent differences.

The particles isolated from plasma of homozygous Tangier subjects demonstrated a capacity to promote the efflux of cell cholesterol, equivalent to that observed in the normolipidemic subjects.

We propose that the apo A-IV containing lipoprotein particles participate in the pathway of return of cholesterol to the liver, and because the fast turnover of apo A-IV (Ghiselli et al., 1986 ; Malmendier et al., 1991) this process could be highly effective in the removal of cholesterol from peripheral tissues, and could explain the lack of severe CHD in Tangier subject.

Moreover, by two dimensional electrophoresis we found that most of the plasma apo A-IV is present in particles with pre-ß mobility (Castro G, unpublished data), close to that of pre-ß-HDL reported to be the primary acceptor of cell cholesterol (Castro et al., 1988).

REFERENCES

Beisiegel U, Utermann G (1979) Eur J Biochem 93: 601-608.

Bisgaier CL, Sachdev OP, Megna L, Glickman RM (1985) J Lipid Res 26: 11-25.

Castro GR, Fielding CJ (1984) J Lipid Res 25: 58-67.

Castro GR, Fielding CJ (1988) Biochemistry 27: 25-29.

Chen CH, Albers JJ (1982) J Lipid Res 23: 680-691.

Chen CH, Albers JJ (1985) Biochim Biophys Acta 836: 279-285.

De Knijff P, Rosseneu M, Beisiegel U, De Keersgieter W, Frants RR, Havekes LM (1988) J Lipid Res 29: 1621-1627.

Dvorin E, Gorder NL, Benson DM, Gotto A (1986) J Biol Chem 261: 15714-15718.

Fruchart JC, Fievet C, Puchois P (1985) in Methods of Enzymatic Analysis Academic Press, New York, pp 126-128.

Ghiselli G, Crump WL III, Gotto A (1986) Biochem Biophys Res Commun 139: 122-128.

Ghiselli G, Krishnan S, Beigel Y, Gotto A (1986) J Lipid Res 27: 813-827.

Green, PHR., Glickman, RM., Saudek, CD., Blum, CB., and Tall, AR (1979) J Clin Invest 64: 233-242.

Green, PHR., Glickman, RM., Riley, JN., and Quinet, E (1980) J Clin Invest 65: 911-919.

Herbert PN, Assmann G, Gotto A, Fredrickson DS (1983) 5th Ed, The metabolic basis of inherited disease, New York, Mc Graw-Hill, pp 589-621.

Kamboh MI, Ferrell RE (1987) Am J Hum Genet 41: 119-127.

Lagrost L, Gambert P, Boquillon M, Lallemant C (1989) J Lipid Res 30: 1525-1534.

Lohse P, Kindt MR, Rader DJ, Brewer HB (1990) J Biol Chem 265: 10061-10064.

Lowry OH, Rosebrough NJ, Farr AL, Rawdall RJ (1951) J Biol Chem 193: 265-275.

Malmendier C, Lontie JF, Lagrost L, Delcroix C, Dubois DY, Gambert P (1991) J Lipid Res 32: 801-808.

Menzel HJ, Kovary PM, Assmann G (1982) Hum Genet 62: 349-351.

Menzel HJ, Boerwinkle E, Schrangl-Will S, Utermann G (1988) Hum Genet 79: 368-372.

Savion N, Gamliel A, Tauber JP, Gospodarowicz D (1987) Eur J Biochem 164: 435-443.

Savion N, Gamliel A (1988) Arteriosclerosis 8: 178-186.

Stein O, Stein Y, Lefebre M, Roheim PS (1986) Biochim Biophys Acta 878: 7-13.

Steinmetz A, Utermann G (1985) J Biol Chem 260: 2258-2264.

Steinmetz A, Barbaras R, Ghalim N, Clavey V, Fruchart JC, Ailhaud G (1990) J Biol Chem 265: 7859-7863.

Tenkanen H, Lukka M, Lauhiainen M, Metso J, Baumann M, Peltonen L, Ehnholm C (1991) Arteriosclerosis 11: 851-856.

Tollefson JH, Albers JJ (1986) Methods Enzymol 129: 797-817.

Weinberg RB, Jordan MK (1990) J Biol Chem 265: 8081-8086.

Weinberg RB, Jordan MK, Steinmetz A (1990) J Biol Chem 265: 18372-18378.

Weinberg RB, Ibdah JA, Phillips M (1992) J Biol Chem 267: 8977-8983.

STRUCTURAL AND FUNCTIONAL PROPERTIES OF APO A-I VARIANTS IN RECONSTITUTED HDL

A. Jonas[a], A. von Eckardstein[b], W. W. Mantulin[c], and G. Assmann[b,d]
[a]Department of Biochemistry, College of Medicine,
University of Illinois at Urbana-Champaign
506 South Mathews Avenue
Urbana, IL 61801
U.S.A.

INTRODUCTION

Apolipoprotein A-I (apo A-I), the major protein component of high density lipoproteins (HDL) defines the structures and functions of these particles. Apo A-I determines the size of HDL subclasses, activates the reaction of lecithin cholesterol acyltransferase (LCAT) with HDL, and may participate in lipid transfers, and in the binding of HDL to cell surfaces. More than 20 natural mutants of apo A-I are known (von Eckardstein et al., 1990); most of these mutants are not associated with detectable disease states nor abnormalities in lipid metabolism; nevertheless, they may provide interesting insights into the structural properties of apo A-I.

The linear sequence of apo A-I contains eight internal repeats of 22 amino acids, six of which start with a Pro residue (Boguski et al., 1986). These segments can potentially form amphipathic α-helices which constitute the lipid binding units of apo A-I. In discoidal complexes with phospholipids, the α-helical segments of apo A-I are thought to be closely packed in an antiparallel arrangement around the periphery of the disc, as shown in Figure 1 (Jonas et al., 1989; Brasseur et al., 1990). The diameter of the complexes is determined by the number of apo A-I molecules per particle and the number of helices (6, 7, or 8) per apo A-I that are in contact with lipid. Since the smallest number of apo A-I molecules in a discoidal complex is 2, the smallest complex

[b]Institute of Clinical Chemistry and Laboratory Medicine, University of Münster, D-4400 Münster, Germany
[c]Department of Physics and Laboratory for Fluorescence Dynamics, University of Illinois at Urbana-Champaign, Urbana, IL 61801, U.S.A.
[d]Institute of Arteriosclerosis Research, D-4400 Münster, Germany

NATO ASI Series, Vol. H 73
Human Apolipoprotein Mutants III
Edited by C. R. Sirtori, G. Franceschini, B.H. Brewer Jr.

has a size determined by 12 α-helices. The ability of apo A-I to adopt different structures may modulate its functions; in fact, the activation of the LCAT reaction decreases by 15-fold when the apo A-I molecule rearranges from the state with 8 helices to that with 6 helices in contact with lipid (Jonas et al., 1989). In spherical HDL particles the helical structure and close packed arrangement of apo A-I appears to be preserved, but the most prevalent particles have 3 instead of 2 apo A-I molecules per particle (Jonas et al., 1990).

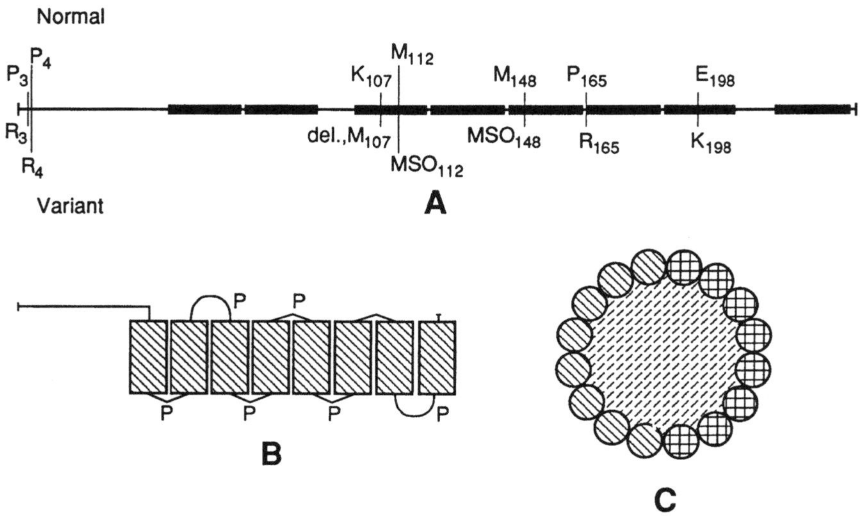

Figure 1. **A** represents the linear sequence of human apo A-I showing the 22 amino acid repeats (black rectangles) separated by Pro residues and the variants investigated by us. **B** shows the proposed antiparallel arrangement of the putative amphipathic α-helical segments. **C** is the top view of a 94–96 Å rHDL disc containing 2 apo A-I molecules each with 8 α-helices in contact with lipid.

Apo A-I variants can be used to address several important questions regarding the general features of the structure of apo A-I and of HDL. Namely, how critical is the integrity of an individual α-helix in the overall structure and stability of HDL, what is the effect of polar groups introduced into the nonpolar faces of a helix, what is the effect of replacing a Pro residue between two helices, how does the substitution of an acidic residue with a basic one affect the structure, is there a structural role for the cluster of Pro residues at the amino terminal end of apo A-I? Figure 1A gives the natural and chemical variants of apo A-I which we used in our studies (Jonas et al., 1991a; Jonas et al., 1992): $Pro_3 \rightarrow Arg$, $Pro_4 \rightarrow Arg$, $Lys_{107} \rightarrow 0$ (deletion), $Lys_{107} \rightarrow Met$, $Pro_{165} \rightarrow Arg$, $Glu_{198} \rightarrow Lys$, and $Met_{112} \rightarrow MetSO$ plus $Met_{148} \rightarrow MetSO$. The latter is a chemical variant which contains oxidized Met, i.e., Met sulfoxide (MetSO) in positions 112 and 148 of the sequence (von Eckardstein et al., 1991).

In order to assess the structural and functional effects of these mutations we reconstituted HDL (rHDL) from the pure apo A-I variants and native apo A-I controls, with palmitoyloleoyl phosphatidylcholine (POPC) and cholesterol, by the sodium cholate-dialysis method, and examined the completeness of the reaction, as well as the sizes and size distributions of the products by nondenaturing gradient gel electrophoresis. Since rHDL, containing native apo A-I, undergo well-defined structural transformations when incubated with LDL or with LDL plus LCAT, we performed the same reactions with the rHDL prepared with the apo A-I variants and examined their products by nondenaturing gradient gel electrophoresis. In addition, we studied the structural properties and stability to denaturation of selected apo A-I variants, in solution and in the discoidal rHDL particles, by means of fluorescence and circular dichroism spectroscopic methods. Finally, we measured the reactivity of all the rHDL particles with LCAT, in order to determine the effects of the apo A-I mutations on the best known function of apo A-I, the activation of the LCAT reaction.

RECONSTITUTED HDL

Figure 2A shows the nondenaturing gradient gel electrophoresis patterns of selected rHDL prepared with native and variant apo A-Is (Jonas et al., 1991a). In all cases, a major particle with a diameter of 94–96 Å is evident, and several minor particle bands with diameters of 78, 87, and 109 Å. In electrophoresis runs stopped before equilibrium was attained, there was no evidence of free apo A-I in any of the rHDL preparations, indicating that all the apo A-I variants we examined are capable of binding lipid. Figure 2B shows the same rHDLs after incubation with LDL for 24 hrs at 37°C. This treatment is known to result in the transfer of phospholipids from the rHDL into LDL, and in the rearrangement of the apo A-I structure and rHDL sizes (Jonas et al., 1988; Jonas et al., 1991b). All the rHDL prepared with native apo A-I and with the

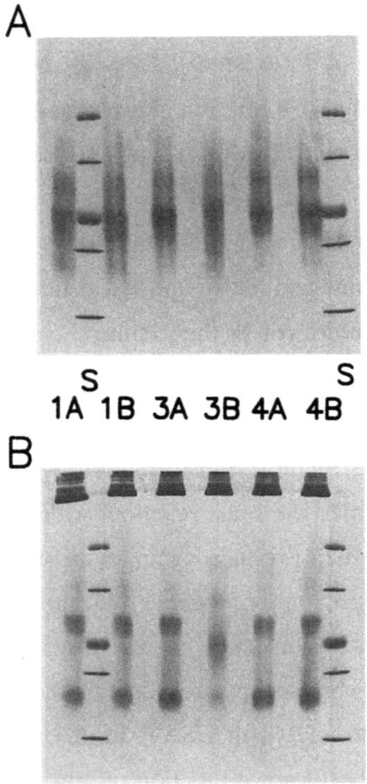

Figure 2

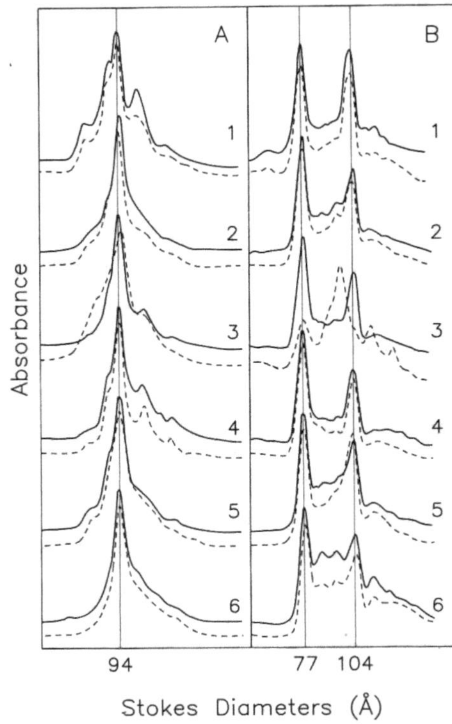

Figure 3

Figure 2. Nondenaturing gradient gel electrophoresis patterns of rHDL particles prepared with apo A-I variants or apo A-I normal controls: **1B**, $Pro_3 \rightarrow Arg$; **3B**, $Lys_{107} \rightarrow 0$; **4B**, $Lys_{107} \rightarrow Met$; **1A**, **3A**, **4A** are the corresponding normal controls; **S** are standard proteins. Panel **A** shows the original rHDL particles. Panel **B** shows the same rHDL particles after incubation with LDL for 24 hrs at 37°C.

Figure 3. Densitometer scans of nondenaturing gradient gel electrophoresis patterns. Panel **A** for the original rHDL particles; and Panel **B** for the rHDL particles incubated with LDL. The full lines correspond to the normal controls and the dashed lines to the apo A-I variants: **1**, $Pro_3 \rightarrow Arg$; **2**, $Pro_4 \rightarrow Arg$; **3**, $Lys_{107} \rightarrow 0$; **4**, $Lys_{107} \rightarrow Met$; **5**, $Pro_{165} \rightarrow Arg$; **6**, $Glu_{198} \rightarrow Lys$.

apo A-I variants show the characteristic particle size transformations that give 77–78 Å and 104–108 Å products, with the exception of the 3B preparation which contains the $Lys_{107} \rightarrow 0$ mutant of apo A-I. Evidently this modification in the structure of apo A-I makes it resistant to the structural rearrangement from the 96 Å particle with 8 α-helices per apo A-I to the 77 Å and 104 Å species which contain 6 α-helices per apo A-I. This observation also implicates the helix between residues 99 and 121 as one of the α-helical segments of apo A-I involved in these structural transformations.

Figure 3 illustrates the size distributions of the rHDL and their rearrangements in the presence of LDL by means of densitometer scans of the nondenaturing gradient gel bands. The rHDL particles containing the $Lys_{107} \rightarrow 0$ mutation, the $MetSO_{112, 148}$ variant, and the corresponding control particles were subjected, in addition, to incubation with LDL plus LCAT for 24 hrs at 37°C; all four preparations gave a major product with a diameter of 92–95 Å (results not shown), which for normal apo A-I is known to be a cholesterol ester-rich, spherical particle, containing 3 molecules of apo A-I (Jonas et al., 1990). The fact that the rHDL containing the apo A-I variants are good substrates for LCAT argues in favor of a similar transformation of their rHDLs into spherical products.

STRUCTURE AND STABILITY OF APO A-I

Four of the rHDL particles, those containing the $Lys_{107} \rightarrow 0$ mutation, the $MetSO_{112, 148}$ variant, the normal allele product from the $Lys_{107} \rightarrow 0$ heterozygote, and the reduced $Met_{112, 148}$ forms of apo A-I, were used in detailed studies of the structure and stability of the apo A-I (Jonas et al., 1992). The free apolipoproteins were included in several of the spectroscopic measurements. The intrinsic fluorescence spectra of the apo A-Is incorporated into the rHDL particles were quite similar in terms of their wavelengths of maximum fluorescence (334–335 nm) and maximum intensities (500–550 arbitrary units). Fluorescence lifetimes were determined by frequency domain methods, and gave average values ranging from 2.7 to 3.5 ns for the free apo A-Is and from 5.0 to 5.6 ns for the rHDLs (see Table 1). These precise measurements show, as expected, that the Trp residue environments change markedly when going from the free apolipoproteins to their lipid-bound states, but indicate quite similar Trp environments in each of these states for the different forms of apo A-I, except for the $MetSO_{112, 148}$ variant in solution. Fluorescence polarization measurements (Table 1) essentially confirm the lifetime results regarding the similarity of Trp residue environments and rotational motions in the lipid-bound forms of the apo A-Is, and a

Table 1. Spectroscopic Properties of Apo A-I Variants in Free Form and in R-HDL Particles

R-HDL	Apo A-I	Ave. Lifetime[a] (ns)	Fl. Polariz.[b]	α-Helix[c] (%)
	Normal	3.06 ± 0.20	0.116 ± 0.003	55 ± 5
	$Lys_{107} \rightarrow 0$	3.47	0.102	49
	MetSO	2.70	0.121	61
	Met	3.25	0.106	52
rHDL-1	Normal	5.56	0.089	73
rHDL-2	$Lys_{107} \rightarrow 0$	5.17	0.094	62
rHDL-3	MetSO	4.96	0.088	74
rHDL-4	Met	5.47	0.089	67

[a] The fluorescence lifetimes are the average of two lifetime values obtained by a double exponential fit of data from frequency domain measurements at 295 nm excitation.
[b] Measured at 24°C using a 280 nm excitation wavelength.
[c] Calculated from ellipticity values at 222 nm, obtained from circular dichroism spectra.

larger structural and dynamic disparity of the free apolipoproteins in solution. The α-helix content estimates from circular dichroism spectra (Table 1) show the expected increases in secondary structure comparing the free apo A-Is and their corresponding rHDLs. The $Lys_{107} \rightarrow 0$ mutant of apo A-I has a somewhat lower α-helix content in both its free and lipid-bound forms. In contrast, the $MetSO_{112,148}$ apo A-I variant appears slightly more helical in solution.

The stability of the four apo A-I forms in rHDL particles was examined by fluorescence and circular dichroism spectroscopic methods as guanidine hydrochloride (GndHCl) concentration was increased from 0 to 6 M. The changes in the wavelengths of maximum fluorescence as a function of GndHCl concentration are shown in Figure 4. Free apo A-Is are completely denatured by 2 M GndHCl, whereas the apo A-Is bound to lipid are markedly stabilized, and are not fully denatured even at 5.5 M GndHCl. The stepwise denaturation of the apo A-Is is strongly suggestive of a domain structure in the rHDL particles. The domain structure and stability appear similar for all the apo A-I

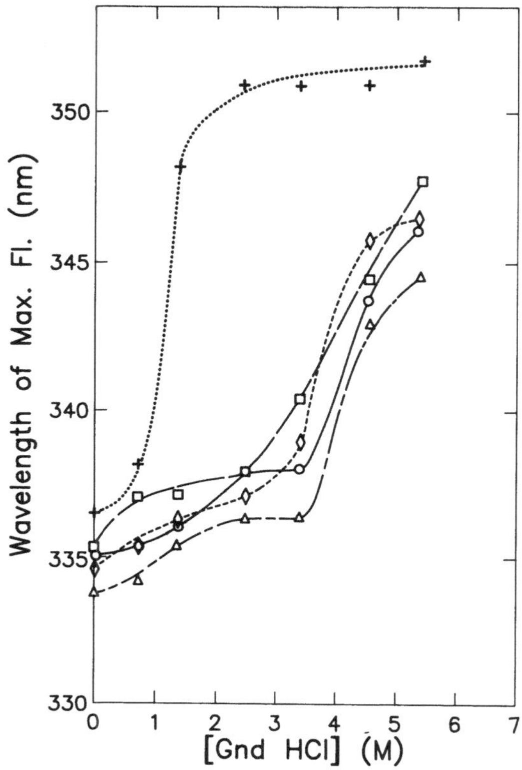

Figure 4. Denaturation of rHDL particles by GndHCl measured by changes in the wavelength of maximum fluorescence of Trp residues in apo A-I. Free, normal apo A-I (++); rHDL-1 with normal apo A-I (○○); rHDL-2 with $Lys_{107} \rightarrow 0$ apo A-I (□□); rHDL-3 with $MetSO_{112,148}$ apo A-I (ΔΔ); rHDL-4 with Met apo A-I (◇◇).

forms; however, at low GndHCl concentrations (1–2 M GndHCl) the $Lys_{107} \rightarrow 0$ mutant particle is the least stable and the $MetSO_{112,148}$ variant particle the most stable in this set of rHDL particles. This order of stability is illustrated more clearly in Figure 5 which gives the kinetics of denaturation of the rHDL particles in 3.4 M GndHCl: here it is clear that the $Lys_{107} \rightarrow 0$ mutant particles are the least stable species, whereas the $MetSO_{112,148}$ variant particles are the most stable ones. All particles exhibit two phases in their denaturation kinetics, a rapid one that occurs in the first ½ hour, and a slow phase that continues past 6 hrs.

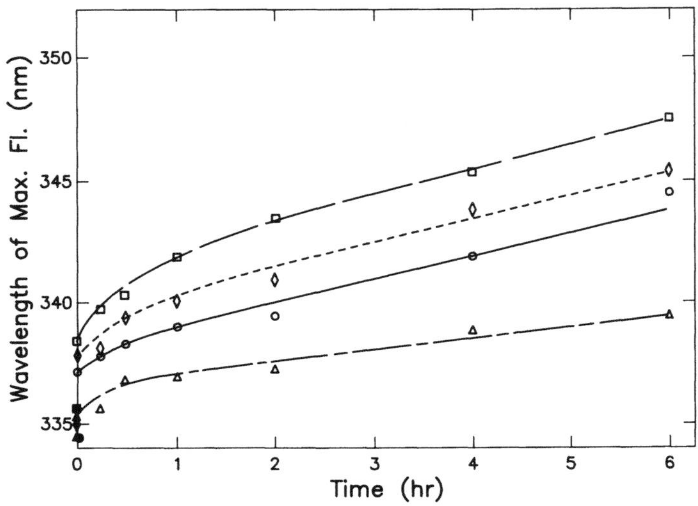

Figure 5. Kinetics of the denaturation of rHDL particles in 3.4 M GndHCl. The same symbols are used as in Figure 4. The closed symbols represent the wavelengths of maximum fluorescence prior to the addition of GndHCl.

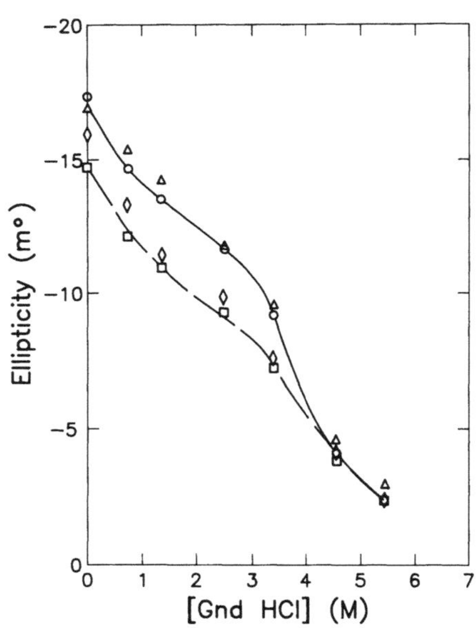

Figure 6. Denaturation of rHDL particles by GndHCl measured by circular dichroism (negative ellipticity) at 222 nm. The same symbols are used as in Figure 4.

The denaturation behavior of the rHDL measured by circular dichroism is depicted in Figure 6. The small differences in α-helix content of the rHDL are evident at the lower GndHCl concentrations. In all particles there appear to be two distinct domains of secondary structure which denature with mid-points at 1.5 or 4.0 M GndHCl. The relative stability of the $MetSO_{112,148}$ rHDL and relative instability of the $Lys_{107} \rightarrow 0$ rHDL are again evident in this experiment.

REACTIVITY WITH LCAT

The kinetics of the reaction of rHDLs, containing all the variant apo A-Is, with pure LCAT were measured at several concentrations of the particles, and were compared with the reaction of the rHDLs containing the corresponding normal, control apo A-Is. The kinetic parameters obtained from a Lineweaver-Burk analysis of the kinetic data are listed in Table 2 (Jonas et al., 1991a; Jonas et al., 1992). Comparing each variant with its control, significant differences in apparent V_{max}/apparent K_m were detected for the $Lys_{107} \rightarrow 0$, the $Glu_{198} \rightarrow$ Lys, and the $MetSO_{112,148}$ variants; however, the differences in reactivity for the pairs were at most 2- to 3-fold. Examination of the kinetic parameters for the "normal" apo A-I reveal comparable differences; therefore, the physiological importance of the differences in the reactivity of LCAT with the rHDL particles containing these apo A-I variants is minimal, especially in view of the fact that rearrangements of the structure of apo A-I can lead to major, 15-fold differences in reactivity with LCAT (Jonas et al., 1989).

CONCLUSIONS

1) All the variants of apo A-I studied by us bound to lipid (POPC) giving rHDL particles with equivalent diameters and overall structural properties. Thus, the partial disruption of an α-helix with a change in its face, as in the $Lys_{107} \rightarrow 0$ mutant, the substitution of a Pro with an Arg between two helices, or the presence of polar MetSO residues in the nonpolar faces of two helices failed to perturb, in a major way, the structural organization of apo A-I in the rHDL complexes. Evidently the extensive

Table 2. Apparent Kinetic Constants for the Reaction of R-HDL Complexes with LCAT[a]

R-HDL	Apo A-I	Apparent V_{max}/Apparent K_m [b] (nmol/h · M)
1A	Normal	6.66×10^6
1B	$Pro_3 \rightarrow Arg$	6.29×10^6
2A	Normal	8.50×10^6
2B	$Pro_4 \rightarrow Arg$	7.82×10^6
3A	Normal	7.80×10^6 [c]
3B	$Lys_{107} \rightarrow 0$	5.40×10^6 [c]
4A	Normal	7.15×10^6
4B	$Lys_{107} \rightarrow Met$	7.74×10^6
5A	Normal	5.97×10^6
5B	$Pro_{165} \rightarrow Arg$	6.30×10^6
6A	Normal	6.42×10^6 [c]
6B	$Glu_{198} \rightarrow Lys$	5.43×10^6 [c]
rHDL-1	Normal	1.13×10^6 [c]
rHDL-2	$Lys_{107} \rightarrow 0$	0.49×10^6 [c]
rHDL-3	$MetSO_{112, 148}$	2.22×10^6 [c]
rHDL-4	$Met_{112, 148}$	1.00×10^6 [c]

[a] The two sets of experiments, those for rHDL 1A to 6B (Jonas et al., 1991a), and rHDL-1 to rHDL-4 (Jonas et al., 1992), are not directly comparable because different levels of LCAT were used in determining the kinetic parameters.

[b] The standard deviations, 0.94×10^6 and 1.07×10^6, were calculated from the means for the six normal (1A through 6A) samples and the six variant (1B through 6B) samples, respectively.

[c] These pairs are significantly different by comparing their ratios to 1.00, and using a S.D. of 0.08.

cooperative interactions of the 6 to 8 amphipathic α-helices of apo A-I with lipid, and perhaps interactions between adjacent helices on the rHDL particles (Rosseneu, 1992) can overcome energetically unfavorable local modifications. For the other mutations in apo A-I, $Pro_3 \rightarrow Arg$, $Pro_4 \rightarrow Arg$, $Lys_{107} \rightarrow Met$, and $Glu_{198} \rightarrow Lys$, the lack of an effect on lipid binding and rHDL properties was less of a surprise since these amino acid residues at the amino terminus or the polar faces of helices were not expected to participate in direct interactions with lipids or in β-turns between helices.

2) In contrast to all the variant and normal rHDL, only the $Lys_{107} \rightarrow 0$ mutant failed to undergo the rearrangement from the 94–96 Å particles with 8 α-helical segments per apo A-I to the 77 Å and 104 Å particles which contain 6 α-helical segments per apo A-I in contact with lipid. Evidently the helical segment of apo A-I encompassing the Lys_{107} residue is involved in these structural rearrangements of apo A-I, which not only change the size of the rHDL particles but also modulate their functions. We have shown that the 77 Å and 104 Å particles are 16- to 12-fold less reactive compared to the 94–96 Å rHDL particles (Jonas et al., 1989). The physiological significance of such structural rearrangements of apo A-I is not yet known, but it is possible that they underlie the diverse properties of HDL subclasses.

3) Spectroscopic methods can reveal minor differences in the structure and stability of the apo A-I variants. The $Lys_{107} \rightarrow 0$ mutant of apo A-I when incorporated into rHDL particles has a lower content of α-helical structure and higher susceptibility to denaturation in GndHCl compared to the other rHDL. This probably reflects the alteration in one amphipathic helix. Surprisingly, the $MetSO_{112, 148}$ variant forms the most stable rHDL particles. A possible explanation for this increased stability is the presence of basic residues within a few Å from the MetSO groups. The basic residues could have strong dipolar interactions with the MetSO groups, extending the nonpolar face of the helices and increasing their affinity for lipid.

4) The absence of any major effects of these apo A-I variants on the overall structure of their rHDL complexes is reflected in their similar ability to activate LCAT. Although the $Lys_{107} \rightarrow 0$ mutant rHDL appears to be only 40 to 70% as reactive as its normal counterpart in agreement with the report by Rall et al. (1984), similar levels of variability in LCAT activation are found among rHDL prepared with normal apo A-I from different donors. This variability among the normal apo A-Is can be accounted, in

part, by the existence of chemical variants of apo A-I as seen for the MetSO variant relative to the reduced Met form, and in part, by experimental variability such as that due to small differences in particle size distributions that result from rather minor differences in the proportions of lipid to protein in the rHDL preparation.

5) The insensitivity of the structure and reactivity with LCAT of rHDL containing the variants of apo A-I studied by us, probably explains the high amino acid variability found in human apo A-I compared to several other proteins (von Eckardstein et al., 1990).

REFERENCES

Boguski MS, Freeman N, Elshourbagy NA, Taylor JM, Gordon JI (1986) On computer-assisted analysis of biological sequences: proline punctuation, consensus sequences, and apolipoprotein repeats. *J Lipid Res* 27:1011-1034

Brasseur R, De Meutter JD, Vanloo B, Goormaghtigh E, Ruysschaert J-M, Rosseneu M (1990) Mode of assembly of amphipathic helical segments in model high-density lipoproteins. *Biochim Biophys Acta* 1043:245-252

Jonas A, Kézdy KE, Williams MI, Rye K-A (1988) Lipid transfers between reconstituted high density lipoprotein complexes and low density lipoproteins: effects of plasma protein factors. *J Lipid Res* 29:1349-1357

Jonas A, Kézdy KE, Hefele Wald J (1989) Defined apolipoprotein A-I conformations in reconstituted high density lipoprotein discs. *J Biol Chem* 264:4818-4824

Jonas A, Hefele Wald J, Harms Toohill K, Krul ES, Kézdy KE (1990) Apolipoprotein A-I structure and lipid properties in homogeneous reconstituted spherical and discoidal high density lipoproteins. *J Biol Chem* 265:22123-22129

Jonas A, von Eckardstein A, Kézdy KE, Steinmetz A, Assmann G (1991a) Structural and functional properties of reconstituted high density lipoprotein discs prepared with six apolipoprotein A-I variants. *J Lipid Res* 32:97-106

Jonas A, Bottum K, Kézdy KE (1991b) Transformations of reconstituted high density lipoprotein subclasses as a function of temperature or LDL concentration. *Biochim Biophys Acta* 1085:71-76

Jonas A, von Eckardstein A, Churgay L, Mantulin WW, Assmann G (1992) Structural and functional properties of natural and chemical variants of apolipoprotein A-I. *J Lipid Res* (submitted)

Rall SC Jr, Weisgraber KH, Mahley RW, Ogawa Y, Fielding CJ, Utermann G, Haas J, Steinmetz A, Menzel H-H, Assmann G (1984) Abnormal lecithin: cholesterol acyltransferase activation by a human apolipoprotein A-I variant in which a single lysine residue is deleted. *J Biol Chem* 259:10063-10070

Rosseneu M (1992) Interactions and stabilization of apoprotein helical segments in high density lipoprotein. *Third International Symposium on Plasma High Density Lipoproteins and Atherosclerosis*, Tall A (ed), Elsevier Science Publishers, Amsterdam (in press) von Eckardstein A, Funke H, Walter M, Altland K, Benninghoven A, Assmann G (1990) Structural analysis of human apolipoprotein A-I variants. Amino acid substitutions are nonrandomly distributed throughout the apolipoprotein A-I primary structure. *J Biol Chem* **265**:8610-8617

von Eckardstein A, Walter M, Holz H, Benninghoven A, Assmann G (1991) Site-specific methionine sulfoxide formation is the structural basis of chromatographic heterogeneity of apolipoproteins A-I, C-II, and C-III. *J Lipid Res* **32**:1465-1476

ACKNOWLEDGEMENTS

The work in A. Jonas' laboratory has been supported by NIH grants HL-16059 and HL-29939; in A. von Eckardstein's laboratory by grant EC 116/1-1 from the Deutsche Forschungsgemeindschaft.

Structural Domains in Human Serum Low Density Lipoprotein Apolipoprotein B100

Lawrence P. Aggerbeck
Centre de Génétique Moléculaire
Centre National de la Recherche Scientifique
91198 Gif-sur-Yvette
France

Introduction

Low density lipoproteins have been studied by a variety of techniques which have provided a global description of the organization of the particle and the location of its components, however the precise details of the organization continue to be explored. It is clear that there is a hydrophobic core in the particle which is composed mainly of the cholesteryl esters and the triglycerides which is surrounded by the more polar phospholipid and cholesterol components and the protein (Mateu et al., 1972; Stuhrmann et al., 1975; Laggner et al., 1976; Tardieu et al., 1976; Atkinson et al., 1977; Deckelbaum et al., 1977; Luzzati et al., 1979). The organization of the core (Deckelbaum et al., 1975) and the surface components (Luzzati et al., 1979) is temperature dependent . The structure of the protein component, apolipoprotein B100 (apo B100), which is present in a single copy per particle, is of particular interest because of its large size (Cladaras et al., 1986; Knott et al., 1986; Law et al., 1986; and Yang et al., 1986), because it is a ligand for the LDL receptor (Brown and Goldstein, 1986) and because the lipid environment (Schonfeld and Patsch, 1980; Tsao et al., 1982; Bradley et al., 1984; Chen et al., 1991) or the temperature may affect its structure and, thus, its function within different VLDL or LDL subclasses.

A Problem in Protein Folding

The structure of apo B100 is also of interest as a problem in protein folding within the endoplasmic reticulum and its relation to lipoprotein assembly and

NATO ASI Series, Vol. H 73
Human Apolipoprotein Mutants III
Edited by C. R. Sirtori, G. Franceschini, B.H. Brewer Jr.

secretion. Through processes which are poorly understood at present, this large protein must, in addition to acquiring its tertiary structure, interact with lipids of several different types and be secreted from the cell. The formation of disulfide bonds and the glycosylation of the protein, events which may be of importance for the proper assembly and secretion of the lipoprotein, probably are processes which are also influenced by the acquisition of the tertiary structure of the protein.

Hereditary Diseases and Factors Involved in Folding, Assembly and Secretion

Factors other than apoB100 itself certainly are involved in the folding, assembly and secretion of lipoproteins. These other factors must play important roles as underlined by two hereditary disorders of lipoprotein deficiency, abetalipoproteinemia and Anderson's disease (Kane and Havel, 1989) These disorders are characterized, among other manifestations, by severe malabsorption which is a consequence of the inability of the affected individuals to assemble and/or secrete apo B containing lipoproteins. Restriction fragment length polymorphism analyses, however, have excluded a causal link to the apo B gene in both disorders (Talmud et al., 1988; Huang et al., 1990; Pessah et al., 1991).

A recent study of several kindreds with abetalipoproteinemia, in which both apo B100 and apo B48 are absent from the plasma, has shown that these patients lack the activity of the microsomal triglyceride transfer protein, MTP (Wetterau et al., 1992). This protein catalyzes the net transfer of lipid (particularly triglyceride) between lipid vesicles in vitro (Wetterau and Zilversmit, 1985). MTP is a heterodimer composed of one molecule of protein disulfide isomerase (PDI) and one molecule of another protein having an apparent molecular weight of 88000 (88kDa) (Wetterau et. al. 1990, 1991a and b). In all cases studied thus far, the 88 kDa subunit has been found to be missing (Wetterau et al., 1992). Although it has been demonstrated in cases of abetalipoproteinemia that apo B is synthesized (Dullaart et al., 1986; Bouma et al., 1990) and that the lipid components are present, lipoprotein assembly does not appear to take place, presumably as a result ot the absence of activity of MTP.

Other factors, as well, could be implicated in the assembly process and defects in these factors could conceivably result in a similar phenotype. Some of these factors may be specific for lipoprotein assembly and may be tissue specific. Others may play a more general role related to protein folding (Gething and Sambrook, 1992; Ellis and van der Vies, 1991; Schmid, 1991; Ellis and Hemmingsen, 1989) such as chaperon proteins which, by transiently binding the protein, prevent aggregation due to incorrect interactions or such as folding catalysts like protein disulfide isomerase or prolyl isomerase, PPlase.

In the case of Anderson's disease, which affects selectively the apo B48 containing lipoproteins, some assembly does appear to take place (Roy et al., 1987; Bouma et al., unpublished results) and, in these cases, the defect appears to be related, instead, to the secretory process.

Protein-Protein Interactions

Although in all cases of abetalipoproteinemia and Anderson's disease studied thus far the apo B gene appears to have been excluded as a cause, it is clear that the other factors which play a role in lipoprotein assembly and secretion must function by recognizing, in some fashion, apo B. This recognition probably depends upon the structure of the apoprotein. Indeed, the importance of apoprotein structure in protein-protein interactions is underscored by the mutations that exist in apo E (Mahley, 1988; Mahley et al., 1990) and apo B (Farese et al., 1992) which affect binding of certain lipoproteins to the LDL receptor or by the effect of the lipid environment which appears to play a role in maintaining a receptor-binding-competent structure for apo E (Innerarity et al., 1979 and 1984; Mims et al., 1990) and apo B (Bradley et al., 1984). As in the case of other protein-protein interactions, the tertiary structure of the proteins must play an important role.

Indications of Structural Domains in Apo B

There exists a considerable amount of information about apo B100, although there is no crystal structure. Apo B100 is a single polypeptide chain of 4536 amino acids with a predicted molecular weight of 512000 (Cladaras et al.,

1986; Law et al., 1986; Knott et al., 1986; Yang et al., 1986). Approximately 8-10% of the mass of apo B is composed of N-linked carbohydrate (Lee and Breckenridge, 1976) resulting in an apparent molecular weight of about 550000 on gel electrophoresis in the presence of SDS. Recent work has established the precise arrangement of most of the disulfide bonds which are situated in the amino-terminal one-fourth of the protein (Yang et al., 1986, 1990). Investigation with proteases has shown the presence of several regions of rather large size which are relatively resistant to proteolytic attack (Ikai and Yagisawa, 1977; Triplett and Fisher, 1978; Chapman et al., 1978; Yang et al., 1986, 1989; Chen, G.C. et al., 1989a and b). Calorimetric studies have shown that the melting of apo B100 proceeds through discrete stages which are associated with changes in the organization of apo B (Walsh and Atkinson, 1990). A portion of this melting is reversible.

This information and the fact that many proteins with molecular weights greater than 20000 are composed of structural domains (Rossmann and Argos, 1981; Kim and Baldwin, 1982) suggests that apo B100 might also contain domains. Structural domains are characterized by spatial compactness and frequently by thermodynamic stability (Privalov, 1979; Wetlaufer, 1973). Specific functions are frequently located within domains or at the interfaces of domains (Janin and Wodak, 1983).

Among the serum apolipoproteins, apoE has been demonstrated to be composed of two structural domains (Wetterau et al., 1988; Aggerbeck et al., 1988a). Discovery of the existence of these domains in apo E led to the crystallization of the amino-terminal domain (Aggerbeck et al., 1988b) and, subsequently, to the determination of its structure (Wilson et al., 1991). As in the case of other proteins composed of domains, the domains in apo E appear, at least superficially, to have distinct functions. The amino-terminal domain appears primarily responsible for binding to the LDL receptor (Innerarity et al., 1983; Weisgraber et al., 1983) whereas the carboxyl-terminal domain, which has a tendency to self-associate (Aggerbeck et al., 1988a) may play a more important role in binding to lipid although both domains are capable of binding lipids. Further, each of these domains appears to be stable in solution separated from the other domain.

The large size of apoB makes it very likely that this protein, too, is composed of several structural domains that might fold and function independently of one another. Demonstrating unequivocally, in apo B, the presence of structural domains and establishing their identity would be of interest for several reasons, as alluded to previously. Knowledge of the domain structure would be important for understanding how the protein folds. Domains may represent regions of the protein which interact with factors residing in the endoplasmic reticulum that are important for the folding of the molecule (chaperon proteins), or which interact with factors that catalyze the formation of disulfide bonds (protein disulfide isomerase) or which catalyze the isomerization of proline residues (prolyl isomerase) or which interact with other factors important for the assembly (such as microsomal triglyceride transfer protein) or secretion of the particle. Elucidation of the structure of these regions of apoB might then aid in understanding how these other factors function. This information could be useful for controlling lipoprotein assembly and secretion.

Demonstration of Structural Domains

In order to demonstrate the presence of structural domains in apoB a number of procedures may be used. A classical approach involves denaturation, either with guanidine hydrochloride or urea or with heat. The unfolding of the protein is followed with techniques such as calorimetry, fluorescence or ultraviolet spectroscopy or circular dichroism. Structural domains within a protein exhibit highly cooperative behavior in folding and unfolding. Demonstration of non-cooperative behavior during denaturation is compatible with the presence of independent domains (Ghelis and Yon, 1982).

In order to define the boundaries of the domains within the sequence of the protein, however, another approach is necessary. A particularly useful approach for defining compact tertiary structures is limited proteolysis (Mihalyi, 1978a and b; Creighton, 1983; Rupley, 1967). This approach employs a number of proteases having different specificities to cleave the protein. Hydrolysis of peptide bonds depends upon the sequence of the protein, that is, the residues forming the bond as well as the residues in the immediate vicinity of the bond. However, the rate of hydrolysis also depends upon the secondary, tertiary and quaternary structure of the protein. Some potentially cleavable bonds are, in fact, not hydrolyzed or hydrolyzed at significantly lower rates than might be expected because the amino acids forming the bonds are inaccessible due to the tertiary structure of the protein or are stabilized due to their involvement with other amino acids in the protein via hydrophobic interactions, hydrogen bonds or salt bridges. By studying the initial cleavages in the protein before the proteolysis markedly disrupts the tertiary structure the stable fragments produced should reflect compact, structured regions in the protein. Thus, the "maps" of the protein defining the regions resistant and accessible to proteases with different specificities provide indications of the boundaries of the presumed stable structural domains. Identification of the boundaries of the domains then permits further functional and structural analysis using model domains that are prepared by limited proteolysis or by expression *in vitro* using bacteria, yeast or mammalian cell lines.

The denaturation with guanidine hydrochloride of apo B100 in human serum low density lipoprotein has been assessed by far ultraviolet circular dichroism (Djavaheri and Aggerbeck, 1990). The loss of the secondary structure with increasing guanidine hydrochloride concentrations appears to occur in a non-cooperative manner consistent with the presence of several structural domains.

Freeze-fracture electron microscopy was used to investigate the effect of denaturing apo B100 upon the low density lipoprotein structure. Remarkably, up to a concentration of 4M guanidine hydrochloride, there appeared to be little effect upon the integrity of the lipoprotein particle. However, beyond 5-6M guanidine hydrochloride, marked disruption of the particle structure occurred. Thus, as assessed by circular dichroism and freeze-fracture electron microscopy, up to 40% of the secondary structure of apo B100 can be disrupted before the lipoprotein particle integrity is altered.

Re-acquisition of the secondary structure as assessed by far ultraviolet circular dichroism was found to be dependent upon the technique used to eliminate the guanidine hydrochloride. At initial concentrations below 4M guanidine hydrochloride, up to 90% of the initial circular dichroic signal could be recovered when dialysis was used to remove the denaturant. However, as the initial concentration exceeded 4M guanidine hydrochloride, then less and less of the secondary structure was recovered and the sample precipitated. Interestingly, the point of irreversibility to denaturation (approximately 4-6M guanidine hydrochloride) corresponded to the disruption of the lipoprotein particle. An alternative approach to remove the guanidine hydrochloride is rapid dilution. Using this technique, a significant portion of the secondary structure (up to 80%) could be recovered even after denaturation with up to 8M guanidine hydrochloride. All these results are consistent with the existence of structural domains in apo B100 in low density lipoprotein.

The results obtained by many authors (Ikai and Yagisawa, 1977; Triplett and Fisher, 1978; Chapman et al., 1978; Yang et al., 1986, 1989; Chen, G.C. et al., 1989a and b; Djavaheri and Aggerbeck, 1990) using several proteases of different specificities clearly shows the presence of a number of stable fragments. Given the large size of apo B100, it is not surprising that, initially, stable fragments of rather large molecular weight are found (greater than 150,000). However, stable fragments of about 100,000 and less are apparent as the proteolysis proceeds. Although it could be suggested that these regions are sequestered by lipid, the sequence of apo B100 does not contain long stretches of very hydrophobic amino acids. It is more likely that these are structured regions of the molecule (which does not mean that they do not interact with lipid).

The location of the stable fragments produced during limited proteolysis can be ascertained by a combination of amino-terminal microsequencing, immunoblotting with monoclonal antibodies and estimation of the molecular weight of the fragment. Using these techniques, we have identified several stable fragments throughout the molecule which could represent domains. Of particular interest are several fragments which are located in the amino-terminal part of the molecule which is rich in disulfide bridges as well as a number of fragments which interact with antibodies which block interaction with the LDL receptor. A number of these fragments are being further characterized as to their stability and function.

Electron Microscopy and Structural Domains

Given the size of the low density lipoprotein particle, one would have expected that electron microscopic techniques would have provided definitive information regarding the organization of the protein in LDL much as they have in the case of the spherical viruses. This has not been the case, however. As opposed to spherical viruses, for which the different electron microscopic techniques (negative staining, freeze-drying, freeze-fracture, freeze-etching, rotary shadowing of molecules dried from glycerol containing solutions) provide a consistant interpretation of the virus structure, the morphology of low density lipoproteins seems particularly sensitive to the technique employed. Almost all studies of low density lipoprotein using the technique of negative staining have shown relatively uniform particles of almost spherical shape with a smooth surface (Forte and Nichols, 1972). Because negative staining is known to induce the formation of artifacts in lipid containing systems, lipoproteins have also been studied by techniques employing cryofixation such as freeze-fracture and freeze-etching electron microscopy (see Aggerbeck and Gulik-Krzywicki, 1986 for review). With these techniques, the sample is visualized by shadowing after fracturing and etching (subliming) away the vitrified water. With these techniques the morphology of the particle surface is completely different from that seen with negative staining. By freeze-fracture or freeze-etching (with short periods of etching), the lipoprotein particle appears to have a rough surface composed of globular domains. However, if the time of etching is prolonged, the particle surface becomes smooth. Further, the technique of freeze-drying, in which the particle is first cryofixed and then the excess water removed at low temperature, also shows a lipoprotein particle with a smooth surface. It would thus appear that the particle surface is particularly sensitive to the removal of the water that surrounds it. Low angle rotary shadowing of low density lipoproteins dried from glycerol containing solutions also shows particules having a rough surface although this technique does not employ any cryofixation step. The glycerol, which is present in high concentration, may have an effect during the removal of the water. Thus, low density lipoprotein may present two distinct particle morphologies depending upon the electron microscopic technique used to visualize it. It must be emphasized that this is in marked contrast to results obtained with spherical viruses where all the techniques used provide a similar morphology for the virus. Cryoelectron microscopy of frozen hydrated low density lipoprotein (a technique in which the sample is not dried) seems to provide a third representation of the particle; a view which might include the internal, as well as, the surface structure. In conclusion, it is clear that low density lipoprotein is particularly sensitive to electron microscopic procedures. Therefore, conclusions derived from the use of these techniques are to be viewed with a certain amount of caution until their effects upon lipoprotein structure are better understood.

References

Aggerbeck, L.P. and Gulik-Krzywicki, T. (1986) Studies of lipoproteins by freeze-fracture and etching electron microscopy. Methods Enzymol. 128: 457-472.

Aggerbeck, L.P., Wetterau, J.R., Weisgraber, K.H., Mahley, R.W. and Agard, D.A. (1988) Crystallization and preliminary X-ray diffraction studies on the amino-terminal (receptor binding) domain of human apolipoprotein E3 from serum very low density lipoproteins. J. Mol. Biol. 202: 179-181.

Aggerbeck, L.P., Wetterau, J.R., Weisgraber, K.H., Wu, C.-S.C. and Lindgren, F.T. (1988) Human apolipoprotein E3 in aqueous solution. II. Properties of the amino- and carboxyl-terminal domains. J. Biol. Chem. 263: 6249-6258.

Atkinson, D., Deckelbaum, R.J., Small, D.M. and Shipley, G.G. (1977) Structure of human plasma low density lipoprotein: Molecular organization of the central core. Proc. Nat. Acad. Sci., U.S.A. 74: 1042-1046.

Bouma, M.E., Beucler, I., Pessah, M., Heinzmann, C., Lusis, A.J., Naim, H.Y., Ducastelle, T., Leluyer, B., Schmitz, J., Infante, R. and Aggerbeck, L.P. (1990) Description of two different patients with abetalipoproteinemia: synthesis of a normal sized apolipoprotein B48 in intestinal organ culture. J. Lipid Res. 31: 1-15.

Bradley, W.A., Hwang, S.-L.C., Karlin, J.B., Lin, A.H.Y., Prasad, S.C., Gotto, A.M. and Gianturco, S.H. (1984) Low density lipoprotein receptor binding determinants switch from apolipoprotein E to apolipoprotein B during conversion of hypertriglyceridemic VLDL to LDL. J. Biol. Chem. 259: 14728-14735.

Brown, M.S. and Goldstein, J.L. (1986) A receptor mediated pathway for cholesterol homeostasis. Science 232: 34-47.

Chapman, M.J., Goldstein, S. and Mills, G.L. (1978) Limited tryptic digestion of human serum low density lipoprotein: Isolation and characterisation of the protein deficient particle and of its apoprotein. Eur. J. Biochem. 87: 475-488.

Chen, G.C., Hardman, D.A., Hamilton, R.L., Mendel, C.M., Schilling, J.W., Zhu, S. Lau, K., Wong, J.S. and Kane, J.P. (1989) Distribution of lipid-binding regions in human apolipoprotein B100. Biochemistry 28: 2477-2484.

Chen, G.C., Lau, K., Hamilton, R.L. and Kane, J.P. (1991) Differences in local conformation in human apolipoprotein B100 of plasma low density and very low density lipoproteins as identified by cathepsin D. J. Biol. Chem. 266: 12581-12587.

Chen, G.C., Zhu, S., Hardman, D.A., Schilling, J.W., Lau, K. and Kane, J.P. (1989) Structural domains of human apolipoprotein B100. Differential accessibility to limited proteolysis of B100 in low density and very low density lipoproteins. J. Biol. Chem. 264: 14369-14375.

Cladaras, C., Hadzopoulou-Cladaras, M., Nolte, R.T., Atkinson, D. and Zannis, V.I. (1986) The complete sequence and structural analysis of human apolipoprotein B100: relationship between apo B100 and apo B48 forms. The EMBO J. 5: 3495-3507.

Creighton, T.E. (1983) Proteins. Structures and molecular principles. Freeman Publications. San Francisco.
Deckelbaum, R.J., Shipley ,G.G. and Small, D.M. (1977) Structure and interactions of lipids in human plasma low density lipoproteins. J. Biol. Chem. 252: 744-754.
Deckelbaum, R.J., Shipley, G.G., Small, D.M., Lees, R.S. and George, P.K. (1975) Thermal transitions in human plasma low density lipoproteins. Science 190: 392-394.
Djavaheri, M. and Aggerbeck, L.P. (1990) Investigation of structural domains in human serum low density lipoprotein apolipoprotein B100. In Hypercholesterolemia, Hypocholesterolemia, Hypertriglyceridemia (Malmendier, C.L., ed.) Plenum Press, New York, p.39-48.
Dullaart, R.P.F., Speelberg, B., Schuurman, H.J., Milne, R.W., Havekes, L.M., Marcel, Y., Geuze, H.J., Hulshof, M.M. and Erkelens, D.W. (1986) Epitopes of apolipoprotein B100 and B48 in both liver and intestine. Expression and evidence for local synthesis in recessive abetalipoproteinemia. J. Clin. Invest. 78: 1397-1404.
Ellis, R.J. and Hemmingsen, S.M. (1989) Molecular chaperones: Proteins essential for the biogenesis of some macromolecular structures. Trends Biochem. Sci. 14: 339-342.
Ellis, R.J. and van der Vies, S.M. (1991) Molecular chaperones. Annu. Rev. Biochem. 60: 321-347.
Farese, R.V., Linton, M.F. and Young, S.G. (1992) Apolipoprotein B gene mutations affecting cholesterol levels. J. Internal Med. 231: 643-652.
Forte, T. and Nichols, A.V. (1972) Application of electron microscopy to the study of plasma lipoprotein structure. Adv. Lipid Res. 10: 1-41.
Gething, M.-J. and Sambrook, J. (1992) Protein folding in the cell. Nature 355: 33-45.
Ghelis, C. and Yon, J. (1982) Protein Folding. Freeman Publications, San Francisco.
Huang, L.-S., Jäune, P.A., deGraaf, J., Cooper, M., Deckelbaum, R.J., Kayden, H.J., and Breslow, J.L. (1990) Exclusion of linkage between the human apolipoprotein B gene and abetalipoproteinemia. Am. J. Hum. Genet. 46: 1141-1148.
Ikai, A. and Yagisawa, H. (1977) Proteolysis of apoproteins in human serum low density lipoprotein. J. Biochem. 81: 955-961.
Innerarity, T.L., Friedlander, E.J., Rall, S.C., Jr., Weisgraber, K.H. and Mahley, R.W. (1983) The receptor-binding domain of human apolipoprotein E. Binding of apolipoprotein E fragments. J. Biol. Chem. 258: 12341-12347.
Innerarity, T.L., Pitas, R.E. and Mahley, R.W. (1979) Binding of arginine rich (E) apoprotein after recombination with phospholipid vesicles to the low density lipoprotein receptors of fibroblasts. J. Biol. Chem. 254: 4186-4190.
Innerarity, T.L., Weisgraber, K.H., Arnold, K.S., Rall, S.C., Jr. and Mahley, R.W. (1984) Normalization of receptor binding of apolipoprotein E2. Evidence for modulation of binding site conformation. J. Biol. Chem. 259: 7261-7267.
Janin, J. and Wodak, S.J. (1983) Structural domains in proteins and their role in the dynamics of protein function. Prog. Biophys. Mol. Biol. 42: 21-78.
Kane, J.P. and Havel, R.J. (1989) In The Metabolic Basis of Inherited Disease

(Scriver, C.R., Beaudet, A.L., Sly, W.S., Valle, D., eds.) McGraw-Hill, New York, ed. 6, pp. 1139-1164.
Kim, P.S. and Baldwin, R.L. (1982) Specific intermediates in the folding reactions of small proteins and the mechanism of protein folding. Annu. Rev. Biochem. 51: 459-489.
Knott, T.J., Pease, R.J., Powell, L.M., Wallis, S.C., Rall, Jr., S.C., Innerarity, T.L., Blackhart, B., Taylor, W.H., Marcel, Y., Milne, R., Johnson, D., Fuller, M., Lusis, A.J., McCarthy, B.J., Mahley, R.W., Levy-Wilson, B. and Scott, J. (1986) Complete protein sequence and identification of structural domains of human apolipoprotein B. Nature 323: 734-738.
Laggner, P., Müller, K., Kratky, O., Kostner, G. and Holasek, A. (1976) J. Coll. Interf. Sci. 55: 102-108.
Law, S.W., Grant, S.M., Higuchi, K., Hospattankar, A., Lackner, K., Lee, N. and Brewer, H.B., Jr. (1986) Human liver apolipoprotein B100 cDNA: Complete nucleic acid and derived amino acid sequence. Proc. Natl. Acad. Sci. U.S.A. 83: 8142-8146.
Lee, P. and Breckenridge, W.C. (1976) Can. J. Biochem. 54: 42-49.
Luzzati, V., Tardieu, A. and Aggerbeck, L.P. (1979) Structure of serum low density lipoprotein. I. A solution x-ray scattering study of a hyperlipidemic monkey low density lipoprotein. J. Mol. Biol. 131: 435-473.
Mahley, R.W. (1988) Apolipoprotein E: Cholesterol transport protein with expanding role in cell biology. Science 240: 622-630.
Mahley, R.W., Innerarity, T.L., Rall, S.C., Jr., Weisgraber, K.H. and Taylor, J.M. (1990) Apolipoprotein E: Genetic variants provide insights into its structure and function. Current Opinion in Lipidology 1: 87-95.
Mateu, L., Tardieu, A., Luzzati, V., Aggerbeck, L. and Scanu, A.M. (1972) On the structure of human serum low density lipoprotein. J. Mol. Biol. 70: 105-116.
Mihalyi, E. (1978) Application of proteolytic enzymes to protein structure-function studies. 2nd edition. Vol. 1. CRC Press, Inc., Boca Raton, FL.
Mihalyi, E. (1978) Application of proteolytic enzymes to protein structure-function studies. 2nd edition. Vol. 2. CRC Press, Inc., Boca Raton, FL.
Mims, M.P., Soma, M.R. and Morrisett, J.D. (1990) Effect of particle size and temperature on the conformation and physiological behavior of apolipoprotein E bound to model lipoprotein particles. Biochemistry 29: 6639-6647.
Pessah, M., Benlian, P., Beucler, I., Loux, N., Schmitz, J., Junien, C. and Infante, R. (1991) Anderson's disease: Genetic exclusion of the apo B gene in two families. J. Clin. Invest. 87: 367-370.
Privalov, P.L. (1979) Stability of proteins. Small globular proteins. Adv. Protein Chem. 33: 167-241.
Rossmann, M.G. and Argos, P. (1981) Protein Folding. Annu. Rev. Biochem. 50: 497-532.
Roy, C.C., Levy, E., Green, P.H.R., Sniderman, A., Letarte, J., Buto, J.P., Orquin, J., Brochu, P., Weber, A.M., Morin, C.L., Marcel, Y. and Deckelbaum, R.J. (1987) Malabsorption, hypocholesterolemia and fat-filled enterocytes with increased intestinal apoprotein B: chylomicron retention disease. Gastroenterology 92: 390-399.
Rupley, J.A. (1967) Susceptibility to attack by proteolytic enzymes. Methods

Enzymol. 11: 905-917.
Schmid, F.X. (1991) Catalysis and assistance of protein folding. Curr. Opin. Struct. Biol. 1: 36-41.
Schonfeld, G. and Patsch, W. (1980) In Drugs Affecting Lipid Metabolism (Fumagalli, R., Kritchevsky, D. and Paoletti, R., eds.) Elsevier/ North Holland Biomedical Press, Amsterdam, p. 37.
Stuhrmann, H.B., Tardieu, A., Mateu, L., Sardet, C. Luzzati, V., Aggerbeck, L. and Scanu, A.M. (1975) Neutron scattering study of human serum low density lipoprotein. Proc. Nat. Acad. Sci., U.S.A. 72: 2270-2273.
Talmud, P.J., Lloyd, J.K., Muller, D.P.R., Collins, D.R., Scott, J. and Humphries, S. (1988) Genetic evidence from two families that the apolipoprotein B gene is not involved in abetalipoproteinemia. J. Clin. Invest. 82: 1803-1806.
Tardieu, A., Mateu, L., Sardet, C., Weiss, B., Luzzati, V., Aggerbeck, L. and Scanu, A.M. (1976) Structure of human serum lipoproteins in solution. II. Small angle X-ray scattering study of HDL3 and LDL. J. Mol. Biol. 101: 129-153.
Triplett, R.B. and Fisher, W.R. (1978) Proteolytic digestion in the elucidation of the structure of low density lipoprotein. J. Lipid Res. 19: 478-488.
Tsao, B.P., Curtiss, L.K. and Edgington, T.S. (1982) Immunochemical heterogeneity of human plasma apolipoprotein B. II. Expression of apolipoprotein B epitopes on native lipoproteins. J. Biol. Chem. 257: 15222-15228.
Walsh, M.T. and Atkinson, D. (1990) Calorimetric and spectroscopic investigation of the unfolding of human apolipoprotein B. J. Lipid Res. 31: 1051-1062.
Weisgraber, K.H., Innerarity, T.L., Harder, K.J., Mahley, R.W., Milne, R.W., Marcel, Y.L. and Sparrow, J.T. (1983) The receptor-binding domain of human apolipoprotein E. Monoclonal antibody inhibition of binding. J. Biol. Chem. 258: 12348-12354.
Wetlaufer, D.B. (1973) Nucleation, rapid folding and globular intrachain regions in proteins. Proc. Natl. Acad. Sci. U.S.A. 70: 697-701.
Wetterau, J.R., Aggerbeck, L.P., Bouma, M.-E., Eisenberg, C., Munck, A., Hermier, M., Schmitz, J., Gay, G., Rader, D.J., Gregg, R.E. (1992) Absence of microsomal triglyceride transfer protein in individuals with abetalipoproteinemia. Science (in press).
Wetterau, J.R., Aggerbeck, L.P., Laplaud, P.M. and McLean, L.R. (1991) Structural properties of the microsomal triglyceride transfer protein complex. Biochemistry 30: 4408-.
Wetterau, J.R., Aggerbeck, L.P., Rall, S.C., Jr. and Weisgraber, K.H. (1988) Human apolipoprotein E3 in aqueous solution. I. Evidence for two structural domains. J. Biol. Chem. 263: 6240-6248.
Wetterau, J.R., Combs, K.A., McLean, L.R., Spinner, S.N. and Aggerbeck, L.P. (1991) Protein disulfide isomerase appears necessary to maintain the catalytically active structure of the microsomal triglyceride transfer protein. Biochemistry 30: 9728-.
Wetterau, J.R., Combs, K.A., Spinner, S.N. and Joiner, B.J. (1990) The 58 kDa subunit of the microsomal triglyceride transfer protein is protein disulfide isomerase. J. Biol. Chem. 255: 9800-9807.
Wetterau, J.R. and Zilversmit, D.B. (1985) Purification and characterization of microsomal triglyceride and cholesteryl ester transfer protein from bovine

liver microsomes. Chem. Phys. Lipids 38: 205-222.

Wilson, C., Wardell, M.R., Weisgraber, K.H., Mahley, R.W. and Agard, D.A. (1991) Three dimensional structure of the LDL receptor-binding domain of human apolipoprotein E. Science 252: 1817-1822.

Yang, C.-Y., Chen, S.-H., Gianturco, S.H., Bradley, W.A., Sparrow, J.T., Tanimura, M., Li, W.-H., Sparrow, D.A., DeLoof, H., Rosseneu, M., Lee, F.-S., Gotto, A.M. and Chan, L. (1986) Sequence, structure, receptor-binding domains and internal repeats of human apolipoprotein B-100. Nature 323: 738-742.

Yang, C.-Y., Kim, T.W., Weng, S.-A., Lee, B., Yang, M. and Gotto, A.M. (1990) Isolation and characterization of sulfhydryl and disulfide peptides of human apolipoprotein B100. Proc. Natl. Acad. Sci. U.S.A. 87: 5523-5527.

Yang, C.-Y., Gu, Z.-W., Weng, S.-A., Kim, T.W., Chem, S.-H., Pownall, H.J., Sharp, P.M., Liu, S.-W., Li, W.-H., Gotto, A.M. and Chan, L. (1989) Structure of apolipoprotein B100 of human low density lipoproteins. Arteriosclerosis 9: 96-108.

Crystal Structures of the Common Apolipoprotein E Variants: Insights into Functional Mechanisms.

Mark R.Wardell,
Department of Haematology
University of Cambridge
MRC Centre, Hills Road
Cambridge CB2 2QH
United Kingdom

Charles Wilson*, David A. Agard#, Robert W. Mahley†, and Karl H. Weisgraber†.

Apolipoprotein (apo) E is found in the plasma on the surfaces of most of the major lipoprotein classes, including chylomicrons, very low density lipoproteins (VLDL), intermediate density lipoproteins and high density lipoproteins (Mahley et al. 1984). One of its major roles is to mediate the delivery of lipids to cells by acting as a specific high affinity ligand for cell surface receptors, in particular the low density lipoprotein (LDL) receptor (Mahley et al. 1984; Mahley 1988).

There are three major isoforms of apoE (apoE2, apoE3, and apoE4) (Utermann et al. 1980) which are coded for by three common alleles at a single genetic locus (Zannis et al. 1981; Zannis and Breslow 1981). The most common, apoE3, is considered the normal isoform and contains one cysteine residue at position 112 of the total of 299 amino acids in its single polypeptide chain (Rall et al. 1982b). The other two isoforms differ from apoE3 by single amino acid substitutions (Weisgraber et al. 1981): apoE4 has arginine rather than cysteine at position 112 (Mahley et al. 1984), and apoE2 has cysteine rather than arginine at position 158 (Rall et al. 1982b). The three different isoforms do not all have the same receptor-binding activities (Weisgraber et al. 1982). Whereas apoE3 and apoE4 both bind with equal affinity to the LDL receptor, apoE2 expresses only ~1% of this normal receptor-binding activity (Weisgraber et al. 1982).

As apoE2 only differs from apoE3 by a single amino acid substitution (Rall et al. 1982b), it is this cysteine-for-arginine substitution at position 158 that is responsible

---

* Department of Molecular Biology, Massachusetts General Hospital, Boston, MA 02114; # Howard Hughes Medical Institute, the Graduate Program in Biophysics, and the Department of Biochemistry and Biophysics, University of California, San Francisco, CA 94143 - 0448; † Gladstone Institute of Cardiovascular Disease, University of California, San Francisco General Hospital, San Francisco, CA 94141 - 9100.

NATO ASI Series, Vol. H 73
Human Apolipoprotein Mutants III
Edited by C. R. Sirtori, G. Franceschini, B.H. Brewer Jr.

for the drammatically decreased receptor-binding activity of this variant (Weisgraber et al. 1982). Subjects homozygous for apoE2, which represent ~1% of the population, all have dysbetalipoproteinemia, where remnant lipoproteins accumulate in the plasma (Utermann et al. 1977; Havel and Kane 1973). Most of these subjects have plasma and LDL cholesterol levels that are slightly lower than normal (Utermann et al. 1979; Davignon et al.1988). However, when they also develop a second factor, which may result either from environmental influences such as obesity or diabetes, or be genetic in origin such as co-inheritance of familial combined hyperlipidemia, they will develop type III hyperlipoproteinemia, which is characterized by severely elevated cholesterol and triglyceride levels and the onset of premature atherosclerosis (Mahley and Rall 1989).

Apolipoprotein E4 also has only a single amino acid substitution relative to apoE3 (Weisgraber et al. 1981), arginine-for-cysteine at position 112 (Mahley et al. 1984). Although apoE4 has normal receptor-binding activity (Weisgraber et al. 1982), it is nevertheless associated with both elevated total plasma and LDL cholesterol levels (Davignon et al. 1988; Bouthillier et al. 1983). The mechanism responsible for this is unknown; however relative to apoE3, apoE4 does preferrentially distribute to triglyceride-rich lipoproteins, particularly chylomicrons and VLDL (Gregg et al. 1986; Steinmetz et al. 1989), and the positively charged arginine at position 112 has been implicated in this process (Weisgraber 1990). This distribution difference leads to a relative enrichment of apoE4 on chylomicron and VLDL particles and is thought to lead to their accelerated uptake and subsequent down-regulation of LDL receptor expression. This in turn would account for the increased levels of LDL (the major cholesterol carrying lipoprotein) in the plasma of apoE4 subjects.

The way in which apoE binds to the LDL receptor has been investigated by a variety of experimental approaches including binding studies of thrombolytic and cyanogen bromide fragments (Innerarity et al. 1983), chemical modification of arginine (Mahley et al. 1977) and lysine (Weisgraber et al. 1978) residues, inhibition of binding with a monoclonal antibody of known epitope (Weisgraber et al. 1983), binding studies with natural genetic variants associated with hyperlipidemia (Rall et al. 1982a; Rall et al. 1983; Wardell et al. 1987; Rall et al. 1989) and site-directed mutagenesis (Lalazar et al. 1988). All investigations indicate that the positively charged amino acids in a small region between residues 135 and 150 probably make direct ionic interactions with negatively charged amino acids in the LDL receptor. These studies paved the way for more detailed investigations which established that apoE has `two structural domains with very different physical properties (Wetterau et al. 1988; Aggerbeck et al. 1988b) and which are responsible for different functions: the amino-terminal domain is responsible for the receptor-binding activity of apoE (Innerarity et al. 1983) while the carboxy-terminal fragment is mainly responsible for the attachment of apoE to lipoprotein particles (Weisgraber 1990). Efforts to crystallize intact apoE have not as yet been successful, probably due to both its multi-domain nature and its tendency to aggregate in the absence of lipid, whereas the amino-terminal receptor-binding domain of apoE remains monomeric up to very high protein concentrations (Aggerbeck et al. 1988b) and did grow single, well ordered crystals that diffracted beyond 2.5Å resolution (Aggerbeck et al. 1988a; Wilson et al. 1991). We therefore first undertook a detailed structural analysis of the amino-terminal domain of apoE3 by X-ray diffraction analysis in order that we might better understand the molecular basis of receptor binding in this protein. We then performed crystallographic analysis of the equivalent domains of apoE2 and apoE4 to directly compare the structural changes that occur in each of these two common mutants, so that we might gain insights into the reasons for the altered functional properties of these variants.

The crystallization conditions for this domain of apoE have been described previously (Aggerbeck et al. 1988a; Wilson et al. 1991); briefly, the crystals were grown by the hanging drop method using 15% polyethylene glycol 400 as precipitant and by raising the pH of the crystallization resevoir from 4.5 to 5.3. The detergent β-*n*-octyl-glucopyranoside was included in the system to improve crystal morphology and the inclusion of β-mercaptoethanol was required for crystal growth as this fragment readily dimerizes through disulfide formation.

Data was collected at liquid nitrogen temperature, because at room temperature the crystals were very susceptible to radiation damage, and three crystals were needed to collect a full data set (Wilson et al. 1991). These crystals belonged to the orthorhombic space group $P2_12_12_1$ and had unit cell dimensions **a** = 41.3Å, **b** = 54.5Å, **c** = 87.0Å, and $\alpha = \beta = \gamma = 90°$ with four molecules in the unit cell (Wilson et al. 1991). The phases were solved using a single isomorphous heavy metal derivative together with accurate anomalous scattering data from the derivative and a structure at 2.5Å resolution was derived (Wilson et al. 1991). The structure consists of residues 23 - 166 and is missing 22 residues from the amino-terminus and the last 25 amino acids from the carboxy-terminus as these residues from each end of the receptor-binding domain appear to be disordered.

Figure 1 is a ribbon diagram of the crystal structure. As can be seen, this domain belongs to the 4-helix bundle family of proteins of which there are about 18 members. Table 1 lists the residue boundaries and lengths of each helix.

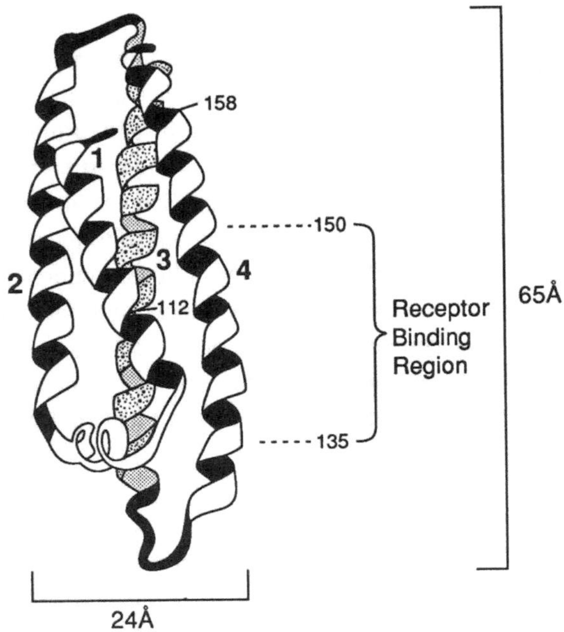

Fig. 1. Ribbon drawing of the amino-terminal, receptor-binding domain of apoE (residues 23 - 166) showing it is an elongated member of the 4-helix bundle family of structures. The four major helices are arranged in an antiparallel fashion. The basic amino acids between residues 135 and 150 on the lower half of helix 4 (indicated) are thought to make direct ionic interactions with acidic residues in the ligand-binding domain of the LDL receptor.

In this structure the connection between the first two helices of the bundle is a short helix (Hc) while there are simple turns between the second and third and between the third and fourth helices. The lower half of the fourth helix contains the basic amino acids between 135 and 150 (Fig. 1) that are thought to make direct ionic interactions with the LDL receptor.

Apolipoprotein E shares a gene structure common to many other apolipoproteins that gives rise to repeat sequences of amino acids. However, the protein sequence identity between the apoE repeats is relatively low compared to other apolipoproteins. Figure 2 shows the distribution of repeats within the helical structure of this domain. There are three 11-mers from exon 3 and five 22-mers from exon 4. There is only one proline residue in this structure (occurring in the loop between helices 2 and 3) whereas prolines exist in many of the putative turns between helices in other apolipoproteins. As can be seen, the boundaries between the repeats in this structure either occur in turns or within the middle of helices. None of the turns in the structure correspond to the middle of a repeat, thus the repeats may function as structural building blocks.

TABLE 1
Residue boundaries and lengths of the helices in the amino-terminal domain of apolipoprotein E.

Helix	Residue Boundaries	Helix Length (# of amino acids)
1	24 - 42	19
Hc	44 - 53	10
2	54 - 81	28
3	87 - 122	36
4	130 - 164	35

As is most common in the structural family to which this apoE structure belongs, the helices are arranged in an antiparallel fashion. However, whereas the average length of the helices in the other members of the family is about 18 residues, three of the helices in this domain of apoE are about twice this length giving rise to an elongated bundle 65Å in height and 24Å wide (Fig. 1). Within the bundle, the side chains of hydrophobic amino acids are tightly packed in the interior of the molecule while the charged amino acids form a ring around the exterior surface. Of the charged residues, 24 are acidic and 24 are basic, together comprising one-third of all the amino acids in this structure and most of these are involved in either intra-helical or inter-helical salt bridges which help to stabilize the folded structure. The basic side chains in the receptor-binding region (residues 135 - 150) on the other hand, are not involved in salt bridges within the molecule. Instead they protrude out into the solvent where they are perfectly placed to make good electrostatic interactions with complementary negatively charged side chains in the ligand-binding domain of the LDL receptor. In support of this, an electrostatic potential map of the surface of this structure indicates that the surface of most of the molecule is relatively neutral but that there is a very large positive potential surrounding residues 135 - 150 (Wilson et al. 1991).

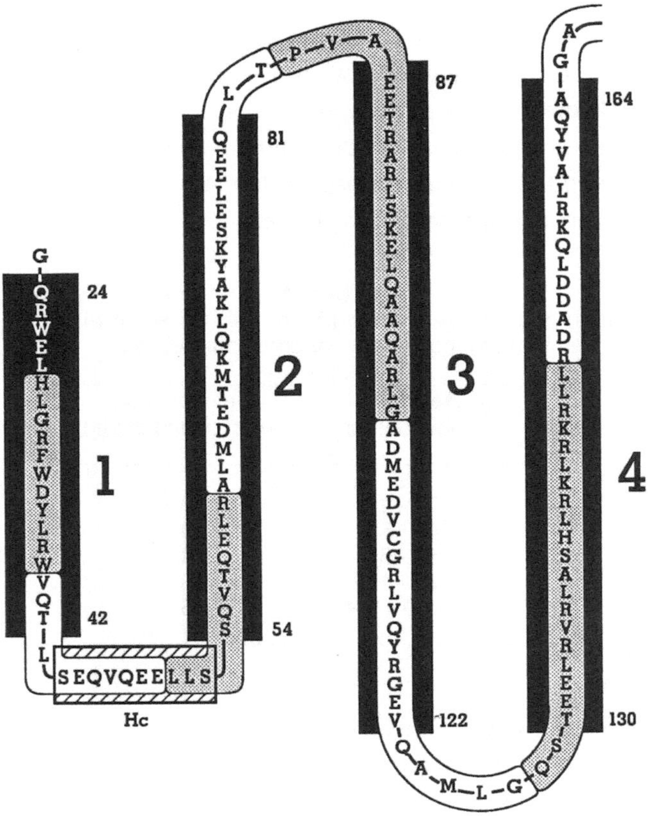

Fig. 2. Amino acid sequence of the amino-terminal, receptor binding domain of apoE showing the distribution of the 11-mer and 22-mer amino acid repeats within the 4-helix bundle.

Arginine-158, which is mutated to cysteine in the common apoE2 variant, is situated 12.5Å away from the cluster of basic amino acids between residues 135 - 150. In addition to being spatially removed from this basic cluster, arginine-158 is also not free to make direct ionic interactions with the LDL receptor because it is involved in two salt bridges; one intra-helical salt bridge with aspartate-154 and one inter-helical salt bridge with glutamate-96 on helix 3 (Fig. 3, A). We performed X-ray crystallographic analysis of the amino-terminal domain of apoE2 in order to investigate how the cysteine-for-arginine substitution at position 158 might have such a drastic effect on receptor-binding activity and thereby contribute to the association of this variant with severe hyperlipoproteinemia in affected individuals.

The most striking difference seen between the amino-terminal structural domains of apoE2 and apoE3 is a markedly altered set of salt bridges in the region of the structure adjacent to the cysteine-for-arginine substitution at position 158 (Fig. 3). In apoE3, the guanidinium group of arginine-158 is salt-bridged with the oppositely

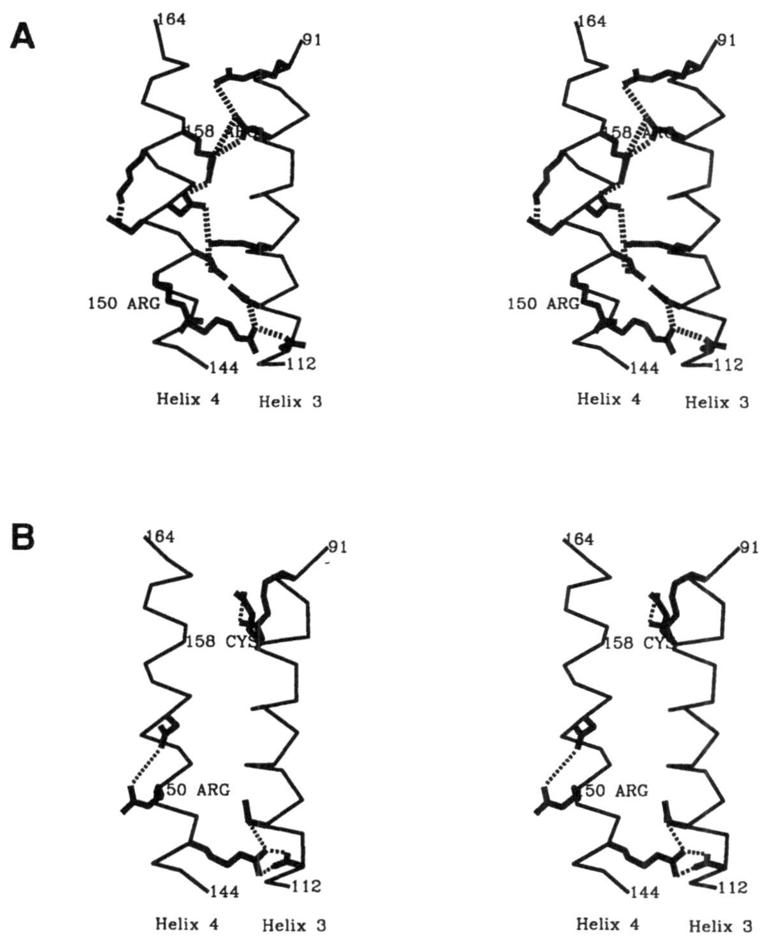

Fig. 3. Stereo diagrams showing that the substantial network of salt bridge interactions present between helices 3 and 4 in apoE3 (A) is markedly reduced in apoE2 (B). Note also that the side chain of arginine-150 is free in the apoE3 structure (A) and available to make direct ionic interactions with acidic residues in the LDL receptor whereas, in the apoE2 structure (B), it forms a salt bridge with aspartate-154 and is no longer free to interact with the receptor.

charged carboxylate groups of both glutamate-96 on helix 3 and aspartate-154 on helix 4. These salt bridges form part of an extended network of charge interactions between helices 3 and 4 (Fig. 3, A) that is completely disrupted when the positively charged arginine-158 on helix 4 is substituted by a neutral cysteine in apoE2 (Fig. 3, B). A consequence of the loss of this tying together of helices 3 and 4 by the salt bridge network is the distortion in the polypeptide backbone of helix 3 in apoE2 at glycine-105 (Fig. 4).

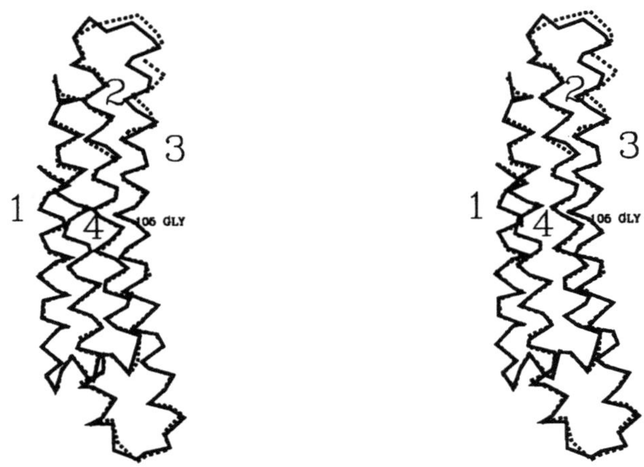

Fig. 4. Stereo diagram showing the superposition of the apoE3 (solid line) and apoE2 (dotted line) structures to illustrate the distortion in the polypeptide backbone that begins at glycine-105 in helix 3 of apoE2 and results from the disruption of the salt bridge network in the vicinity of residue 158 which normally ties helices 3 and 4 together (see Fig. 3A).

Of particular interest is arginine-150 which, when changed to alanine by site-directed mutagenesis, reduces LDL receptor binding activity to only 25% of normal (Lalazar et al. 1988). In the apoE3 structure, arginine-150 is free and solvent exposed (Fig. 3, A) and therefore accessible for direct interaction with acidic residues in the LDL receptor. In the apoE2 structure however, this residue is no longer free but forms a new salt bridge with aspartate-154 (Fig. 3, B). The result of this is a significant reduction in the positive electrostatic surface potential surrounding the receptor-binding site on helix 4 (Wilson et al. 1992). Thus, the loss of arginine-158 which was forming a salt bridge with aspartate-154 in the normal apoE3 structure, means that the side chain of arginine-150 is now recruited into a salt bridge with aspartate-154 and no longer contributes to the positive potential over residues 135 - 150. We speculate that both the marked reduction in positive surface potential at the putative receptor-binding site on helix 4, and the disruption of the salt bridge network between helices 3 and 4, together account for the drammatically decreased receptor-binding activity of apoE2.

Apolipoprotein E4 has normal LDL receptor-binding activity (Weisgraber et al. 1982) but is nevertheless associated with elevated total plasma and LDL cholesterol levels (Davignon et al. 1988; Bouthillier et al. 1983) most probably due to its preferrential association with VLDL-sized particles relative to apoE3 (Gregg et al. 1986; Steinmetz et al. 1989; Weisgraber 1990). The single amino acid substitution at position 112 within the amino-terminal domain of apoE is the only difference between apoE3 and apoE4 and recent biochemical experiments have implicated the positive charge at position 112 as the agent responsible for directing apoE4 to VLDL-sized particles (Weisgraber 1990). However, when only the amino-terminal domains of apoE3 and apoE4 were incubated with plasma, this difference was no longer seen. Neither fragment bound to any of the major lipoprotein classes thus establishing the carboxy-terminal domain of apoE as the most important lipid-binding domain (Weisgraber 1990). However, the cysteine-arginine interchange at position 112 in the amino-terminal domain determines the lipoprotein class to which the carboxy-terminal domain will bind (Weisgraber 1990). Thus, the lipid-binding function of the carboxy-terminal domain of apoE is modulated by the amino-terminal, receptor-binding domain.

In the crystal structure of the amino-terminal domain of apoE4 only small distortions in the polypeptide backbone were seen. However, there were two significant movements of charged side chains, one of which involved the formation of a new salt bridge (Fig. 5). In the first of these, glutamate-109, which was free in the apoE3 structure (Fig. 5, A), swings down to form a new salt bridge with the introduced arginine-112 in apoE4 (Fig. 5, B). This causes arginine-61, which is normally partially filling the space above cysteine-112, to move out of the way. It is possible that one or both of these differences within the amino-terminal domain of apoE4 are important influences of the conformation of the lipid-binding domain, and as such may play a role in determining the lipoprotein preferrence of this variant. In this connection it has been predicted by Segrest et al. (1992) that the carboxy-terminal, lipid-binding domain of apoE contains a continuous amphipathic helix that is 64 amino acids in length (residues 203 - 266), the longest helical domain predicted for any of the apolipoproteins investigated. These authors speculated that longer helices may preferrentially target larger, less curved particles such as triglyceride-rich lipoproteins (Segrest et al. 1992). It is possible that the mutation in the amino-

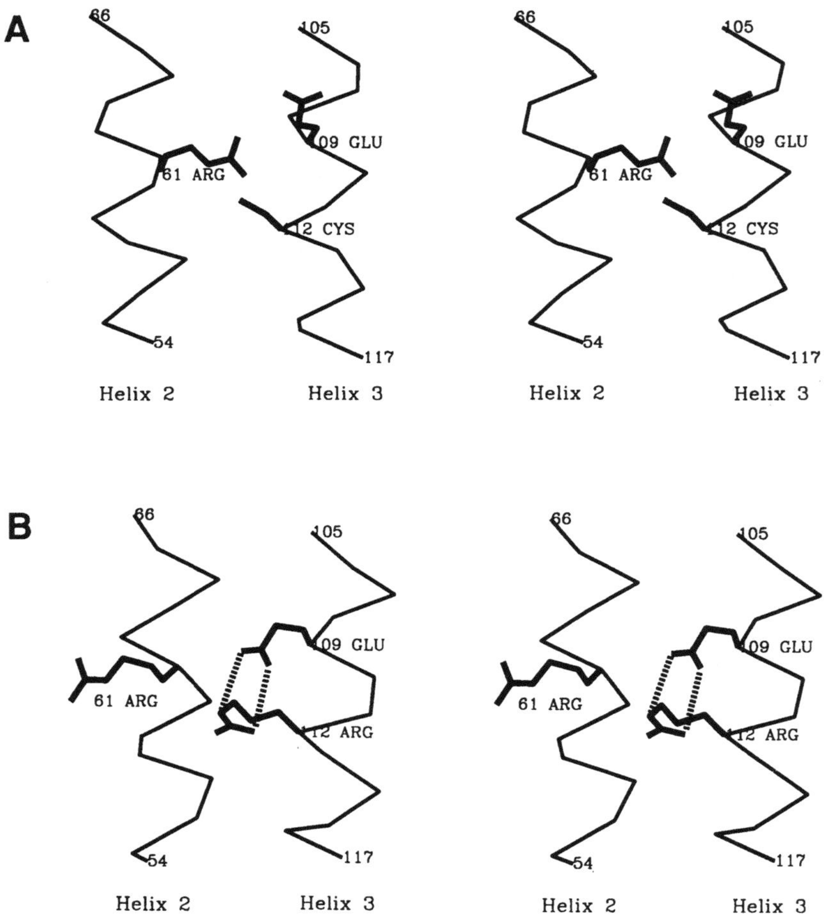

Fig. 5. Stereo diagrams showing the major charged side chain differences between apoE3 (A) and apoE4 (B) in the regions of helices 2 and 3 that are adjacent to residue 112, the site of substitution in apoE4. In apoE3 which has cysteine at position 112, the side chain of arginine-61 lies just above cysteine-112 and partially fills the space between helices 2 and 3, and glutamate-109 lies free in an external location (A). In apoE4, a new salt bridge is formed between the introduced arginine-112 and glutamate-109 causing the side chain of arginine-61 to be displaced from the pocket between helices 2 and 3 to a more exterior location.

terminal domain of apoE4 acts in some way to stabilize the formation of longer helices in this variant relative to apoE3, thus promoting the preferrential distribution of apoE4 to larger VLDL-sized particles.

Two possibilities as to how this might happen are suggested from the crystal structure. First, the new salt bridge that forms between the mutant arginine at position 112 and glutamate-109 perhaps causes a previously existing salt bridge between glutamate-109 and a basic residue in the carboxy-terminal, lipid-binding domain to be broken as shown schemmatically in Figure 6, A. Alternatively, to accommodate the new arginine-112, the side chain of arginine-61 has to move from a position partially included within the structure to a more exterior position. It is possible that this now forms a new salt bridge with an acidic residue in the carboxy-terminal, lipid-binding domain as depicted in Figure 6, B. Both of these hypotheses are presently being tested by site-directed mutagenesis of glutamate-109 and arginine-61 in the amino-terminal domain and by substitution and truncation mutagenesis of the carboxy-terminal domain. The exact mechanism of interaction between the two structural domains of apoE will be better understood once a crystal structure of the intact protein is available.

A second apolipoprotein structure has recently been reported by Breiter et al. (1991), that of apolipophorin III from the African migratory locust. The X-ray crystallographic structure of apolipophorin III shows that it has the same tertiary fold as the amino-terminal domain of apoE, with five amphipathic helices connected by short loops (Breiter et al. 1991). Interestingly, apolipophorin III exists both free as a monomer in the hemolymph of the resting insect, and bound to lipid as it functions to deliver lipid to the active flight muscles (Wells et al. 1987). Consequently, it must readily convert from one state to the other to carry out its biological function. Using air-water interface studies Kawooya et al. (1986) determined that the helical bundle of apolipophorin III probably was open when in the lipid-associated conformation. Breiter et al. (1991) discussed that for energetic reasons it was unlikely that apolipophorin III underwent major changes in either secondary structural elements or their relative orientations with respect to one another when it associated with lipid.

Air-water interface studies of the amino-terminal domain of apoE would indicate that this fragment could unfold in a similar fashion when bound to lipid (Weisgraber et al. 1991). Whereas the surface area calculated for the long face of this domain when in the 4-helix bundle conformation of the crystal structure is ~1560Å^2, the

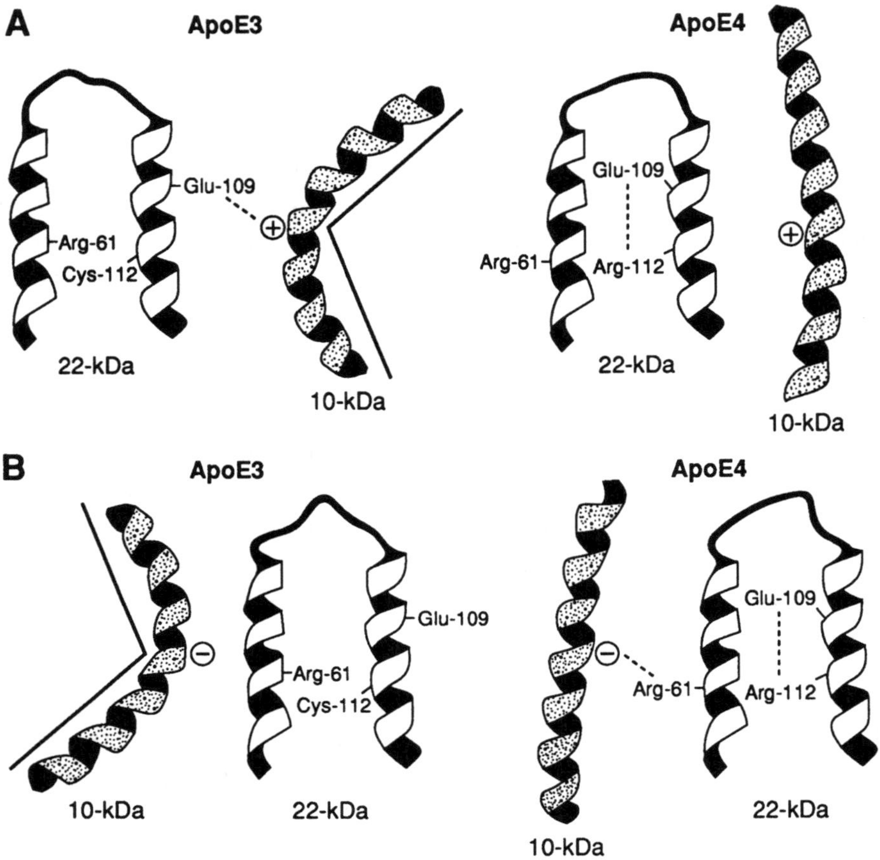

Fig. 6. Schemmatic representations of how the arginine-for-cysteine mutation at residue 112 in the amino-terminal domain (22-kDa) might influence the lipoprotein binding preferrence of the carboxy-terminal domain (10-kDa). The side chain of glutamate-109 occupies a very exterior position in apoE3 and perhaps salt bridges with a basic residue in the carboxy-terminal domain. This salt bridge contact may be lost in apoE4 when glutamate-109 forms a new salt bridge with the introduced arginine-112 leading to a reorganization of the carboxy-terminal domain into a conformation that prefers triglyceride-rich lipoproteins (A). Alternatively, the displacement to a more exterior position of arginine-61 in apoE4 may allow it to form a new salt bridge with an acidic residue in the carboxy-terminal domain thus changing the conformation of the lipid-binding region (B).

area this fragment occupied at the air-water interface was found to be ~3000Å^2 (Weisgraber et al. 1991). This is very close to that calculated (~2860Å^2) if the bundle simply opened while maintaining the 4 α-helices in the same orientation with respect to each other. Certainly, there is some difference between the lipid-free and lipid-associated states of this structure as only in the lipid-associated conformation does it bind to the LDL receptor with a physiologically significant association rate. Furthermore, it is possible that *in vivo,* two conformations of the amino-terminal domain of apoE may be in equilibrium; one lipid-associated and the other the lipid-free conformation of the crystal structure, where the ratio of each perhaps depends upon the available phospholipid which is apparently the major lipid class to which it will bind (Weisgraber 1990). The observation several years ago by Gianturco et al. (1983) that there were two populations of apoE present on hypertriglyceridemic VLDL, one receptor-active and one receptor-inactive is perhaps consistent with this possibility although other mechanisms involving protein-lipid or protein-protein interactions could also account for this observation. An additional feature of the crystal structure may suggest how the lipid-associated

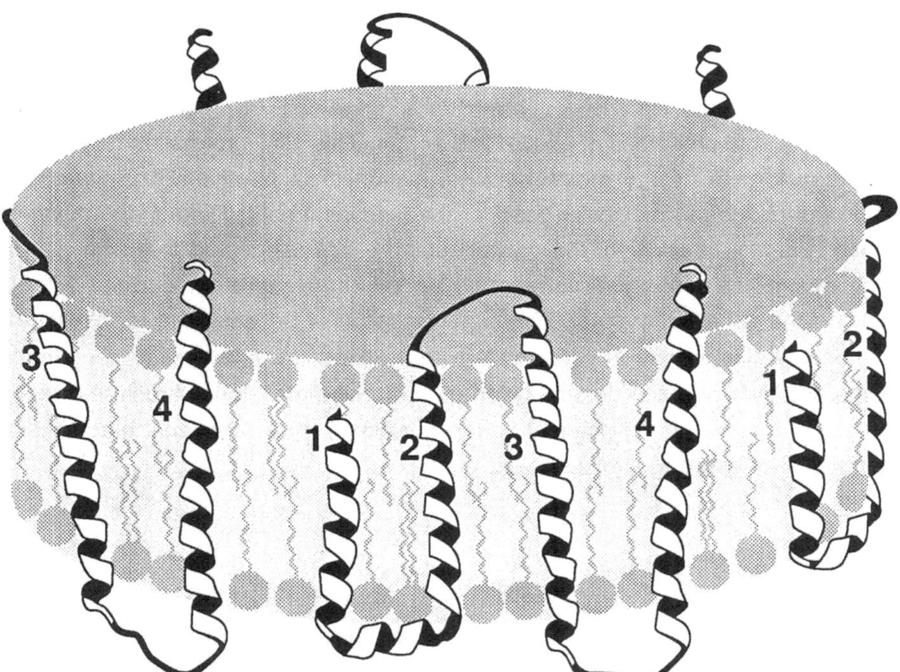

Fig. 7. Schemmatic representation of how the 4-helical bundle of the amino-terminal domain of apoE may open up and align on the edge of a phospholipid disc such that the helical axes run parallel to the phospholipid acyl chains. The lower dielectric constant of the lipid environment would increase the net positive potential of the region around residues 135 - 150 and perhaps it is this rather than a major conformational change which increases the binding affinity of apoE for the LDL receptor.

structure might differ from the lipid-free 4-helix bundle. The turn between helices 2 and 3 of the bundle is poorly defined in the electron density map because the atoms in this loop are very mobile. This loop may therefore normally be in a different conformation when apoE is bound to a lipoprotein surface such that the bundle opens. It has been shown by Hefele Wald et al. (1990) that the α-helical axes of apolipoproteins associated with phospholipid discs run parallel to the phospholipid acyl chains. Therefore, the receptor-active conformation of the amino-terminal domain of apoE may exist as depicted in Figure 7. Experiments are ongoing to test this hypothesis, for example by locking the protein in the closed conformation via a disulfide bond between helices 1 and 4 and then determining whether it can bind lipid and become receptor active.

In conclusion, the structures of the amino-terminal domains of the three common genetic variants of apolipoprotein E, apoE2, apoE3, and apoE4 have been determined by X-ray crystallography. Although the structures were necessarily determined in the absence of lipid and therefore may not represent the receptor-active conformations, they nevertheless do suggest possible mechanisms for the altered functional properties of apoE2 and apoE4. This new information has laid the foundation for the design of the next generation of biochemical experiments to better understand the molecular basis of the functional aberrations in the common genetic variants. In addition, the crystal structure has suggested a mechanism by which this domain of apoE may open up on the lipoprotein surface without changing the α-helical content. Experiments can now be designed to investigate whether it is this structural conformation that allows the expression of receptor-binding activity. Thus, the determination of the three-dimensional structure of apoE represents a new starting point to better understand how functional mechanisms are intrinsically linked to structure.

References:

Aggerbeck LP, Wetterau JR, Weisgraber KH, Mahley RW, and Agard DA (1988a) Crystallization and preliminary X-ray diffraction studies on the amino-terminal (receptor-binding) domain of human apolipoprotein E3 from serum very low density lipoproteins. *J Mol Biol* **202:**179-181

Aggerbeck LP, Wetterau JR, Weisgraber KH, Wu C-SC, and Lindgren FT (1988b) Human apolipoprotein E3 in aqueous solution. II. Properties of the amino- and carboxyl-terminal domains. *J Biol Chem* **263:**6249-6258

Bouthillier D, Sing CF, and Davignon J (1983) Apolipoprotein E phenotyping with a single gel method: application to the study of informative matings. *J Lipid Res* **24:**1060-1069

Breiter DR, Kanost MR, Benning MM, Wesenberg G, Law JH, Wells MA, Rayment I, and Holden HM (1991) Molecular structure of an apolipoprotein determined at 2.5-Å resolution. *Biochemistry* **30**:603-608

Davignon J, Gregg RE, and Sing CF (1988) Apolipoprotein E polymorphism and atherosclerosis. *Arteriosclerosis* **8**:1-21

Gianturco SH, Gotto AM Jr, Hwang S-LC, Karlin JB, Lin AHY, Prasad SC, and Bradley WA (1983) Apolipoprotein E mediates uptake of S_f 100-400 hypertriglyceridemic very low density lipoproteins by the low density lipoprotein receptor pathway in normal human fibroblasts. *J Biol Chem* **258**:4526-4533

Gregg RE, Zech LA, Schaefer EJ, Stark D, Wilson D, and Brewer HB Jr (1986) Abnormal *in vivo* metabolism of apolipoprotein E_4 in humans. *J Clin Invest* **78**:815-821

Havel RJ, and Kane JP (1973) Primary dysbetalipoproteinemia: predominance of a specific apoprotein species in triglyceride-rich lipoproteins. *Proc Natl Acad Sci USA* **70**:2015-2019

Hefele Wald J, Goormaghtigh E, De Meutter J, Ruysschaert J-M, and Jonas A (1990) Investigation of the lipid domains and apolipoprotein orientation in reconstituted high density lipoproteins by fluorescence and IR methods. *J Biol Chem* **265**:20044-20050

Innerarity TL, Friedlander EJ, Rall SC Jr, Weisgraber KH, and Mahley RW (1983) The receptor-binding domain of human apolipoprotein E. Binding of apolipoprotein E fragments. *J Biol Chem* **258**:12341-12347

Kawooya JK, Meredith SC, Wells MA, Kézdy FT, and Law JH (1986) Physical and surface properties of insect apolipophorin III. *J Biol Chem* **261**:13588-13591

Lalazar A, Weisgraber KH, Rall SC Jr, Giladi H, Innerarity TL, Levanon AZ, Boyles JK, Amit B, Gorecki M, Mahley RW, and Vogel T (1988) Site-specific mutagenesis of human apolipoprotein E. Receptor binding activity of variants with single amino acid substitutions. *J Biol Chem* **263**:3542-3545

Mahley RW (1988) Apolipoprotein E: cholesterol transport protein with expanding role in cell biology. *Science* **240**:622-630

Mahley RW, Innerarity TL, Pitas RE, Weisgraber KH, Brown JH, and Gross E (1977) Inhibition of lipoprotein binding to cell surface receptors of fibroblasts following selective modification of arginyl residues in arginine-rich and B apoproteins. *J Biol Chem* **252**:7279-7287

Mahley RW, Innerarity TL, Rall SC Jr, and Weisgraber KH (1984) Plasma lipoproteins: apolipoprotein structure and function. *J Lipid Res* **25**:1277-1294

Mahley RW, and Rall SC Jr (1989) Type III hyperlipoproteinemia (dysbetalipoproteinemia): the role of apolipoprotein E in normal and abnormal lipoprotein metabolism. *In* The Metabolic Basis of Inherited Disease, 6th edition, chapter 47, Scriver CR, Beaudet AL, Sly WS, and Valle D (eds). Mcgraw-Hill, New York 1195-1213

Rall SC Jr, Newhouse YM, Clarke HRG, Weisgraber KH, McCarthy BJ, Mahley RW, and Bersot TP (1989) Type III hyperlipoproteinemia associated with apolipoprotein E phenotype E3/3. Structure and genetics of an apolipoprotein E3 variant. *J Clin Invest* **83**:1095-1101

Rall SC Jr, Weisgraber KH, Innerarity TL, Bersot TP, Mahley RW, and Blum CB (1983) Identification of a new structural variant of human apolipoprotein E, E2($Lys_{146} \rightarrow$Gln), in a type III hyperlipoproteinemic subject with the E3/2 phenotype. *J Clin Invest* **72**:1288-1297

Rall SC Jr, Weisgraber KH, Innerarity TL, and Mahley RW (1982a) Structural basis for receptor binding heterogeneity of apolipoprotein E from type III hyperlipoproteinemic subjects. *Proc Natl Acad Sci USA* **79**:4696-4700

Rall SC Jr, Weisgraber KH, and Mahley RW (1982b) Human apolipoprotein E. The complete amino acid sequence. *J Biol Chem* **257:**4171-417

Segrest JP, Jones MK, De Loof H, Brouillette CG, Venkatachalapathi YV, and Anantharamaiah GM (1992) The amphipathic helix in the exchangeable apolipoproteins: a review of secondary structure and function. *J Lipid Res* **33:**141-166.

Steinmetz A, Jakobs C, Motzny S, and Kaffarnik H (1989) Differential distribution of apolipoprotein E isoforms in human plasma lipoproteins. *Arteriosclerosis* **9:**405-411

Utermann G, Hees M, and Steinmetz A (1977) Polymorphism of apolipoprotein E and occurrence of dysbetalipoproteinemia in man. *Nature* **269:**604-607

Utermann G, Langenbeck U, Beisiegel U, and Weber W (1980) Genetics of the apolipoprotein E system in man. *Am J Hum Genet* **32:**339-347

Utermann G, Pruin N, and Steinmetz A (1979) Polymorphism of apolipoprotein E: III. Effect of a single polymorphic gene locus on plasma lipid levels in man. *Clin Genet* **15:**63-72

Wardell MR, Brennan SO, Janus ED, Fraser R, and Carrell RW (1987) Apolipoprotein E2-Christchurch (136Arg→Ser). New variant of human apolipoprotein E in a patient with type III hyperlipoproteinemia. *J Clin Invest* **80:**483-490

Weisgraber KH (1990) Apolipoprotein E distribution among human plasma lipoproteins: role of the cysteine-arginine interchange at residue 112. *J Lipid Res* **31:**1503-1511

Weisgraber KH, Innerarity TL, Harder KJ, Mahley RW, Milne RW, Marcel YL, and Sparrow JT (1983) The receptor-binding domain of human apolipoprotein E. Monoclonal antibody inhibition of binding. *J Biol Chem* **258:**12348-12354

Weisgraber KH, Innerarity TL, and Mahley RW (1982) Abnormal lipoprotein receptor-binding activity of the human E apoprotein due to cysteine-arginine interchange at a single site. *J Biol Chem* **257:**2518-2521

Weisgraber KH, Innerarity TL, and Mahley RW (1978) Role of lysine residues of plasma lipoproteins in high affinity binding to cell surface receptors on human fibroblasts. *J Biol Chem* **253:**9053-9062

Weisgraber KH, Lund-Katz S, Mahley RW, and Phillips M (1991) Modulation of apolipoprotein E structure by lipid. *Arterioscler Thromb* **11:**1419a

Weisgraber KH, Rall SC Jr, and Mahley RW (1981) Human E apoprotein heterogeneity. Cysteine-arginine interchanges in the amino acid sequence of the apoE isoforms. *J Biol Chem* **256:**9077-9083

Wells MA, Ryan RO, Kawooya JK, and Law JH (1987) The role of apolipophorin III in *in vivo* lipoprotein interconversions in adult Manduca sexta. *J Biol Chem* **262:**4172-4176

Wetterau JR, Aggerbeck LP, Rall SC Jr, and Weisgraber KH (1988) Human apolipoprotein E3 in aqueous solution. I. Evidence for two structural domains. *J Biol Chem* **263:**6240-6248

Wilson C, Wardell MR, Weisgraber KH, Mahley RW, and Agard DA (1991) Three-dimensional structure of the LDL receptor-binding domain of human apolipoprotein E. *Science* **252:**1817-1822

Wilson C, Weisgraber KH, Wardell MR, Mahley RW, and Agard DA (1992) Salt bridge relay triggers defective LDL receptor binding by mutant apolipoprotein E. *Submitted for publication*

Zannis VI, and Breslow JL (1981) Human very low density lipoprotein apolipoprotein E isoprotein polymorphism is explained by genetic variation and post translational modification. *Biochemistry* **20:**1033-1041

Zannis VI, Just PW, and Breslow JL (1981) Human apolipoprotein E isoprotein subclasses are genetically determined. *Am J Hum Genet* **33:**11-24

ANALYSIS OF APOLIPOPROTEIN E FUNCTION IN ADRENAL CELLS: EFFECTS ON CHOLESTEROL METABOLISM AND SIGNAL TRANSDUCTION PATHWAYS

David L. Williams, Margaret M. Prack, M. Nicosia, J.T. Gwynne[1], and Mary E. Reyland
Department of Pharacological Sciences, Health Sciences Center, State University of New York at Stony Brook, Stony Brook, NY 11794

Introduction

Apolipoprotein E (apoE) is atypical of the other plasma apolipoproteins in that it is synthesized in a wide variety of peripheral tissues and in diverse cell types and not just in liver and small intestine (Blue et al.1983; Williams et al.1985b). Although considerable information about the regulation of apoE expression in various cell types has accumulated over the past decade, little is known about the function of apoE that is synthesized locally in many tissues. Among the potential functions in lipid metabolism that have been suggested are 1) a role in reverse cholesterol transport to target interstitial lipoproteins for hepatic clearance, 2) as a component of a local mechanism for removing cholesterol from cells or shuttling lipid between cells in a tissue, 3) as a means to enhance targeting of lipoproteins for uptake by cells expressing apoE, and 4) as a regulator of intracellular cholesterol utilization (Blue et al.1983; Williams et al.1985b; Williams et al.1985a). ApoE may also have autocrine or paracrine activities not directly related to lipid metabolism since apoE can modify hormonal or mitogenic stimuli in ovarian theca/interstitial cells and T lymphocytes, respectively (Dyer & Curtiss, 1988; Curtiss et al.1977).

Although as much as 20-40 % of apoE synthesis occurs in peripheral tissues (Newman et al.1985), a primary role in reverse cholesterol transport is made less likely by two factors. First, apoE is expressed in the central nervous system (Williams et al.1985b; Elshourbagy et al.1985) into which there is little or no traffic of lipoproteins because of the blood-brain barrier (Serougne et al.1976; Pardridge & Mietus, 1980). Expression of apoE in a closed compartment such as the central nervous system is more suggestive of a local rather than a systemic function. Second, studies in liver transplant patients indicate that 90% or more of plasma apoE is derived from the liver (Linton et al., 1991). This argues that little of the apoE synthesized in peripheral tissues remains in the plasma compartment. Nevertheless, it has been demonstrated that apoE synthesized in dog limb is recovered in lymph drainage (Dory et al., 1986) and that apoE made by grafted epidermal keratinocytes reaches the systemic circulation in the mouse (Fenjves et al.1989). Thus, it is possible that interstitial

[1] Department of Endocrinology, University of North Carolina at Chapel Hill, Chapel Hill, NC 27514

NATO ASI Series, Vol. H 73
Human Apolipoprotein Mutants III
Edited by C. R. Sirtori, G. Franceschini, B.H. Brewer Jr.

lipoproteins are targeted for hepatic removal by peripheral apoE, although the steady state concentation of such particles in plasma must be quite low.

The possibility that locally-derived apoE functions in lipid shuttling among cells in a tissue is strengthened by the observation that apoE expression is enhanced in optic nerve during Wallarian degeneration (Dawson et al. 1986) and in injured peripheral nerve during both degeneration and regeneration phases (Boyles et al. 1989). ApoE is also expressed at very high levels in steroidogenic tissues such as adrenal gland, testis, and ovary (Blue et al. 1983; Williams et al. 1985b; Driscoll & Getz, 1984) that have a major cholesterol requirement for hormone synthesis. These observations, of course, also are consistent with possibilities 3 and 4 noted above or with a role for apoE as an autocrine or paracrine factor. As regards the question of function, the difficulty with many studies on apoE expression is that the experiments are purely observational and, hence, lead to deductions about function but do not provide direct evidence for function. In order to take a more direct approach to function, we have recently focused on apoE expression in adrenal cells. Two experimental models have been used. The first is the rat adrenal gland as an *in vivo* model in which cellular cholesterol metabolism can be readily altered by agents that stimulate or inhibit steroid production. The second is the murine Y1 adrenal cell line in which the transfected human apoE gene was expressed constitutively. This system permits one to ask more directly how apoE may alter cholesterol metabolism whereas the *in vivo* model provides the relationship between cholesterol and apoE expression when cholesterol metabolism is perturbed. The expectation was that by using both approaches definitive evidence for apoE function could be obtained.

The *In Vivo* Model

The relationship between apoE mRNA levels and adrenal gland cholesterol was assessed by treating rats with agents known to alter adrenal steroidogenesis and/or cholesterol content (Prack et al. 1991a). Treatment with the synthetic glucocorticoid, Dexamethasone (DEX), for example, inhibits steroidogenesis by blocking the release of adrenocorticotropin (ACTH) from the pituitary. This suppression of steroid production also leads to accumulation of free and esterified cholesterol in adrenocortical cells. As shown in Table 1, DEX treatment increased apoE mRNA concentration in parallel to the increase in tissue cholestrol and inversely to steroid production as judged by serum corticosteroids.

Rats were also treated with 4-aminopyrazolopyrimidine (4APP), an agent that decreases plasma cholesterol by blocking lipoprotein secretion from the liver (Schiff et al. 1971). Such animals also are stressed leading to hyperstimulation of adrenal steroidogenesis and depletion of adrenal gland cholesteryl ester (Brecher & Hyun, 1978). ApoE mRNA levels were reduced in 4APP-treated animals in parallel with the depletion of adrenal gland cholesteryl ester content. A similar reduction in apoE mRNA and cholesteryl ester content was seen when ACTH was administered to DEX-treated animals. The correlation in Figure 1 shows a strong relationship between apoE mRNA levels and adrenal gland total cholesterol content ($r = 0.75$), suggesting that the regulation of apoE expression closely parallels tissue cholesterol content.

Table 1. Serum corticosteroids, adrenal cholesterol, and adrenal apoE mRNA concentrations in untreated and dexamethasone-treated rats

Treatment	Adrenal ApoE mRNA	Adrenal Total Cholesterol	Adrenal Free Cholesterol	Serum Corticosteroids
	pg/μg RNA	μg/mg protein	μg/mg protein	ng/ml
Control	14.6 ± 2 (8)	178 ± 18 (7)	28.3 ± 5 (3)	560 ± 100 (4)
Dexamethasone	45.9 ± 4 (10)	517 ± 43 (13)	69.5 ± 9 (10)	42.7 ± 7 (10)

Data shown are the average ± SEM for untreated (control) rats or rats given three daily 1 mg dexamethasone injections. Number of rats indicated in parentheses. Reprinted with permission from (Prack et al. 1991a).

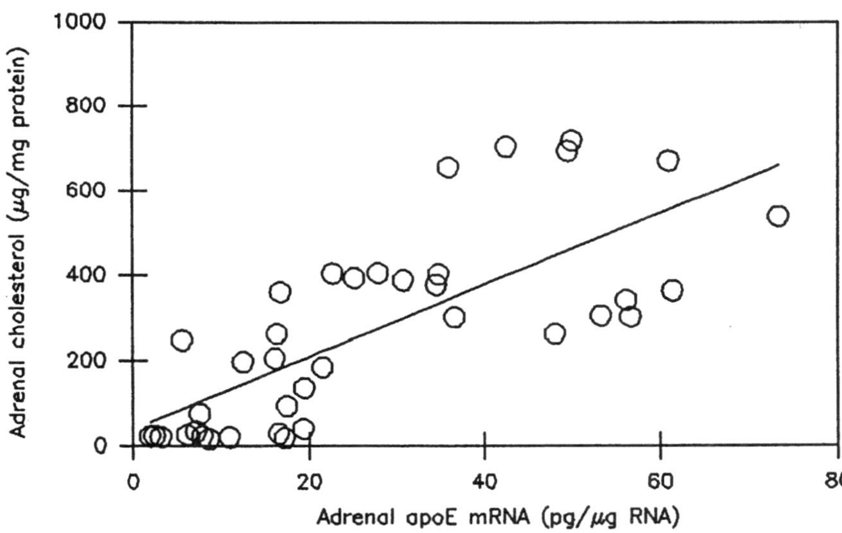

Figure 1 Correlation between adrenal apoE mRNA and adrenal cholesterol in untreated rats and rats treated with DEX, 4APP, ACTH, or a combination of these agents in 6 independent experiments. Adapted from (Prack et al. 1991a).

These results may indicate that apoE expression is positively regulated by tissue cholesterol as has been suggested in the macrophage (Mazzone et al. 1987; Mazzone et al. 1989; Basu et al. 1981). However, when rats were treated with aminogluthethimide, an inhibitor of adrenal steroidogenesis, apoE mRNA concentration was unchanged whereas tissue cholesterol content was greatly increased (Prack et al. 1991a). Similarly, ACTH decreased adrenal gland apoE mRNA concentration in animals treated with DEX and aminoglutethimide while tissue cholesterol content was increased. These findings argue against cholesterol as a primary positive regulator of apoE expression in adrenal cells and suggest, instead, that ACTH is a primary negative regulator. The important point with regard to function is that apoE expression is high when cholesterol content is high and steroidogenesis is low, and apoE expression is low when cholesterol content is low and steroidogenesis is high. Since the demand for cholesterol for steroid production is high when apoE expression is low, it seems unlikely that apoE is made by adrenal cells to enhance lipoprotein uptake by either the LDL receptor pathway (Brown et al.1979) or the selective uptake pathway (Rinninger & Pittman, 1987).

The cell type distribution of apoE mRNA expression in rat adrenal gland was determined by *in situ* hybridization (Nicosia et al.1992b). The results showed that apoE mRNA is 3-4 fold more abundant in z. fasciculata and z. reticularis cells as compared to z. glomerulosa or medulla in control adrenal glands. After treatment with DEX, apoE mRNA content was selectively increased in in the outer z. fasciculata (Table 2). As compared to control animals or 4APP-treated animals, apoE mRNA increased 3-4 fold in z. fasciculata cells from DEX-treated animals. These results indicate that the increase in total adrenal gland mRNA in DEX-treated animals is due to cell-type specific regulation within the cell population that is responsible for glucocorticoid synthesis.

Table 2. Distribution of ApoE mRNA in Zones of the Rat Adrenal Gland

Group	Glom	FascI	FascII	FascIII	Retic	Medulla
CONTROL n=7	4±2	9±5	12±5	16±10	12±10	3±2
4APP n=4	5±4	8±6	8±4	14±8	11±8	3±3
DEX n=5	9±4	28±3	33±6	28±6	21±8	5±2
DEX vs CONTROL	ns	p<0.001	p<0.001	ns	ns	ns
DEX vs 4APP	ns	p<0.01	p<0.001	ns	ns	ns

The table shows the mean net hybridization signal for apoE mRNA within the indicated regions, expressed as autoradiographic grains per 10 nuclei. Statistical analyses via one way analysis of variance were performed with the GraphPad INSTAT program. Reproduced with permission from (Nicosia et al.1992a).

The Y1 Adrenal Cell Model

The murine Y1 adrenal cell has many properties of normal adrenocortical cells *in vivo* (Kowal & Fiedler, 1968) although Y1 cells do not express apoE mRNA or protein. To investigate the influence of apoE on cholesterol metabolism, we transfected Y1 cells with a human genomic apoE expression vector, and isolated stable cell lines which constitutively express the human apoE protein (Reyland et al.1991). The most striking phenotype of the Y1E lines is a marked suppression of basal and cAMP- or ACTH-stimulated steroidogenesis (Reyland et al. 1991). Further analysis showed that Y1E cells are blocked in a number of cAMP-mediated responses including changes in cell morphology, the induction of ornithine decarboxylase mRNA, and the maintenance of basal levels of mRNAs encoding two of the P450 steroid hydroxylases (Reyland & Williams, 1991). The reduction in the basal level of the P450-side chain cleavage enzyme ($P450_{scc}$) mRNA ranged from 5-10 fold in the various Y1E cell lines and paralleled the reduction in basal steroid production in the apoE-expressing cells (Table 3). This result suggests that the reduction in basal steroid production is due to the reduced level of expression of the side chain cleavage enzyme. It is clear from the data in Table 3, however, that reduced cAMP-stimulated steroidogenesis in the Y1E cell lines can not be explained by the 5-10 fold reduction in $P450_{scc}$ mRNA since cAMP-stimulated steroidogenesis is inhibited as much as 100 fold.

Table 3. Comparison of ApoE mRNA, SCC mRNA and steroid secretion in the Y1-E cell lines

Cell Line	apoE mRNA pg/μg	SCC mRNA % of Y1-neo	Steroid (μg/mg cell protein) Basal	+Bt_2cAMP
Y1-neo	<0.1	100	1.15	24.60
Y1-E15	0.8	61	0.90	18.40
Y1-E2	2.9	19	0.11	0.33
Y1-E10	63.4	16	0.06	0.15
Y1-E12	120.0	16	0.19	0.25
Y1-E16	57.2	14	0.18	0.27
Y1-E1	17.2	9	0.09	0.18

ApoE mRNA was measured by DNA-excess solution hybridization using a human cDNA probe. SCC mRNA was quantified by dot blot hybridization with a mouse cDNA probe and normalized to the amount of GAPDH mRNA hydridized. Secreted steroids were measured using a fluorogenic assay as described (Gwynne & Hess, 1980). Bt_2cAMP stimulated cells were incubated for 24 hrs in complete medium containing one mM Bt_2cAMP. Steroid values represent μg fluorogenic steroid secreted/24 hrs/mg total cell protein. All values represent the average of two or more experiments with the exception of Y1-E10 SCC and apoE mRNA, which are from a single experiment. Reprinted with permission from (Reyland & Williams, 1991).

The block in the cAMP - protein kinase A (PKA) pathway in Y1E cells suggests that apoE expression can modulate a specific signal transduction pathway. Several points are relevant to this finding. First, although several cAMP responses are blocked in the Y1E cells, cAMP stimulation still yields a modest 2-3 fold increase in steroid production (Table 3). Analysis of $P450_{scc}$ mRNA levels in Y1E cells showed a similar increase in response to cAMP (Reyland & Williams, 1991) suggesting that the increase in steroidogenesis is due to increased expression of the $P450_{scc}$ enzyme. Of the various cAMP responses that were examined in the Y1E cells, induction of the $P450_{scc}$ mRNA is the only response that is not blocked. This result argues that the block in the PKA signalling pathway in Y1E cells is selective for some but not all responses. Second, previous studies have shown that maintenance of the basal level of $P450_{scc}$ mRNA in the parent Y1 cell requires a functional PKA but not exogenous cAMP (Wong et al. 1989). Thus, much of the observed phenotype in the Y1E cells could arise if PKA were absent or nonfunctional. To test this point directly, PKA was assayed in extracts from Y1 and Y1E cells. As shown in Figure 2, PKA activity is present in Y1E cells at levels comparable to Y1 cells, and activation by cAMP appears to be normal. PKA activation by cAMP also appears to be normal *in vivo* as judged by similar activity ratios in Y1 and Y1E cells (Reyland & Williams, 1991). These results support that idea that the PKA signalling pathway is blocked in Y1E cells at a point distal to activation of the kinase itself. However, these data do not eliminate the possibility of a PKA inhibitor in Y1E cells that is inactive in the PKA assays due to dilution in the cell free extracts.

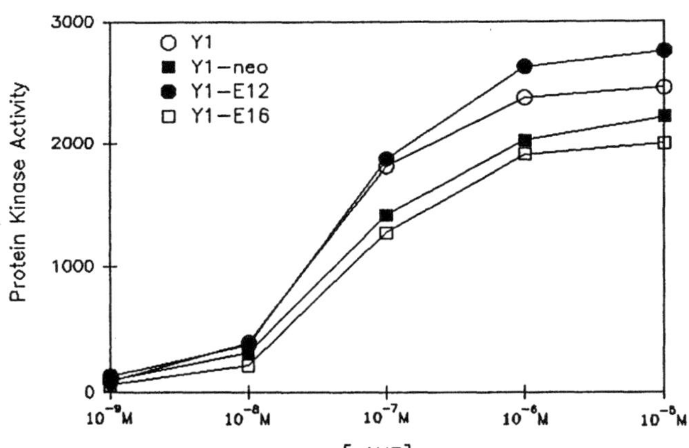

Figure 2. cAMP-dependent protein kinase activity in Y1, Y1-neo, and Y1E cell lines. Reproduced with permission from Reyland & Williams, 1991).

In more recent studies we have examined the protein kinase C (PKC) pathway in Y1E cells since PKC appears to be a negative regulator of $P450_{scc}$ mRNA levels in Y1 cells (Moore et al. 1990). The results show that Y1E cell lines have 3-5 fold increased levels of PKC activity and PKC_α isoform protein (Reyland et al. 1992). Interestingly, Y1E cells also show 2-3 fold increased concentrations of *sn*-1,2-diacylglycerol, an *in vivo* activator of PKC. These results suggest that Y1E cells are chronically exposed to elevated levels of activated PKC. To test the role of the increased PKC activity in the suppression of steroidogeneis in Y1E cells, Y1 or Y1E cells were cultured with calphostin C, an inhibitor of PKC. The result showed that treatment with calphostin C increased steroid production by Y1E cells up to the basal level of steroid production in the Y1 control cell line. Accompanying the increase in steroidogenesis was a similar increase in the concentration of $P450_{scc}$ mRNA. These data indicate that reduced basal steroidogenesis and reduced basal expression of $P450_{scc}$ mRNA are due, in part, to elevated PKC activity in Y1E cells (Reyland et al. 1992). Restoration of basal steroidogenesis by inhibition of PKC, however, did not restore cAMP-stimulated steroidogenesis in Y1E cells.

Y1E cells also show two distinct alterations in cholesterol metabolism in comparison to the parental Y1 cell (Prack et al. 1991b). First, Y1E cell lines have 2-5 fold increased content of cholesteryl ester, and Y1E cells expressing higher levels of apoE also have increased free cholesterol levels. Second, Y1E cells show a reduced efflux of free cholesterol from the cell. Thus, despite having elevated levels of free cholesterol, Y1E cells show a reduced efflux of cholesterol from the cell membrane. Since it is the cell membrane pool of free cholesterol that is available for efflux (Johnson et al. 1991; Mahlberg & Rothblat, 1992), reduced efflux in Y1E cells suggests that movement of free cholesterol to the cell membrane is hindered in apoE-expressing adrenal cells. These data suggest that apoE expression promotes cholesterol accumulation and cholesteryl ester storage at the expense of cholesterol availability for efflux from the cell or for steroidogenesis. These alterations may reflect an influence of apoE on the trafficking of cholesterol within the cell.

Summary and Perspective

With regard to cholesterol metabolism in adrenal cells, the *in vivo* studies with the rat adrenal gland show that apoE expression changes in parallel with cholesteryl ester storage and inversely with the rate of steroid production (Prack et al. 1991a; Nicosia et al. 1992b). ApoE mRNA is present at the highest levels in z. fasciculata cells, and is markedly increased in these cells when steroidogenesis is suppressed by administration of exogenous glucocorticoid (Nicosia et al. 1992b). These results are consistent with the findings in Y1E cells in which constitutive expression of apoE leads to suppression of steroidogenesis and enhanced cholesteryl ester storage. Taken together, the results with the *in vivo* model and Y1E cell model support the idea that apoE acts to regulate cholesterol utilization for steroid production in adrenal cells (Blue et al.1983). One interpretation of these findings is that, in the rat adrenal gland, apoE expression is reduced when steroidogenesis is stimulated thereby permitting utilization of stored cholesteryl ester for steroid production. In the Y1E cell, this pattern of regulation is disrupted due to the constitutive expression of apoE with the result that cholesterol is not available for steroidogenesis.

The mechanism through which apoE expression alters cholesterol metabolism in Y1E cells is not known, but two possibilities may be suggested. The first is that apoE may act directly to alter some aspect of cholesterol trafficking in the cell to promote cholesteryl ester storage and reduce free cholesterol availability for steroid production. The recent finding that apoE is found in rat hepatocytes in locations other than in the secretory or endocytic pathways (Hamilton et al.1990) raises the possibility that apoE may be present in the cytoplasm where it could have direct effects on cholesterol movement in the cell. The second possibility is that the effects of apoE expression on cholesterol metabolism indirectly reflect alterations in signal transduction pathways. A focus of future studies will be to distinguish between these possibilities.

A surprising result of the Y1E studies is that apoE expression alters both the PKA and PKC signal transduction pathways. Previous work by Curtiss and colleagues (Dyer & Curtiss, 1988; Curtiss et al.1977) showed that apoE can modifiy cell specific responses. Mitogen-stimulated T-cell proliferation, for example, is blocked by exogenous apoE (Curtiss et al.1977). Similarly, in cultured theca/interstitial cells from rat ovary, apoE blocks the induction of androgen synthesis by luteinizing hormone (LH) when apoE is present at or prior to the time of hormone treatment (Dyer & Curtiss, 1988). This result may be explained if apoE blocks the LH-mediated induction of the C17-C20 lyase that is responsible for the conversion of progesterone to androgen, although this has not been tested directly. Interestingly, *in situ* hybridization analysis showed that apoE mRNA is expressed at high levels in both theca and intestitial cells in rat ovary (Nicosia et al., 1992a). The hypothesis has been advanced that apoE is expressed as an autocrine regulator to suppress androgen production in theca cells of follicles destined for atresia. Similarly, apoE expression in interstitial cells may serve to maintain the interstitial cells as a progesterone-producing compartment (Nicosia et al., 1992a). These findings in theca/interstitial cells and in T-cells are suggestive that apoE acts to alter signal transduction pathways. Our findings with the Y1E cells show directly that signal transduction pathways can be altered by apoE expression and raise the possibility that apoE acts to modulate signal transduction pathways in a wide variety of cells. Such actions could have diverse effects on cellular function and may be the reason why apoE is expressed in such a wide variety of tissues and cell types. A major goal of future work is to explore the role of apoE as a signal transduction modulator protein.

Acknowledgement

This work was supported by NIH Grants HL 32868 (DLW) and HL 28306 (JG), National Research Service Award 1T32-DK-07521 (MMP), and American Heart Association New York Affiliate Grant 90-111G (MER)

References

Basu, S.K., Brown, M.S., Ho, Y.K., Havel, R.J., and Goldstein, J.L. (1981). Mouse macrophages synthesize and secrete a protein resembling apolipoprotein E. Proc. Natl. Acad. Sci. USA *78*, 7545-7549.

Blue, M.-L., Williams, D.L., Zucker, S., Khan, S.A., and Blum, C.B. (1983). Apolipoprotein E synthesis in human kidney, adrenal gland, and liver. Proc. Natl. Acad. Sci. USA *80*, 283-287.

Boyles, J.K., Zoellner, C.D., Anderson, L.J., Kosik, L.M., Pitas, R.E., Weisgraber, K.H., Hui, D.Y., Mahley, R.W., Gebicke-Haerter, P.J., Ignatius, M.J., and Shooter, E.M. (1989). A role for apolipoprotein E, Apolipoprotein A-I, and low density lipoprotein receptors in cholesterol transport during regeneration and remyelination of the rat sciatic nerve. J. Clin. Invest. *83*, 1015-1031.

Brecher, P.I. and Hyun, Y. (1978). Effect of 4-aminopyrazolopyrimidine and aminoglutethimide on cholesterol metabolism and steroidogenesis in the rat adrenal. Endocrinology *102*, 1404-1413.

Brown, M., Kovanen, P., and Goldstein, J. (1979). Receptor-mediated uptake of lipoprotein-cholesterol and its utilization for steroid synthesis in the adrenal cortex. Recent Progress in Hormone Research *35*, 215-256.

Curtiss, L.K., DeHeer, D.H., and Edgington, T.S. (1977). In vivo suppression of the primary immune response by a species of low density serum lipoprotein. J. Immunol. *118*, 648-652.

Dawson, P.A., Schechter, N., and Williams, D.L. (1986). Induction of rat E and chicken A-1 apolipoproteins and mRNAs during optic nerve degeneration. J. Biol. Chem. *261*, 5681-5684.

Dory, L., Boquet, L.M., Tate, C.R., and Sloop, C.H. (1986). Peripheral synthesis and isoform distribution of dog apoprotein E - an *in vivo* approach. J. Biol. Chem. *261*, 811-816.

Driscoll, D.M. and Getz, G.S. (1984). Extrahepatic synthesis of apolipoprotein E. J. Lipid Res. *25*, 1368-1379.

Dyer, C.A. and Curtiss, L.K. (1988). Apoprotein E-rich high density lipoprotein inhibit ovarian androgen synthesis. J. Biol. Chem. *263*, 10965-10973.

Elshourbagy, N.A., Liao, W.S., Mahley, R.W., and Taylor, J.M. (1985). Apolipoprotein E mRNA is abundant in the brain and adrenal, as well as in the liver and is present in other peripheral tissues of rats and marmosets. Proc. Natl. Acad. Sci. (U. S. A.) *82*, 203-207.

Fenjves, E.S., Gordon, D.A., Pershing, L.K., Williams, D.L., and Taichman, L.B. (1989). Systemic distribution of apolipoprotein E secreted by grafts of epidermal keratinocytes: Implications for epidermal function and gene therapy. Proc. Natl. Acad. Sci. U. S. A. *86*, 8803-8807.

Gwynne, J. and Hess, B. (1980). The role of high density lipoproteins in rat adrenal cholesterol metabolism and steroidogenesis. J. Biol. Chem. *255 (22)*, 10875-10883.

Hamilton, R.L., Wong, J.S., Guo, L.S.S., Krisans, S., and Havel, R.J. (1990). Apolipoprotein E localization in rat hepatocytes by immunogold labeling of cryothin sections. J. Lipid Res. *31*, 1589-1603.

Johnson, W.J., Mahlberg, F.H., Rothblat, G.H., and Phillips, M.C. (1991). Cholesterol transport between cells and high-density lipoproteins. Biochim. Biophys. Acta Lipids Lipid Metab. *1085*, 273-298.

Kowal, J. and Fiedler, R. (1968). Adrenal cells in tissue culture. I. Assay of steroid products. Arch. Biochem. Biophys. *128*, 406-421.

Linton, M.F., Gish, R., Hubl, S.T., Butler, E., Esquivel, C., Biy, W.I., Boyles, J.K., Wardell, M.R., & Young, S.G. (1991). Phenotypes of apolipoprotein B and apolipoprotein E after liver transplantation. J. Clin. Invest. *88*, 270-281.

Mahlberg, F.H. and Rothblat, G.H. (1992). Cellular cholesterol efflux. Role of cell membrane kinetic pools and interaction with apolipoproteins AI, AII, and Cs. J. Biol. Chem. *267*, 4541-4550.

Mazzone, T., Gump, H., Diller, P., and Getz, G.S. (1987). Macrophage free cholesterol content regulates apolipoprotein E synthesis. J. Biol. Chem. *262*, 11657-11662.

Mazzone, T., Basheeruddin, K., and Poulos, C. (1989). Regulation of macrophage apolipoprotein E gene expression by cholesterol. J. Lipid Res. *30*, 1055-1064.

Moore, C.C.D., Brentano, S.T., and Miller, W.L. (1990). Human P450scc gene transcription is induced by cyclic AMP and repressed by 12-*O*-tetradecanoylphorbol-13-acetate and A23187 through independent *cis* elements. Mol. Cell. Biol. *10*, 6013-6023.

Newman, T.C., Dawson, P.A., Rudel, L.L., and Williams, D.L. (1985). Quantitation of apolipoprotein E messenger RNA in the liver and peripheral tissues of nonhuman primates. J. Biol. Chem. *260*, 2452-2457.

Nicosia, M., Moger, W.H., Dyer, C.A., Prack, M.M., and Williams, D.L. (1992a). Apolipoprotein E messenger RNA in rat ovary is expressed in theca and interstitial cells and presumptive macrophage, but not in granulosa cells. Mol. Endocrinol. *6*, 978-988.

Nicosia, M., Prack, M.M., and Williams, D.L. (1992b). Differential regulation of apolipoprotein E mRNA in z. fasciculata cells of rat adrenal gland determined by in situ hybridization. Mol. Endocrinol. *6*, 288-298.

Pardridge, W.M. and Mietus, L.J. (1980). Palmitate and cholesterol transprot through the blood-brain barrier. J. Neurochem. *34*, 463-466.

Prack, M.M., Nicosia, M., Williams, D.L., and Gwynne, J. (1991a). Relationship between apolipoprotein E mRNA expression and tissue cholesterol content in rat adrenal gland. J. Lipid Res. *32*, 1611-1618.

Prack, M.M., Rothblat, G.H., Glick, J.M., Reyland, M.E., and Williams, D (1991b). Altered cholesterol metabolism in adrenal cells secreting apoE. Circulation *84*, 2259.

Reyland, M.E., Gwynne, J.T., Forgez, P., Prack, M.M., and Williams, D.L. (1991). Expression of the human apolipoprotein E gene suppresses steroidogenesis in mouse Y1 adrenal cells. Proc. Natl. Acad. Sci. USA *88*, 2375-2379.

Reyland, M.E., Prack, M.M., & Williams, D.L. (1992). Increased expression of protein kinase C in Y1 cellls which express apolipoprotein E decreases basal steroidogenesis by inhibiting expression of P450-cholesterol side chain cleavage enzyme mRNA. J. Biol. Chem. *267, 17935-17942.*

Reyland, M.E. and Williams, D.L. (1991). Suppression of cAMP-mediated signal transduction in mouse adrenocortical cells which express apolipoprotein E. J. Biol. Chem. *266*, 21099-21194.

Rinninger, F. and Pittman, R. (1987). Regulation of the selective uptake of high density lipoprotein-associated cholesteryl esters. J. Lipid Res. *28*, 1313-1325.

Schiff, T.S., Roheim, P.S., and Eder, H.A. (1971). Effects of high sucrose diets and 4-aminopyrazolopyrimidine on serum lipids and lipoproteins in rats. J. Lipid Res. *12*, 596-603.

Serougne, C., Lefevre, C., and Chevallier, F. (1976). Cholesterol transfer between brain and plasma in the rat: a model for the turnover of cerebral cholesterol. Exp. Neurol. *51*, 229-240.

Williams, D.L., Dawson, P.A., Newman, T.C., and Rudel, L.L. (1985a). Apolipoprotein E synthesis in peripheral tissues of nonhuman primates. J. Biol. Chem. *260*, 2444-2451.

Williams, D.L., Dawson, P.A., Newman, T.C., and Rudel, L.L. (1985b). Synthesis of apolipoprotein E by peripheral tissues. Potential functions in reverse cholesterol transport and cellular cholesterol metabolism. Ann. N. Y. Acad. Sci. *454*, 222-229.

Wong, M., Rice, D.A., Parker, K.L., and Schimmer, B.P. (1989). The roles of cAMP and cAMP-dependent protein kinase in the expression of cholesterol side chain cleavage and steroid 11β-hydroxylase genes in mouse adrenocortical tumor cells. J. Biol. Chem. *264*, 12867-12871.

SPONTANEOUS HYPERCHOLESTEROLEMIA IN PIGS

Alan D. Attie, Robert J. Aiello, Scott T. Cooper, and Patricia J. Uelmen
Departments of Biochemistry and Comparative Biosciences
University of Wisconsin-Madison
420 Henry Mall
Madison, WI 53706-1569

ApoB genetic heterogeneity and plasma cholesterol

Genetic factors contribute approximately 50% to the variability in plasma cholesterol. The best-characterized cholesterol-elevating mutations are those of the LDL receptor locus. Together, these mutations account for approximately 1% of the incidence of hypercholesterolemia in the human population. Since the estimated heritability coefficient for hypercholesterolemia is approximately 0.5, the quest is on to identify other gene loci that contribute to hypercholesterolemia.

As LDL receptor mutations dramatically elevate plasma LDL, it is easy to envision the same consequence emanating from a dysfunctional ligand. Surprisingly, direct evidence for such a phenomenon did not surface until relatively recently.

Indirect evidence that variation in apoB might be common in humans stemmed from the observation that patients receiving blood transfusions frequently had antibodies to apoB (Blumberg *et al.* 1962). The implication was that apoB might be somewhat more variable in structure than most other plasma proteins.

Additional evidence of apoB variability came from alloimmunization experiments in pigs. In an attempt to identify highly polymorphic proteins in pig blood, Rapacz cross-immunized pigs and showed that the predominant antigen recognized by the alloantisera was associated with plasma lipoproteins, particularly LDL (Rapacz *et al.* 1970). The most frequent alloantigen was subsequently shown to be apoB (Rapacz *et al.* 1986).

The first indication that apoB variability might contribute to hypercholesterolemia was the discovery that pigs carrying a particular immunologically-defined haplotype of apoB had higher than normal plasma cholesterol concentrations (Rapacz 1978). The haplotype, designated "Lpb5", was shown to segregate in a manner consistent with an autosomal co-dominant gene locus (Rapacz 1978). Seven additional pig apo-B haplotypes were described, all segregating as alleles at the same

NATO ASI Series, Vol. H 73
Human Apolipoprotein Mutants III
Edited by C. R. Sirtori, G. Franceschini, B.H. Brewer Jr.

gene locus. Southern blot analysis of genomic DNA from pigs with each haplotype revealed unique restriction fragment length polymorphisms, indicating that each haplotype corresponded to an allele at the apoB locus (Maeda *et al.* 1988).

Cultured skin fibroblasts from the Lpb5 pigs were shown to have normal LDL receptor activity (Rapacz *et al.* 1986), hence the hypercholesterolemia in these animals could not be attributed to a mutation in the LDL receptor gene. However, when LDL isolated from Lpb5 and non-Lpb5 pigs were simultaneously injected into non-Lpb5 pigs, Lpb5 LDL was cleared with a fractional clearance rate one-third lower than that of non-Lpb5 LDL (Checovich *et al.* 1988).

Consistent with a slower *in vivo* clearance rate, LDL from Lpb5 pigs binds to the LDL receptor with one-sixth the affinity of LDL from non-Lpb5 pigs (Lowe *et al.* 1988). When LDL subspecies with identical lipid compositions were isolated from Lpb5 and non-Lpb5 pigs, the LDL receptor binding defect was still evident, suggesting that defective receptor binding was attributable to a difference in apoB rather than a differences in particle lipid composition (Lowe *et al.* 1988).

LDL heterogeneity

Analysis of the size and density distribution of LDL particles from normal and Lpb5 pigs revealed considerable structural heterogeneity. In addition to a higher concentration of LDL, Lpb5 pigs had a dramatic increase in a buoyant, cholesterol ester-enriched LDL subspecies. In fact, virtually all of the LDL elevation in these animals was in the buoyant LDL fraction.

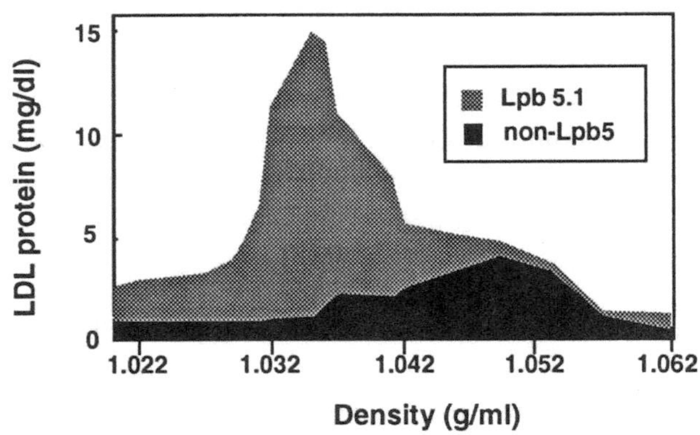

Figure 1. Density distribution of LDL from pigs lacking the *Lpb⁵* apo-B allele and from pigs with the *Lpb⁵* allele showing the Lpb5.1 phenotype.

During the course of our studies on the Lpb5 pigs, we noted phenotypic heterogeneity in the animals with this apoB allele. Some had a severe hypercholesterolemia and a ratio of buoyant to dense LDL of approximately 3. Others showed only moderate hypercholesterolemia and a much reduced ratio of buoyant-to-dense LDL. We therefore designated the severely hypercholesterolemic Lpb5 pigs "Lpb5.1" and the moderately hypercholesterolemic Lpb5 pigs "Lpb5.2".

Receptor binding studies showed that buoyant LDL binds more poorly to the LDL receptor than does dense LDL. In addition, the difference in binding affinity between Lpb5 and non-Lpb5 LDL was only demonstrable within the dense LDL subspecies. These results raised the question, how would a defect that affects dense LDL lead to an increase in buoyant LDL?

Metabolic basis for LDL heterogeneity

One possible answer to this question is that delayed clearance of dense LDL would lead to increased conversion to buoyant LDL. This could occur through the action of cholesterol ester transfer protein. To determine the metabolic origin of buoyant LDL, radiolabeled dense and buoyant LDL preparations were simultaneously injected into Lpb5 and non-Lpb5 pigs. The rate of disappearance from each subspecies pool and the flux between each subspecies was determined. In non-Lpb5 pigs, most of the buoyant LDL subspecies was formed from dense LDL. However, most of the input was into the dense LDL subspecies. By contrast, in Lpb5.1 pigs,

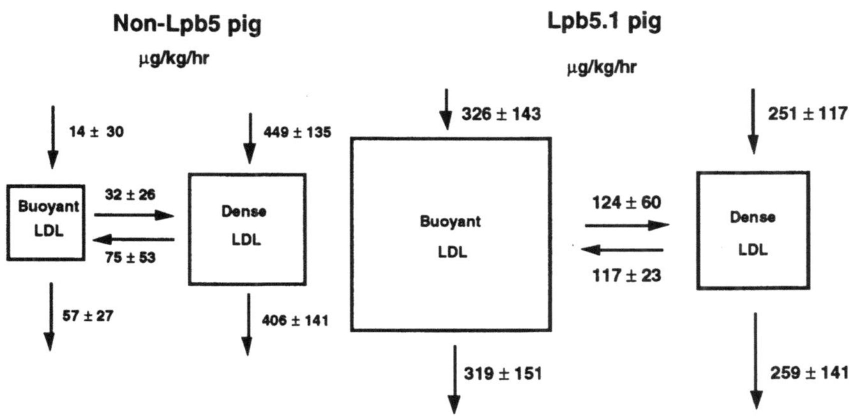

Figure 2. Average production, interconversion, and catabolism fluxes of buoyant and dense LDL in non-Lpb5 and Lpb5.1 pigs. The size of each box is proportional to the LDL pool size. (Reprinted, with permission, from (Checovich *et al.* 1991))

there was relatively little buoyant LDL formed from dense LDL; the increase in buoyant LDL was primarily due to increased *de novo* production. We therefore concluded that the buoyant LDL increase in the Lpb5.1 pigs could not be the result of delayed dense LDL clearance (Checovich *et al.* 1991).

Previous experiments have shown that one consequence of delayed IDL clearance is increased conversion of IDL to LDL. It therefore seemed plausible that the overproduction of buoyant LDL might result from increased flux through the VLDL cascade to buoyant LDL.

To measure the production of LDL via the VLDL cascade, ^{125}I-VLDL and ^{131}I-LDL were injected into Lpb5.1 and Lpb5.2 pigs. The proportion of buoyant and dense LDL derived from VLDL was calculated. The study showed that the great majority of both LDL subspecies were not apparently derived from VLDL (Cooper *et al.* 1992) in both Lpb5.1 and Lpb5.2 pigs. This result is in agreement with turnover studies of Huff and co-workers in minipigs (Huff 1985), which showed that most LDL in the pig does not appear to be formed via the VLDL cascade. As recently explained by Shames and Havel (Shames and Havel 1991), the *in vivo* VLDL turnover experiments have some potential pitfalls. For example, there may be a subfraction of VLDL with a fast turnover rate that comprises a small portion of the VLDL pool. This pool would be labeled in proportion to its steady-state mass rather than its flux rate. Its contribution to LDL production would therefore be underestimated in a turnover study.

The VLDL turnover data, together with the studies of LDL subspecies turnover, suggested that the LDL receptor binding defect alone could not explain the abundance of buoyant LDL in the Lpb5 pigs. One potential explanation is that the two phenotypes are caused by two separate intragenic or extragenic defects.

Analysis of the density distribution of LDL in the Lpb5.2 pigs showed an average ratio of the two LDL subspecies of 0.7. Accordingly, *in vivo* turnover studies showed a much smaller increase in the buoyant LDL production rate than was observed in the Lpb5.1 pigs. However, LDL particles isolated from Lpb5.1 and Lpb5.2 pigs, when injected into normal pigs, disappeared at identical rates. Their ability to compete with non-Lpb5 LDL for LDL receptor binding was similarly defective. Thus, the catabolic defect associated with Lpb5 LDL particles is identical in pigs with the moderate and severe hypercholesterolemia phenotypes. What distinguishes the two animals is the degree of buoyant LDL overproduction.

Table 1. Phenotypic Profile of Pig Variants

Phenotype	non-Lpb5	Lpb5.2	Lpb5.1
Plasma cholesterol	70-80 mg/dl	<180 mg/dl	>180 mg/dl
Ratio of buoyant-to-dense LDL	0.4	0.7	3.0
LDL receptor activity in cultured fibroblasts	normal	(not determined)	normal
% of *in vivo* LDL clearance mediated by the LDL receptor	50	50	0
K_d of LDL for the LDL receptor	1.9×10^{-9} M	11.6×10^{-9} M	11.6×10^{-9} M

The *Lpb*^{5} allele is normally distinguished from all other pig apoB alleles by the presence of a 283-bp insertion within intron 28 (Maeda *et al.* 1988). By this analysis, the apoB gene in animals with the Lpb5.1 and the Lpb5.2 phenotypes are identical. If these animals share the same apoB allele, is there an additional gene locus responsible for the phenotypic differences between them? Recent data from our laboratory suggests that a second locus might be involved. First, we have observed that breedings of animals with the Lpb5.2 phenotype occasionally yield animals with the Lpb5.1 phenotype. Second, preliminary quantitative pedigree analysis (carried out by Drs. Jean MacCluer and John Blangero) suggest that two gene loci (apoB and another locus) contribute to the elevated cholesterol in the pigs. The data are consistent with a putative second locus being transmitted in a recessive fashion.

Speculation and conclusion

If we infer that animals with the Lpb5.2 phenotype have the *Lpb*^{5} apoB allele but are wild type at a putative second locus, then we can see that a six-fold lower binding affinity of LDL for its receptor produces moderate hypercholesterolemia. Interestingly, patients heterozygous for an $Arg_{3500} \rightarrow$ Gln mutation, which completely abolishes receptor binding, have, on average, moderate hypercholesterolemia (Innerarity *et al.* 1990). Therefore, screening human subjects for partially deleterious apoB mutations, such as that of the Lpb5 pigs, will require a very sensitive bioassay or more specific information about the functional domains of apoB.

The hypercholesterolemic pigs are a great resource for exploring unsuspected regions of the apoB molecule that are essential for proper binding to the LDL receptor. In addition, the recent information suggesting the presence of a second cholesterol-elevating mutation means that these animals may lead to the identification of a new mechanism for genetically-caused hypercholesterolemia.

References

Blumberg, B.S., Dray, S. & Robinson, J.C. (1962) Antigen polymorphism of a low density lipoprotein. *Nature* 194: 656-658.

Checovich, W.J., Aiello, R.J. & Attie, A.D. (1991) Overproduction of a buoyant low density lipoprotein in spontaneously hypercholesterolemic mutant pigs. *Arteriosclerosis & Thrombosis* 11: 351-361.

Checovich, W.J., Fitch, W.L., Krauss, R.M., Smith, M.P., Rapacz, J., Smith, C.L. & Attie, A.D. (1988) Defective catabolism and abnormal composition of low-density lipoproteins from mutant pigs with hypercholesterolemia. *Biochemistry* 27: 1934-1941.

Cooper, S.T., Aiello, R.J., Checovich, R.J. & Attie, A.D. (1992) Low density lipoprotein heterogeneity in spontaneously hypercholesterolemic pigs. *Mol. Cell. Biochem.* (in press).

Huff, M.W., Telford, D.E. (1985) Direct synthesis of low-density lipoprotein apoprotein B in the miniature pig. *Metabolism* 34: 36-42.

Innerarity, T.L., Mahley, R.W., Weisgraber, K.H., Bersot, T.P., Krauss, R.M., Vega, G.L., Grundy, S.M., Friedl, W., Davignon, J. & McCarthy, B.J. (1990) Familial defective apolipoprotein B-100: A mutation of apolipoprotein B that causes hypercholesterolemia. *Journal of Lipid Research* 31: 1337-1349.

Lowe, S.W., Checovich, W.J., Rapacz, J. & Attie, A.D. (1988) Defective receptor binding of low density lipoprotein from pigs possessing mutant apolipoprotein B alleles. *Journal of Biological Chemistry* 263: 15467-15473.

Maeda, N., Ebert, D.L., Doers, T.M., Newman, M., Hasler-Rapacz, J., Attie, A.D., Rapacz, J. & Smithies, O. (1988) Molecular genetics of the apolipoprotein B gene in pigs in relation to atherosclerosis. *Gene* 70: 213-218.

Rapacz, J. (1978) Lipoprotein immunogenetics and atherosclerosis. *Am. J. Med. Genetics* 1: 377-405.

Rapacz, J., Grummer, R.H., Hasler, J. & Shackelford, R.M. (1970) Allotype polymorphism of low density ß-lipoproteins in pig serum (LDLpp1, LDLpp2). *Nature* 225: 941-942.

Rapacz, J., Hasler-Rapacz, J., Taylor, K.M., Checovich, W.J. & Attie, A.D. (1986) Lipoprotein mutations in pigs are associated with elevated plasma cholesterol and atherosclerosis. *Science* 234: 1573-1577.

THE APOLIPOPROTEIN E CYS-142 MUTANT: ROLE IN DOMINANT INHERITANCE OF TYPE III HYPERLIPOPROTEINEMIA AND EXPRESSION IN TRANSGENIC MICE

Stanley C. Rall, Jr., Yukio Horie, and Sergio Fazio
Gladstone Institute of Cardiovascular Disease
Cardiovascular Research Institute
University of California
P.O. Box 419100
San Francisco, California 94141-9100
U.S.A.

Type III hyperlipoproteinemia (HLP) is a genetic disorder of lipid metabolism in humans in which the primary molecular defect is a mutation(s) in apolipoprotein (apo) E that causes defective interaction of apoE with lipoprotein receptors and an accumulation of cholesterol-rich β-VLDL in the blood (Mahley and Rall, 1989). Most often, type III HLP is associated with homozygosity for an apoE mutant that has cysteine instead of arginine at residue 158, *i.e.*, with recessive inheritance. Many rare mutants of apoE have now been described (Fig. 1), some of which are also associated with type III HLP. In most of these rare instances, type III HLP is transmitted in a dominant fashion, *i.e.*, heterozygosity for the apoE mutant is sufficient for expression of the disorder. One of these rare apoE variants (cysteine instead of arginine at residue 142) has been found in a single family in which seven members spanning four generations are heterozygous for this apoE variant and all seven have type III HLP (Havel *et al.* 1983; Rall *et al.* 1989). Because of the unusual nature of the disorder in this family, we undertook investigations of the properties of the apoE Cys-142 mutant to determine its role in the expression of type III HLP.

Functional Properties of the ApoE Cys-142 Mutant Expressed in Bacteria

Because the physical and chemical properties of the apoE Cys-142 mutant are identical to those of normal apoE3, the other allele expressed in the family, it was

NATO ASI Series, Vol. H 73
Human Apolipoprotein Mutants III
Edited by C. R. Sirtori, G. Franceschini, B.H. Brewer Jr.

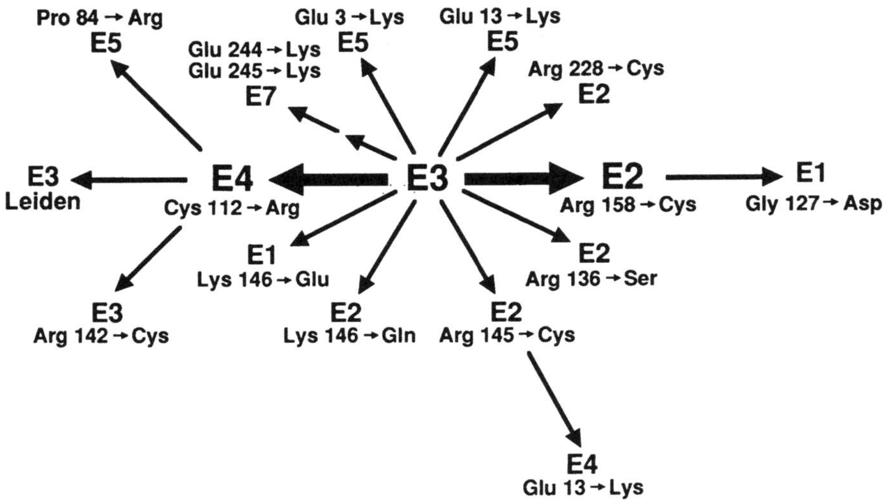

Figure 1. Naturally occurring mutants of human apoE. Each mutant is defined relative to the most commonly occurring isoform, apoE3. For each variant, the isoelectric focusing position and amino acid substitution(s) are given. Apolipoprotein E3-Leiden has a seven-amino acid insertion that is a direct repeat of the preceding seven residues (121-127). The mutants with substitutions at residues 142, 146, 145, 136, or 158, in addition to apoE3-Leiden, are all associated with type III HLP.

impossible to prepare the purified mutant from the plasma of the affected human subjects. Therefore, we expressed and isolated this apoE mutant in bacteria using a previously described vector (Vogel *et al.* 1985). We investigated four functional properties of the apoE Cys-142 mutant (Horie *et al.* 1992): receptor binding, lipoprotein association preference, heparin affinity, and interaction with the monoclonal antibody 1D7, which blocks the binding of apoE to lipoprotein receptors.

We found that the apoE Cys-142 mutant had 15-20% of normal apoE3 receptor binding activity, based on a competition assay using iodinated LDL as the competitor. We also proved that the substitution at residue 142, and not the simultaneous arginine for cysteine substitution at residue 112, was responsible for the binding defect.

We determined the lipoprotein distribution of the apoE Cys-142 mutant using an *in vitro* assay and separation of lipoproteins on Superose 6 columns. This method demonstrated that the mutant, similar to apoE4, has a preference for triglyceride-rich lipoproteins compared with apoE3: 2.0 times as much of the radioiodinated mutant (and 1.7 times as much apoE4) appeared in the VLDL fraction compared with apoE3. Since both the mutant and apoE4 have arginine at residue 112, whereas apoE3 has cysteine at this position, it is likely that it is this arginine that determines the preference for VLDL.

We then studied the interaction of the mutant apoE with heparin and the monoclonal antibody 1D7, which blocks the binding of apoE to lipoprotein receptors. We predicted that both interactions would be affected by the presence of the substitution at residue 142 (cysteine for arginine) in the mutant, because it has been demonstrated that the high-affinity heparin-binding site is coincident with the receptor-binding site, occurring in a cluster of basic residues between positions 135 and 150 of apoE, and that the 1D7 epitope is determined by this same cluster of basic residues (Weisgraber *et al.* 1983; Weisgraber *et al.* 1986). This prediction was confirmed in both instances. By heparin affinity chromatography, we found that the apoE mutant eluted in a salt gradient at about half the salt concentration required to elute normal apoE3, demonstrating the reduced affinity of the mutant for heparin. In a solid-phase competition assay, we found that the apoE mutant had 25-fold less affinity for antibody 1D7 than did normal apoE3. Furthermore, in immunoblot assays, antibody 1D7 failed to detect the apoE mutant under conditions where normal apoE3 was readily detected.

We took advantage of this differential antibody reactivity to determine the natural distribution of the apoE Cys-142 mutant in the plasma lipoproteins of a type III HLP subject heterozygous for this mutant. The plasma lipoproteins of the subject were separated by Superose 6 chromatography, delipidated, and electrophoresed on sodium dodecyl sulfate (SDS) gels. The gel was first immunoblotted with 1D7 and then with a polyclonal anti-apoE (Fig. 2). Antibody 1D7 detected only the normal apoE in each lipoprotein fraction, whereas the polyclonal antibody detected both the mutant and normal apoE3. Densitometry of the autoradiograms demonstrated that normal apoE3 accounted for 27% of the total apoE in VLDL, 50% in IDL/LDL, and 55% in HDL. Therefore, the apoE mutant is overrepresented in the VLDL fraction by almost a 3:1 ratio. This result may explain the apparent discrepancy in binding activity of the total apoE from the VLDL fraction of the type III HLP subjects previously reported. We had found that their total apoE had about 25% normal binding, and we thought that this represented about a 1:1 mixture of mutant apoE and normal apoE3 (Rall *et al.* 1989). It is now clear that this is not the case, and the finding that the isolated mutant has 15-20% of normal binding is now more consistent with the previous result, because that previous result was actually obtained from a mixture in which the mutant apoE predominated 3:1.

The ApoE Cys-142 Mutant and Dominant Inheritance of Type III HLP

The functional characteristics of the apoE Cys-142 mutant provide a possible explanation for its association with dominant inheritance of type III HLP. The receptor binding defect alone cannot be the explanation for dominant transmission. The receptor binding defect of an apoE mutant appears to be a necessary, but not sufficient, property for the expression of type III HLP. Part of the explanation for dominant inheritance may lie with the lipoprotein distribution of the mutant apoE. Because the presence of arginine at residue 112 confers upon apoE a preference for association with triglyceride-rich lipoproteins, this could account for the observed selective accumulation of the apoE Cys-142 mutant in the triglyceride-rich lipoproteins of affected subjects. Furthermore, it is likely that the natural preference for triglyceride-rich lipoproteins of any apoE with Arg-112, in conjunction with a receptor binding defect, will be a major part of the explanation for the accumulation of β-VLDL.

Likewise, if the interaction of apoE with heparin-like molecules is important for lipoprotein processing and/or catabolism *in vivo*, then the lower affinity of the apoE Cys-142 mutant for heparin may contribute to the accumulation of β-VLDL. The

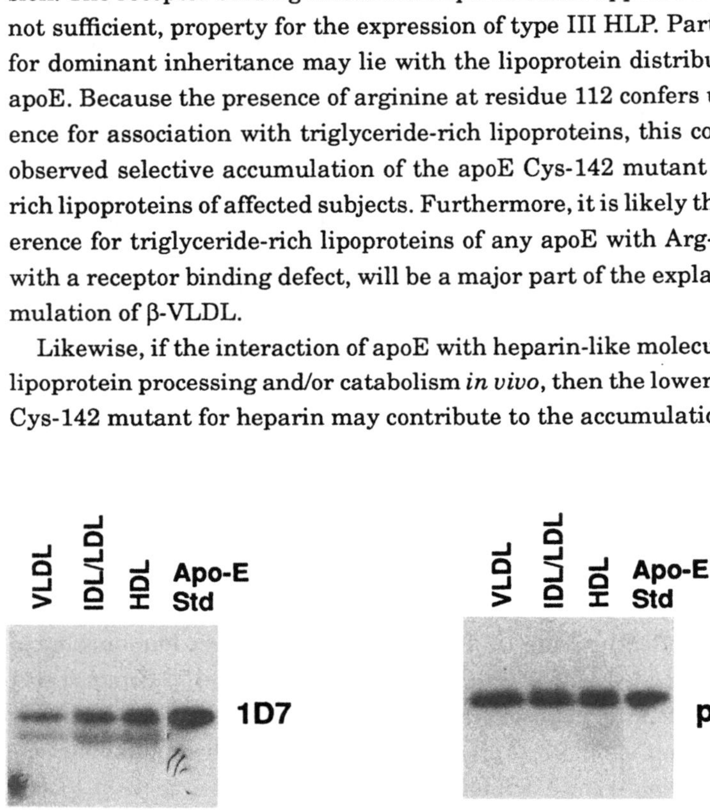

Figure 2. Distribution of the apoE Cys-142 mutant among an affected subject's plasma lipoproteins. Plasma was fractionated on Superose 6 and the pooled VLDL, IDL/LDL, and HDL fraction apoproteins were electrophoresed on an SDS gel and immunoblotted with antibody 1D7 (left panel) and then an anti-apoE polyclonal antibody (right panel). From Horie *et al.* (1992); reproduced with permission of *The Journal of Biological Chemistry*.

physiological significance of the heparin interaction with apoE is uncertain, but heparin is known to interact with lipoproteins, and heparin-like molecules on the cell surface of the vascular endothelium may play a role in lipoprotein processing and even in the development of atherosclerosis. If this is the case, then the reduced affinity of the apoE Cys-142 mutant for heparin-like molecules may interfere with normal processing of triglyceride-rich lipoproteins and contribute to the accumulation of these lipoproteins in the plasma of affected subjects. This is probably not the case for apoE2 Cys-158, which binds normally to heparin, but might be the case for any other apoE mutant that has a substitution for a basic residue in the region of the heparin-/receptor-binding site (residues 135-150). It is already known that some of these latter mutants are associated with dominant expression of type III HLP.

In summary, we believe that the unique combination of functional properties of the apoE Cys-142 mutant, *i.e.*, defective receptor binding, reduced affinity for heparin, and especially its preference for association with triglyceride-rich lipoproteins, leads to excessive accumulation of β-VLDL, which in turn is manifested in dominant expression of type III HLP.

Expression of ApoE Cys-142 in Transgenic Mice

In order to determine the impact of the apoE Cys-142 mutant on lipoprotein metabolism in an animal model, we expressed the protein in transgenic mice. For this purpose, we used a vector known to express normal human apoE3 primarily in the kidney of transgenic mice (Simonet *et al.* 1990). Because previous studies showed that human apoE from this construct co-distributed with endogenous mouse apoE on the lipoproteins of transgenic mice (Simonet *et al.* 1990), and because apoE is readily exchangeable among lipoproteins, we believed that the site of synthesis would not affect the ability to generate β-VLDL. For apoE Cys-142, four founders from two litters were identified, and two of these that expressed the highest level of protein in the plasma were expanded into the F1 generation. The ratio of transgene to endogenous apoE expression was about 0.5 in these two lines. For both lines, the major site of synthesis was the kidney, as expected, as assessed by RNase protection assays.

Superose 6 chromatography of lipoproteins from transgenic mice and non-transgenic littermates showed that there was neither a significant difference in total cholesterol between the two groups, nor major differences in the lipoprotein cholesterol profile. Thus, the transgenic mice did not spontaneously develop a significant increase in VLDL cholesterol that would be indicative of β-VLDL formation and accumulation that is seen in type III HLP. The apoE distribution across the lipoprotein profile was examined by immunoblotting of SDS gels. Unexpectedly, the apoE Cys-142 distributed dramatically differently from both normal human and endogenous mouse apoE, which distributed similarly to one another. In contrast to normal apoE3, which appeared primarily in VLDL and the typical HDL, the apoE Cys-142 accumulated selectively in the region just preceding the main HDL peak.

Because the region between VLDL and the typical HDL contains several types of lipoproteins, including VLDL remnants of both apoB100 and apoB48, LDL (apoB100 only), and apoE-containing HDL_1, agarose gel electrophoresis and immunoblotting were performed to determine on which type of lipoprotein particle the transgenic apoE resided. The transgenic apoE was found primarily on HDL_1 particles, with little on the apoB-containing lipoproteins. In contrast, the endogenous mouse apoE, although also on HDL_1, was more prevalent on the VLDL and their remnants and the typical HDL. These results indicate that the transgenic apoE selectively accumulates on large HDL particles rather than on apoB-containing lipoproteins. The results further suggest that the transgenic apoE accumulation on HDL_1 particles in the steady-state plasma of transgenic mice is due to a metabolic perturbation of HDL metabolism. Since the same phenomenon does not occur in transgenic mice with normal human apoE3, we believe that the mutant apoE, which is defective in receptor binding, is interfering with apoE-mediated catabolism of the large HDL particles. The perturbation that has occurred in these animals is analogous to what occurs in type III HLP, except that HDL are affected, not apoB-containing lipoproteins.

The apparent lack of preference for association of the mutant apoE with apoB-containing lipoproteins led us to investigate whether there was a circumstance under which this preference could be altered. Fat-feeding greatly increases the apoB-containing lipoproteins in mice, a finding we confirmed when we placed transgenic mice on a high-fat, high-cholesterol diet for 5 weeks. This diet caused a dramatic increase in total cholesterol in both transgenic and non-transgenic mice. All of the increase could be accounted for in the VLDL and remnant lipoproteins, with little change in HDL cholesterol. Agarose electrophoresis of Superose 6 fractions demonstrated that fat-feeding had indeed induced the transgenic apoE to vacate almost completely the HDL in preference for apoB-containing lipoproteins, although the typical HDL still persisted after fat-feeding. These results demonstrate that there are conditions under which the transgenic apoE will associate with the apoB-containing lipoproteins.

Comparative analysis of fat-fed apoE Cys-142 animals with fat-fed non-transgenic littermates revealed that the transgenic animals had a significantly greater increase in cholesterol (Table 1), and that all of the increase could be attributed to VLDL cholesterol. The fat-fed transgenic animals had about 50% higher total and VLDL cholesterol than their fat-fed non-transgenic littermates. This result is what would be expected if the mutant apoE were interfering with apoE-mediated catabolism of apoB-containing lipoproteins, which is the metabolic defect in human type III HLP subjects. However, agarose electrophoresis of the VLDL revealed no discernible differences between the transgenic and non-transgenic animals, *i.e.*, there was no clearly distinguishable β-migrating VLDL species in the transgenic animal VLDL fraction. Both transgenic and non-transgenic animals had a broad VLDL band that extended into the β-migrating position. It is probable that even the non-transgenic animals have a diet-induced β-VLDL-like spectrum similar to what is observed in other animal species upon fat-feeding. In these instances, β-VLDL are produced in the absence of a defective apoE and the particles are characteristically cholesterol-rich and triglyceride-deficient, similar to what we find with our mice, whose triglycerides are extremely low (Table 1). Thus, the diet-induced β-VLDL species may be confounding the interpretation of the results in this case. Nevertheless, our results to date suggest that perturbations of the kind needed to yield animals with the type III HLP phenotype are possible to obtain. We are currently investigating other factors and parameters necessary to induce the type III HLP phenotype in mice so that we can gain further insights into the pathogenesis of this disorder caused by the apoE Cys-142 mutant.

Table 1. Plasma Lipids of Non-Transgenic and ApoE Cys-142 Transgenic Mice

	Cholesterol (mg/dl)		Triglycerides (mg/dl)	
	Non-transgenic	Transgenic	Non-transgenic	Transgenic
Chow diet	106±18	100±25	83±50	67±38
(n)	(7)	(9)	(7)	(9)
High-fat diet	192±64	290±81[a]	22±19	33±13
(n)	(7)	(9)	(7)	(9)

[a]P=0.01 (pooled t test) compared with non-transgenic mice on high-fat diet.

Literature References

Havel, RJ, Kotite, L, Kane, JP, Tun, P, and Bersot, T (1983) Atypical familial dysbetalipoproteinemia associated with apolipoprotein phenotype E3/3. J Clin Invest 72:379-387

Horie, YS, Fazio, S, Westerlund, JR, Weisgraber, KH, and Rall, SC, Jr. (1992) The functional characteristics of a human apolipoprotein E variant (cysteine at residue 142) may explain its association with dominant expression of type III hyperlipoproteinemia. J Biol Chem 267:1962-1968

Mahley, RW, and Rall, SC, Jr. in *The Metabolic Basis of Inherited Disease, 6th Edition* (eds Scriver, CR, Beaudet, AL, Sly, WS, and Valle, D) 1195-1213 (McGraw-Hill, New York, 1989)

Rall, SC, Jr., Newhouse, YM, Clarke, HRG, Weisgraber, KH, McCarthy, BJ, Mahley, RW, and Bersot, TP (1989) Type III hyperlipoproteinemia associated with apolipoprotein E phenotype E3/3: structure and genetics of an apolipoprotein E3 variant. J Clin Invest 83:1095-1101

Simonet, WS, Bucay, N, Lauer, SJ, Wirak, DO, Stevens, ME, Weisgraber, KH, Pitas, RE, and Taylor, JM (1990) In the absence of a downstream element, the apolipoprotein E gene is expressed at high levels in kidneys of transgenic mice. J Biol Chem 265:10809-10812

Vogel, T, Weisgraber, KH, Zeevi, MI, Ben-Artzi, H, Levanon, AZ, Rall, SC, Jr., Innerarity, TL, Hui, DY, Taylor, JM, Kanner, D, Yavin, Z, Amit, B, Aviv, H, Gorecki, M, and Mahley, RW (1985) Human apolipoprotein E expression in *Escherichia coli*: structural and functional identity of the bacterially produced protein with plasma apolipoprotein E. Proc Natl Acad Sci USA 82:8696-8700

Weisgraber, KH, Innerarity, TL, Harder, KJ, Mahley, RW, Milne, RW, Marcel, YL, and Sparrow, JT (1983) The receptor-binding domain of human apolipoprotein E. Monoclonal antibody inhibition of binding. J Biol Chem 258:12348-12354

Weisgraber, KH, Rall, SC, Jr., Mahley, RW, Milne, RW, Marcel, YL, and Sparrow, JT (1986) Human apolipoprotein E. Determination of the heparin binding sites of apolipoprotein E3. J Biol Chem 261:2068-2076

APOLIPOPROTEINS AS "DRUGS": CURRENT STATUS

Cesare R. Sirtori
Center E. Grossi Paoletti, Institute of Pharmacological Sciences
University of Milano
Via Balzaretti 9
20133 Milano
Italy

Apolipoproteins are attractive molecules for a potential use in therapeutics. They are generally quite stable, capable of binding at the same time lipids and other chemicals, water soluble and in some cases presenting with a dominant positive or negative charge. In addition, some apolipoproteins can interact with cell receptors, either regulating cholesterol biosynthesis, or possibly cholesterol removal.

It should be noted that a therapeutic use of apolipoproteins may not be strictly linked to their lipid binding properties. In some cases, apolipoprotein segments have been proposed for clinical use; in other, complexes of apolipoproteins with other circulating proteins (eg. immunoglobulins) may provide a potentially useful therapeutic tool. The present review examines some of the possible or futuristic uses of apolipoproteins in a variety of clinical fields, also including the prevention/treatment of arterial disease.

Activation of sperm motility

Studies on in vitro fertilization have clearly indicated that human serum is superior both to seminal plasma and to follicular fluid in supporting motility of human spermatozoa. The major sperm activating capacity in serum was previously found to be mediated by a fraction with a MW of about 250 kD (Akerlöf et al, 1989).

Very recently, the same authors, by isolating the active fraction of serum, could describe this complex as containing albumin, apolipoprotein AI and immunoglobulin heavy and light chains (Akerlöf et al, 1991). Boiling destroys the activity, thus suggesting that the complex acts as a macromolecule. The biological activity of the sperm activating protein (SPAP) is described in Fig. 1 and indicates that, at specific

NATO ASI Series, Vol. H 73
Human Apolipoprotein Mutants III
Edited by C. R. Sirtori, G. Franceschini, B.H. Brewer Jr.

concentrations, sperm motility (SPM) is increased, and so is the ATP content of sperm.

The mechanism by which SPAP induces sperm activation is unknown. Judging from the small amounts of SPAP required for activation, a likely hypothesis may be that of a receptor-mediated or enzyme linked mechanism. Since the effect on ATP content is not of a similar extent as that on SPM, it is not clear whether SPAP affects motility via an increase of ATP content. The clinical use of SPAP is postulated both for diagnostic and therapeutic purposes in human fertilization.

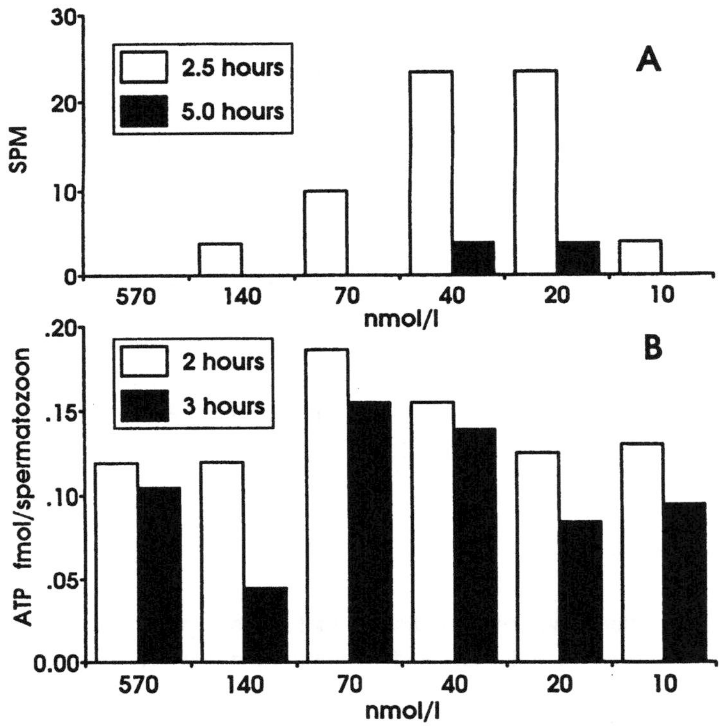

Fig. 1. Effect of purified sperm activating protein (SPAP) on sperm motility (SPM) and ATP content of human spermatozoa (Akerlöf et al, 1991).

Gout

Gout is a metabolic disorder, characterized by increased levels of uric acid in plasma. This is consequent either to excess production or to reduced excretion. The pathological condition resulting from gout is arthritis, mainly at the lower extremities (toe) and characterized by crystal formation with abundant inflow of polymorphonuclear leukocytes (PMN) (Levinson, 1989). Treatment of gout is of a preventive nature, either using drugs increasing urate excretion or, more often, inhibiting biosynthesis (allopurinol).

Interest in the possible therapeutic use of apolipoproteins has been stimulated by the knowledge of the properties of apolipoprotein E (apo E). This relatively basic protein can bind urate crystals in vitro and, in experiments involving cultured PMNs, it was shown that apo E is the secreted protein responsible for urate crystal binding (Terkeltaub et al, 1991) (Fig. 2).

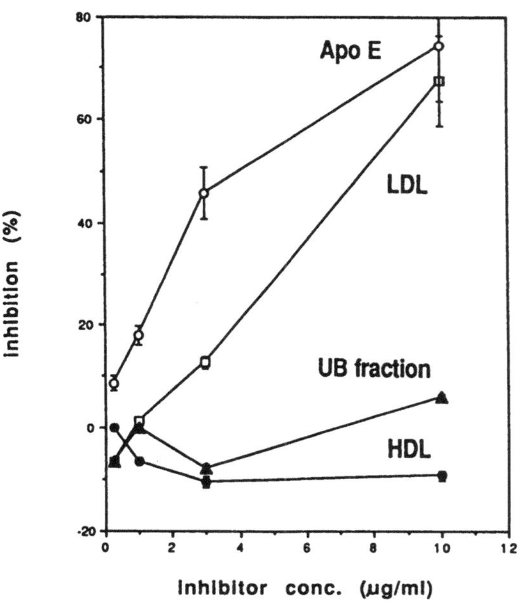

Fig. 2. Inhibitory affect by apo E on neutrophil responsiveness to urate crystals. The effects of apo E, isolated by affinity chromatography, and of the unbound (UB) fraction are indicated (Terkeltaub et al, 1991).

These authors suggest that the effect of apo E is not only of a chemical nature, but it may also affect lymphocyte maturation, up to the formation of active cells. This conclusion stems from prior observations, linking a subfraction of LDL to an immunoregulatory property of these lipoproteins. The so called LDL_{in} are capable of inhibiting lymphocyte immortalization, a pre-requisite for lymphoid tumor development (Chisari et al, 1981). The suppression of pre-inductive events involved in mitogen- and antigen-induced lymphocyte proliferation is independent of the fatty acid content of the lipoprotein (Curtiss and Edgington, 1981). As a result of later studies, apo E, or possibly a modified form of this apolipoprotein, was identified as being responsible for these very interesting immunoregulatory properties (Pepe and Curtiss, 1986).

Whatever the case, these observations, pertaining to the immunoregulatory effect of apo E, may be pertinent to its possible use in inflammatory disorders, such as gouty arthritis (where the lipid binding properties would not be in play), or else in immunopathological conditions. Among these, progressive lupus erythematosus, rheumatoid artritis or even the control of graft rejection may be indicated.

Neural degenerative disorders

Degenerative disorders of the nervous system, both at the central or peripheral level, may be the result of trauma, infectious diseases, or part of a general condition of the central nervous system. Injured mammalian peripheral nerves have been known for many years to release a soluble protein of MW 37,000 (Skene and Shooter, 1983). The rate of secretion of this protein increases dramatically after injury to the peripheral nerves of adult animals, accounting for up to 5% of the total secretory protein three weeks after injury. This protein was later on identified as apo E and found to be produced by the macrophages that sorround the nerve, within 3 days of injury (Snipes et al, 1986).

Isolation of apo E the from injured nervous tissues has led to a number of investigations, carried out in order to examine the possibility that the release of apo E might not be only the consequence of tissue damage, but also a mechanism for tissue repair. In particular, it could be shown that released apo E is recognized by high affinity receptors, with characteristics of the apo B, E receptors, capable of binding apo E-containing lipoproteins (Pitas et al, 1987). This receptor/mediated uptake is elavated in growing neuronal cells, in particular growth cones, thus possibly providing an additional, local source of lipid for myelin regeneration (Ignatius et al,

1987). Interestingly, quantitative evaluation of accumulated apolipoproteins in regenerating rat peripheral nerves shows that the concentrations of apo D and E are increased over 500 and 250-fold, respectively, three weeks after injury, vs very low increases of apo AI and apo AIV (both, probably, reaching the nerve from plasma) (Boyles et al, 1990). Very recently, it has been confirmed that cholesterol released from injured nerves becomes associated with apo E-containing lipoproteins, thus becoming available for local reutilization (Goodrum, 1991). This finding is of particular significance, in view of the fact that regenerating nerves synthesize very little cholesterol.

In other conditions, marked increases of apo E expression has been described. This is particularly the case of astrocytes from brains of Alzheimer's patients and scrapie-infected mice (Diedrich et al, 1991). Both of these diseases are characterized by focal or diffuse neural degeneration, accompanied by extensive myelin changes (Fig. 3).

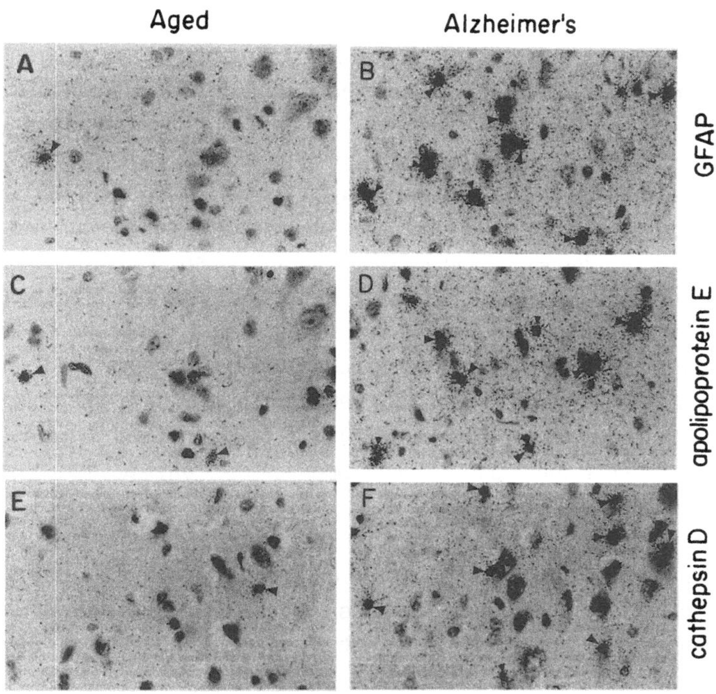

Fig. 3. In situ hybridization of glial fibrillar acid protein (GFAP), apo E and cathepsin D (CD) mRNAs in brains from control and Alzheimer's patients (Diedrich et al, 1991)

All these data provide, therefore, clear and substantial indication that apo E is produced in large quantities after nerve injury (be it acute or chronic, as in Alzheimer's disease) and that apo E is stored in lipoprotein particles interacting with high affinity receptors in regenerating nerve. It is, therefore, unfortunate that the hypothesis of a therapeutic use of apo E in cases of neuronal injury, has never been tested either in experimental animals or in man.

Acquired immunodeficiency syndrome (AIDS)

AIDS is a rapidly diffusing immunological disorder, related to an infection with a retrovirus (HIV), specifically affecting helper T-lymphocytes (T4) (Robey et al, 1985). This infection is characterized by an asymptomatic period of variable length, followed by a full blown disease, associated with diffuse tissue damage, high frequency of infectious complications and, in some cases, development of characteristic tumors (Kaposi sarcomas).

Among the typical pathological features of HIV infection is the presence, in different tissues, of cytopathic effects, characterized by the formation of multinucleated syncytia following cell fusion (McCune et al, 1988). This effect is mediated by the intracellular cleavage of the envelope glycoprotein gp160 to a gp120-gp41 complex, transported to the plasma membrane. After cleavage, a new hydrophobic amino terminus is exposed on gp41, mediating fusion of the envelope with cell membranes, also leading to the release of the viral nucleocapsids into the cytosol. The process of virus-induced cell fusion is analogous to the process of fusion of viral membranes or cell membranes: interruption of this process can reduce viral replication and syncytium formation in a number of virus systems (Richardson and Choppin, 1983). In model membrane systems, agents stabilizing the membrane bilayer have been also shown to inhibit membrane fusion (Epand, 1986). Among agents with a higher affinity for membrane bilayers, amphipathic helical peptides may offer an attractive option for the prevention of syncytium formation.

Recent reports have, in fact, indicated the presence of potential amphipatic helical domains in the HIV envelope glycoprotein. It is thus suggested that proteins with a high content of amphipathic helices may interact with the system, thus in some way affecting cell fusion.

Candidate proteins possessing amphipathic a-helices are numerous. Among them, lipid associating polypeptide hormones (β-endorphin and calcitonin), novel antimicrobial agents, such as magainins and cecropins, as well as type I and type II (IFN-γ) interferons (Owens et al, 1990). Apolipoprotein AI was suggested to possibly play a role in HIV infections, due to a clear effect on the Herpes simplex virus-induced cell fusion at physiological (1 mM) concentrations (Srinivas et al, 1990). Later on, in fact, a similar preventive effect on HIV induced syncytiun formation was shown for apo AI (Owens et al, 1990). The possible mechanism/s of inhibition of HIV-induced membrane fusion by the amphipathic helices of the apolipoprotein are shown in Fig. 4.

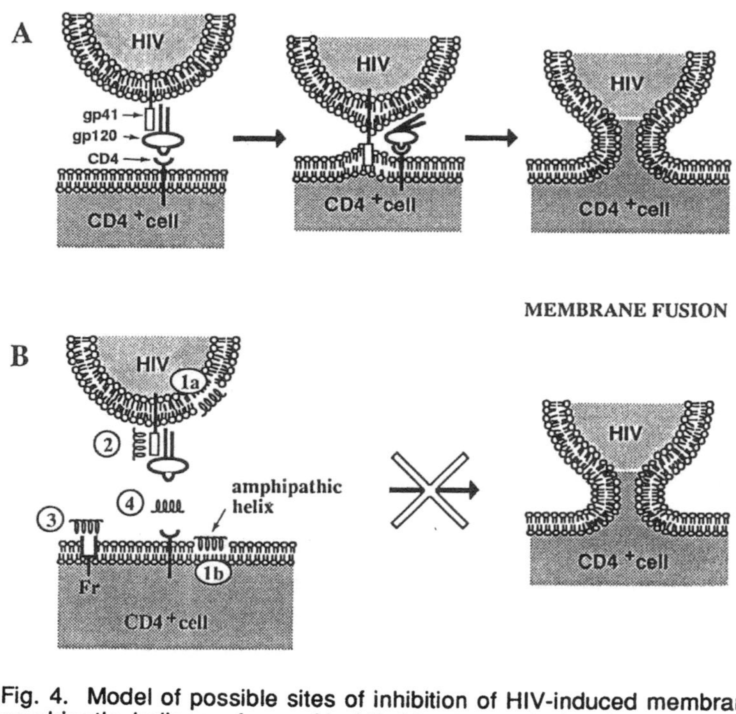

Fig. 4. Model of possible sites of inhibition of HIV-induced membrane fusion by the amphipatic helices of apo AI. (*A*) Membrane fusion in the absence of peptide treatment. After interaction between gp120 and the CD4 receptor, gp41 probably perturbs the membranes bilayers, resulting in fusion of the two membranes. (*B*) The amphipatic helix of apo AI could be acting at different sites: 1a and/or 1b, perturbation of the lipid bilayer; 2, interaction with the fusogenic domain in gp41; 3, interaction with a specific fusion receptor (Fr); 4, interference with gp120-CD4 binding (Owens et al, 1990).

A similar activity has been more recently shown for synthetic amphipathic peptide analogues of apo AI, namely 22-aminoacid residue long synthetic peptides, corresponding to the helical domains of apo AI or to 18-residue long model amphipathic peptides (Srinivas et al, 1991). Although it seems premature to hypothesize the clinical use of these molecules, still the severity of the disease and the relative simplicity of clinical protocols evaluating activity and safety of these molecules, encourage further work in the area.

Atherosclerosis treatment

In addition to being certainly involved in the determination of atherosclerotic lesions, apolipoproteins may possibly serve as useful therapeutic tools for arterial plaque removal. Two apolipoproteins have been designated as potentially effective in atherosclerosis prevention/reversal. Studies with apo E have been focused on the clearance of VLDL or remnant particles, and on the regression of arterial lesions in the Watanabe rabbit. The hypothetical beneficial effect of apo AI essentially stems from studies on atherosclerosis regression, induced by infusions of HDL.

Apo E

The use of apo E for improving lipoprotein metabolism was suggested, based on the affinity of apo E containing lipoproteins for liver cell membranes, responsible for the uptake of dietary cholesterol by way of chylomicron remnants (Hussain et al, 1989). The chylomicron-remnant receptor is believed to be distinct from the LDL receptor, because remnants are efficiently removed from the blood in human type II patients, as well as in the Watanabe heritable hyperlipidemic rabbits (WHHL), both deficient in LDL receptors.

Yamada et al (1989) tested the hypothesis that injections of apo E in WHHL rabbits could improve remnant clearance, in this way reducing cholesterolemia. These authors, by infusing approximately 30 mg of apo E per rabbit, obtained a plasma cholesterol reduction of 8.3% by 1 h and of 19% after 3 h. The reduced levels were maintained for up to 8 h after the injection. Cholesterol reduction occurred primarily in VLDL and intermediate density lipoproteins during the first 2 h after injection, followed by a reduction of LDL cholesterol later on.

A similar approach was tested by Mahley et al. (1989) in cholesterol fed rabbits, which received an infusion of apo E (70 mg) determining a 20-40% acute reduction of plasma cholesterol levels within 2-3 h. Furthermore, the addition of apo E to ^{14}C-cholesterol labelled chylomicrons resulted in an accelerated clearance of these lipoproteins, with an enhanced uptake, occurring primarily in the liver. The major findings of the studies by Yamada et al (1989) and Mahley et al (1989) are summarized in Fig. 5.

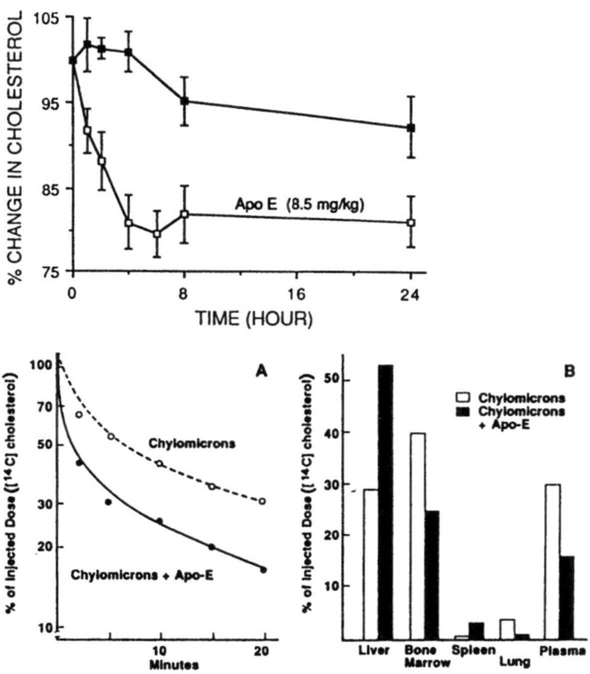

Fig. 5. Major findings in studies on the acute infusions of apo E. In the upper panel (Yamada et al, 1989), the acute infusion of apo E in WHHL rabbits determined a significant, although transient, reduction of cholesterolemia. In the study reported in the lower panel, (Mahley et al, 1989), the infusion of apo E (70 mg) to cholesterol fed rabbits resulted in an accelerated clearance of chylomicrons (A) and also in an enhanced uptake of chylomicron cholesterol in the liver (B).

Very recently, Yamada et al. (1992) showed the results of a chronic treatment in WHHL rabbits with thrice-weekly injections of apo E (30 mg per rabbit). This study clearly indicates that repeated injections of the apolipoprotein for 8.5 months may result in a dramatic reduction (-40%) of the lesions in treated animals.

Apo AI/HDL

The beneficial effects of elevated of HDL concentrations against arterial lipid deposition, well substantiated in tissue culture systems and in vivo, have suggested testing the activity of HDL plasma fractions on arterial lipid lesions in cholesterol-fed rabbits.

In a first study by Badimon et al. (1989) the lipoprotein plasma fraction, isolated at d 1.063 to 1.25 gml (HDL-VHDL) was administered to rabbits receiving a 0.5% cholesterol rich diet for 8 weeks. The isolated lipoproteins were given in saline (50 mg of protein in a final volume of 5 ml) once weekly i.v. throughout the period of cholesterol feeding. The control group was administered the same volume of saline. The authors could recognize an almost identical kinetic behavior of the isolated lipoproteins in normolipidemic and cholesterol-fed rabbits. Treatment did not lead to any marked changes in plasma lipid/lipoprotein distribution of treated rabbits. However, aortic sudanophilia was markedly reduced in the HDL-VHDL treated group (15.2% of fatty streak involvement, vs 38.6% in controls), and the cholesteryl ester content of aortas from treated animals went down from 10.9 ±2.6 mg/g to 4.5±0.6 mg/g of tissue ($p < 0.03$).

These findings prompted a follow-up study, this time investigating the potential of the d>1.125 lipoproteins to induce atherosclerosis regression in rabbits with established lesions. Badimon et al. (1990) again induced arterial disease by feeding the animals a 0.5% cholesterol-rich diet for 60 d (group 1). Another group of animals was maintained on the same diet for 90 d (group 2), whereas a third group was fed these cholesterol diet for 90 d, but received in addition 50 mg of HDL-VHDL protein per week during the last 30 d (group 3). At the end of the study aortic atherosclerosis involvement was 34±4% in group 1, 38.8±5% in group 2 and 17.8±4% in group 3 ($p < 0.005$) (Fig. 6). Aortic lipid deposition was also significantly reduced in group 3, compared to group 1 (studied only at 60 days) and group 2. These studies suggest that HDL may not only inhibit disease progression but also reduce established atherosclerotic lesions.

At this stage, it seems necessary to identify the responsible component of HDL. Due to the mixed apoprotein composition of the isolated fraction used by Badimon et al, it cannot be ruled out that components outside of apo AI may be responsible (apo E, the cholesteryl ester transfer protein, or possibly others). Also the mechanism is in

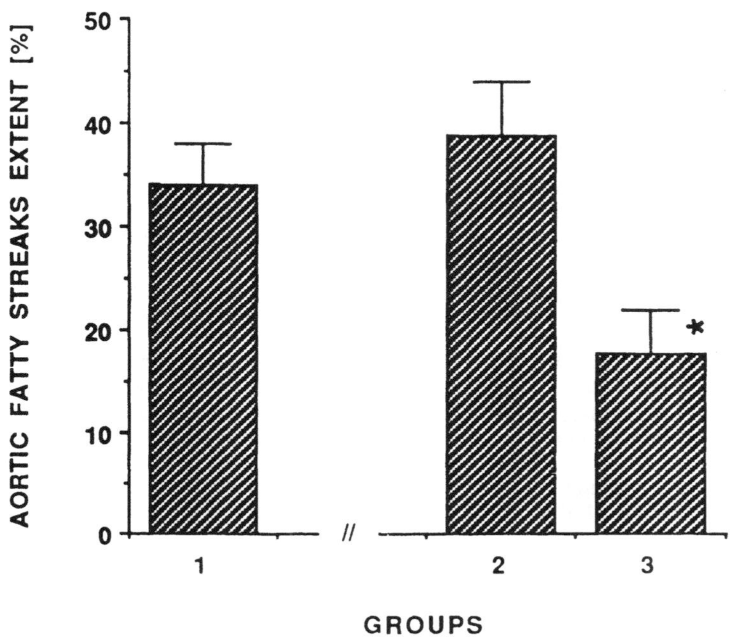

Fig. 6. Aortic atherosclerotic involvement (% ± SE of aortic surface covered by fatty streaks) in the three groups of rabbits. 1: cholesterol-fed for 60 d; 2: cholesterol-fed for 90 d; 3: cholesterol-fed for 90 d, receiving in addition 50 mg of HDL-VHDL protein per week during the last 30 d (Badimon et al, 1990).

discussion. HDL has been shown to exert a prostacyclin stabilizing property, as well as to stimulate prostacyclin release for the arterial wall (Aoyama et al, 1990). Apo AI shows a very potent fibrinolytic activity in <u>in vitro</u> models (Saku et al, 1985).

A possible clue to the identification of apo AI as the responsible apolipoprotein comes from a study in mice, transgenic for human apo AI (Rubin et al, 1991). These animals acquire an HDL particle size pattern similar to humans and, in this condition, become extremely resistant to the induction of atherosclerotic lesions by a high fat diet. The authors concluded that apo AI from humans has a direct inhibitor effect at an early stage of atherogenesis, but it also seems necessary to assess the impact of apo AI on the progression of mature plaques.

Part of the reported research work was supported by a Grant from the Consiglio Nazionale delle Ricerche of Italy, Progetto Finalizzato Ingegneria Genetica

References

Akerlöf E, Fredricsson B, Gustafson O, Lunell N-O, Nylund L, Rosenborg L, Pousette A (1987) Serum factors stimulate the motility of human spermatozoa. Int J Androl 10: 124-130

Akerlöf E, Jörnvall H, Slotte H, Pousette A (1991) Identification of apolipoprotein AI and immunoglobulin as components of a serum complex that mediates activation of human sperm motility. Biochemistry 30: 8996-8990

Aoyama T, Yui Y, Morishita H, Kawai C (1990) Prostaglandin I_2 half-life regulated by high density lipoprotein is decreased in acute myocardial infarction and unstable angina pectoris. Circulation 81: 1784-1791

Badimon JJ, Badimon L, Galvez A, Dische A, Fuster V (1989) High density lipoprotein plasma fractions inhibit aortic fatty streaks in cholesterol-fed rabbits. Lab Invest 60: 455-461

Badimon JJ, Badimon L, Fuster V (1990) Regression of atherosclerotic lesions by high density lipoprotein plasma fraction in rabbits. J Clin Invest 85: 1234-1241

Boyles JK, Notterpek LM, Anderson LJ (1990) Accumulation of apolipoprotein in the regenerating and remyelinating mammalian peripheral nerve. J Biol Chem 265: 17805-17815

Chisari FV, Curtiss LK, Jensen FC (1981) Physiologic concentrations of normal human plasma lipoproteins inhibit the immortalization of peripheral B lympocytes by Epstein-Barr virus. J Clin Invest 68: 329-336

Curtiss LK, Edgington TS (1981) The biologic activity of the immunoregulatory lipoprotein, LDL-In, is independent of its free fatty acid content. J Immunol 126: 1382-1386

Diedrich JF, Minnigan H, Carp RI, Whitaker JN, Race R, Frey W II, Haase AT (1991) Neuropathological changes in scrapie and Alzheimer's disease are associated with increased expression of apolipoprotein E and cathepsin D in astrocytes. J Virol 65: 4759-4768

Epand RM (1986) Virus replication inhibitory peptides inhibit the conversion of phospholipid bilayers to the hexagonal phase. Biosci Rep 6: 647-653

Goodrum JF (1991) Cholesterol from degenerating nerve myelin becomes associated with lipoproteins containing apolipoprotein E. J Neurochem 56: 2082-2086

Hussain MM, Mahley RW, Boyles JK, Fainaru M, Brecht WJ, Lindquist PA (1989) Chylomicron-chylomicron remnant clearance by liver and bone marrow in rabbits. Factors that modify tissue-specific uptake. J Biol Chem 264: 9571-9582

Ignatius MJ, Shooter EM, Pitas RE, Mahley RW (1987) Lipoprotein uptake by neuronal growth cones in vitro. Science 236: 959-962

Levinson DJ (1989) Clinical gout and the pathogenesis of hyperuricemia. In: McCarthy DJ (ed) Arthritis and Allied Conditions. Lea and Fibiger, Philadelphia, p 1691

Mahley RW, Weisgraber KH, Hussain MM, Greenman B, Fisher T, Vogel T, Gorecki M (1989) Intravenous infusion of apolipoprotein E accelerates clearance of plasma lipoproteins in rabbits. J Clin Invest 83: 2125-2130

Owens RJ, Anantharamaiah GM, Kahlon JB, Srinivas RV, Compans RW, Segrest JP (1990) Apolipoprotein A-I and its amphipathic helix peptide analogues inhibit human immuno-deficiency virus-induced syncytium formation. J Clin Invest 86: 1142-1150

Pepe MG, Curtiss LK (1986) Apolipoprotein E is a biologically active constituent of the normal immunoregulatory lipoprotein, LDL-In. J Immunol 136: 3716-3723

Pitas RE, Boyles JK, Lee SH, Foss D, Mahley RW (1987) Astrocytes synthesize apolipoprotein E and metabolize apolipoprotein E-containing lipoproteins. Biochim Biophys Acta 917: 148-161
Richardson CD, Choppin PW (1983) Oligopeptides that specifically inhibit membrane fusion by paramyxoviruses: studies on the site of action. Virology 131: 518-532
Robey WG, Safai B, Oroszlan S, Arthur LO, Gonda MA, Gallo RC, Fishinger PJ (1985) Characterization of envelope and structural gene products of HTLV-III with sera from AIDS patients. Science 228: 1402-1405
Rubin EM, Krauss RM, Spangler EA, Verstuyft JG, Clift SM (1991) Inhibition of early atherogenesis in transgenic mice by human apolipoprotein AI. Nature 353: 265-267
Saku K, Ahmad M, Glas-Greenwalt P, Kashyap M (1985) Activation of fibrinolysis by apolipoproteins of high density lipoproteins in man. Thromb Res 39: 1-8
Skene JHP, Shooter EM (1983) Denervated sheath cells secrete a new protein after nerve injury. Proc Natl Acad Sci USA 80: 4169-4173
Snipes GJ, McGuire CB, Norden JJ, Freeman JA (1986) Nerve injury stimulates the secretion of apolipoprotein E by nonneuronal cells. Proc Natl Acad Sci USA 83: 1130-1134
Srinivas RV, Birkedal B, Owens RJ, Anantharamaiah GM, Segrest JP, Compans RW (1990) Antiviral effects of apolipoprotein A-I and its synthetic amphipathic peptide analogs. Virology 176: 48-57
Srinivas RV, Venkatachalapathi YV, Rui Z, Owens RJ, Gupta KB, Srinivas SK, Anantharamaiah GM, Segrest JP, Compans RW (1991) Inhibition of virus-induced cell fusion by apolipoprotein A-I and its amphipathic peptide analogs. J Cell Biochem 45: 224-237
Terkeltaub RA, Dyer CA, Martin J, Curtiss LK (1991) Apolipoprotein (Apo) E inhibits the capacity of monosodium urate crystals to stimulate neutrophils. J Clin Invest 87: 20-26
Yamada N, Shimano H, Mokuno H, et al (1989) Increased clearance of plasma cholesterol after injection of apolipoprotein E into Watanabe heritable hypolipidemic rabbits. Proc Natl Acad Sci USA 86: 665-669
Yamada N, Inoue I, Kawamura M, et al. (1992) Apolipoprotein E prevents the progression of atherosclerosis in Watanabe heritable hyperlipidemic rabbits. J Clin Invest 89: 706-711

NON-MAMMALIAN HOST SYSTEMS FOR THE EXPRESSION OF MAMMALIAN APOLIPOPROTEINS

S.L. Sturley, D.G. Gretch, P.D. Friesen*, M.R. Culbertson$^{\Phi}$, N. Beckage# and A.D. Attie.
Departments of Biochemistry & Comparative Biosciences
University of Wisconsin-Madison
Madison, Wisconsin 53706

Recombinant mammalian apolipoproteins have been expressed in a variety of cultured mammalian cell lines (Reardon et al. 1986; Mallory et al. 1987; Blackhart et al. 1990; Yao et al. 1991), transgenic animals (Simonet et al. 1991), and bacteria (Vogel et al. 1985; Lalazar et al. 1988). These approaches have yielded considerable insight into the structure and function of several apolipoproteins but are rather limited for research or pharmaceutical purposes due to laborious technology or low yield often associated with these systems. This is particularly problematic if large scale production of biologically active lipoproteins for therapeutic purposes is to be attempted.

Described here are three non-mammalian expression systems (Figure 1, and Gretch et al. 1991; Sturley et al. 1991). Budding yeast (*Saccharomyces cerevisiae*), baculovirus expressed proteins in cultured *Spodoptera frugiperda* insect cells and the same baculovirus constructs in *Manduca sexta* insect larvae. All three systems have different advantages that are exemplified here in the expression of human apolipoprotein E3 (apo-E3). The hosts described are well suited for the production of active mammalian proteins. The presence of a vesicular secretion pathway should ameliorate toxicity previously observed upon expression of apo-E in *Escherichia coli.* (Vogel et al. 1985). In addition, secondary modifications to the apolipoprotein of interest, such as glycosylation, and more importantly, lipid association to produce functional particles, have the potential to occur in all these hosts. All three systems lack endogenous proteins with similar activity to mammalian lipoproteins. This renders assessment of

*Institute of Molecular Virology and $^{\Phi}$Departments of Genetics and Molecular Biology, University of Wisconsin-Madison, Madison, WI 53706, USA and #Department of Entomology, University of California-Riverside, Riverside, California 92521, USA.

NATO ASI Series, Vol. H 73
Human Apolipoprotein Mutants III
Edited by C. R. Sirtori, G. Franceschini, B.H. Brewer Jr.

function more feasible than, for example, the same proteins expressed from.human hepatoma cells. More specific advantages of these systems are itemized in Figure 1.

Non-mammalian expression systems for apolipoproteins:

Advantages

Spodoptera frugiperda	*Manduca sexta*	*Saccharomyces cerevisiae*
high yield	high yield	moderate yield
virus tolerant of large inserts	virus tolerant of large inserts	inexpensive scale-up
	inexpensive to maintain	molecular genetics
	lipoprotein production capability	classical genetics

Figure 1. Three eukaryotic expression systems for apolipoproteins and their associated advantages. The scanning electron micrograph of budding yeast was kindly supplied by Dr. M. Gustin.

Expression of Human apo-E in Insect Cells and Larvae.

The polyhedrin gene promoter of *Autographa californica* nuclear polyhedrosis virus (AcNPV) has been used extensively in recent years for high level expression of mammalian genes in cultured cells of the fall armyworm (*Spodoptera frugiperda*, Luckow and Summers 1988). Due to the genome size of AcNPV (~128 kbp) and the non-essential nature of the polyhedrin gene, the incorporation of large DNA segments is likely to be tolerated. Furthermore, the strength of the polyhedrin promoter is such that yields as high as 500 mg/liter of recombinant protein, representing 50% of total cell protein, have been reported (Smith et al. 1985). Moreover, nuclear polyhedrosis viruses have the capacity to infect and be expressed in insect larvae such as the silkworm, *Bombyx mori*, or the tobacco hornworm, *Manduca sexta* (Luckow 1990). These organisms possess the fat body, an organ which is physiologically equivalent to the mammalian liver and adipose tissue. In particular, this organ is responsible for endogenous lipoprotein production (Shapiro et al. 1988; Ryan 1990). It therefore seems likely that insect larvae might be particularly well-suited for the heterologous expression of proteins that require lipid association for activity.

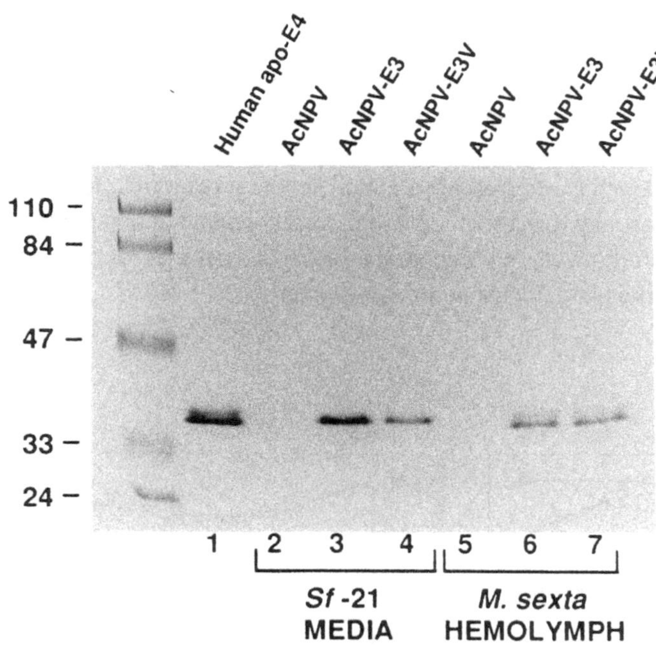

Figure 2. Immunoblot of apo-E3 and apo-E3V produced in *Sf-21* cells *in vitro* and in *M. sexta* larvae *in vivo*. Lane 1, 50 ng human apo-E4 (kindly provided by Dr. R.D. Gregg); lanes 2-4, media from *Sf-21* cells infected with AcNPV (10 µl), AcNPV-E3 (5 µl), AcNPV-E3V (10 µl); lanes 5-7, 1 µl of a 1:4 dilution of hemolymph from larvae infected with AcNPV, AcNPV-E3, AcNPV-E3V. Reproduced from Gretch et al. 1991.

To facilitate this study, cDNA sequences encoding human apo-E3 and a receptor binding-defective variant of apo-E3 (apo-E3V, Ala_{152}-Pro and Ala_{99}-Leu) were fused to the polyhedrin gene promoter in pAcYM1 (Matsuura et al. 1987). These transfer vectors were used to produce recombinant viruses by homologous recombination with wild-type AcNPV by established methods (Summers and Smith 1987).

Upon infection of *S. frugiperda* (*Sf*-21) cells in culture, these recombinant viruses directed the extracellular production of approximately 30 μg/ml of apo-E3 and 2 μg/ml apo-E3V. These proteins were of the predicted molecular weight as shown in Figure 2. A minor proportion of this protein was lipid associated as indicated by its buoyant density (<1.21g/ml) in a sodium bromide density gradient. By contrast, considerably higher levels of expression were obtained upon infection of *M.sexta* larva of the 4th-5th instar. Approximately 220 μg of apo-E3 per ml of insect hemolymph was observed with a molecular weight of the predicted size (Figure 2). As much as 33% of this protein was associated with a surprisingly buoyant lipoprotein (density<1.02g/ml NaBr) indicating a high lipid to protein ratio for these particles. These particles may result from either *de novo* secretion due to infection of the fat body or alternatively an exchange process between apo-E and endogenous insect lipoprotein particles in the hemolymph.

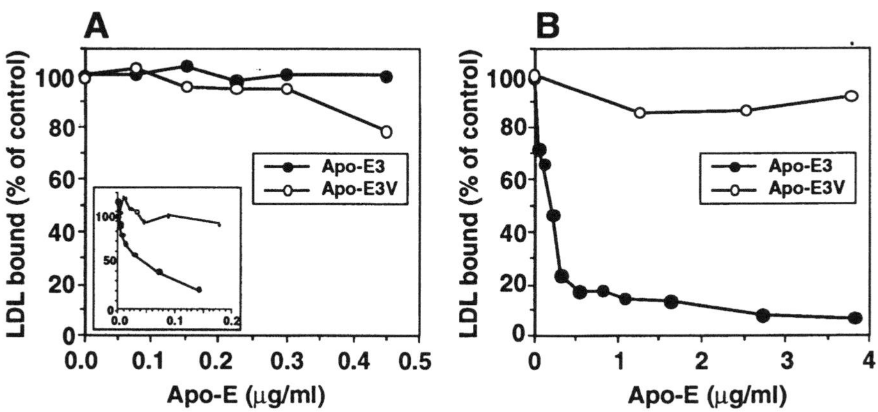

Figure 3. Competition of apo-E3 and apo-E3V with [^{125}I]-LDL for binding to the human LDL receptor at 4°C. **A**. Recombinant apo-E derived from infected *Sf*21 cell culture. The inset panel represents a competition with *in vitro* DMPC associated apo-E. **B**. Recombinant apo-E produced by infected larvae of *M. sexta*. Reproduced from Gretch et al. 1991.

Sf-21 cell culture supernatants and *M. sexta* hemolymph samples containing apo-E3 and apo-E3V were compared for their ability to compete with ^{125}I-labeled Low Density Lipoprotein (LDL) for binding to the human LDL receptor at 4°C. Neither apo-E3 or apo-E3V produced by *Sf*-21 cells was able to effectively compete for binding to the LDL receptor. By contrast, the apo-E3-containing lipoprotein particles in *M. sexta* hemolymph were excellent ligands for the LDL receptor (50% displacement at approximately 0.20 µg/ml). The apo-E3V produced in *M. sexta* had little detectable receptor binding activity, consistent with a previously described receptor binding defect associated with the Ala_{152} to Pro substitution (Lalazar et al. 1988). LDL receptor binding activity of apo-E3 from *Sf*-21 cell media was regenerated upon association with dimyristoyl phosphatidylcholine (DMPC) liposomes *in vitro*. Conversely, apo-E3V from *Sf*-21 cell media remained a poor ligand for the LDL receptor even after lipid association (Figure 3, inset).

Expression of Human apo-E in Yeast.

In a parallel study (Sturley et al. 1991), we expressed human apo-E in the microbial eukaryote *Saccharomyces cerevisiae* to investigate sterol regulation of apo-E expression in an organism amenable to genetic manipulation. In contrast to bacterial expression systems, *S. cerevisiae* utilizes a vesicular secretion pathway and performs many secondary modifications to proteins typically seen in higher eukaryotic cells (Nevalainen and Makarow 1988). Although the major yeast sterol is ergosterol and not cholesterol, other aspects of lipid metabolism in yeast are similar to these processes in animal cells and are genetically well-defined (Henry 1982). Furthermore, the cholesterol and sterol ester status of the yeast cell can be artificially modulated by the use of cholesterol uptake mutants (*upc; uptake control* see Figure 4; Lewis et al. 1985; Lewis et al. 1988).

The cDNA sequences encoding human apo-E3 and apo-E3V were fused to the promoter of the repressible acid phosphatase gene (*PHO5*) such that expression was only achieved under culture conditions of phosphate limitation. The majority of yeast strains transformed with autonomously replicating vectors containing this chimera expressed apo-E protein only at the intracellular level; secretion of apo-E into the culture supernatant could not be detected by immunoblotting. Under normal culture conditions, wild-type strains of *S. cerevisiae* lack the capacity to absorb cholesterol from the medium. However, *upc2* yeast mutants (Lewis et al. 1988) are deregulated for cholesterol uptake control and absorb approximately 150-fold higher levels of sterol (Figure 4). The majority of this sterol is then esterified.

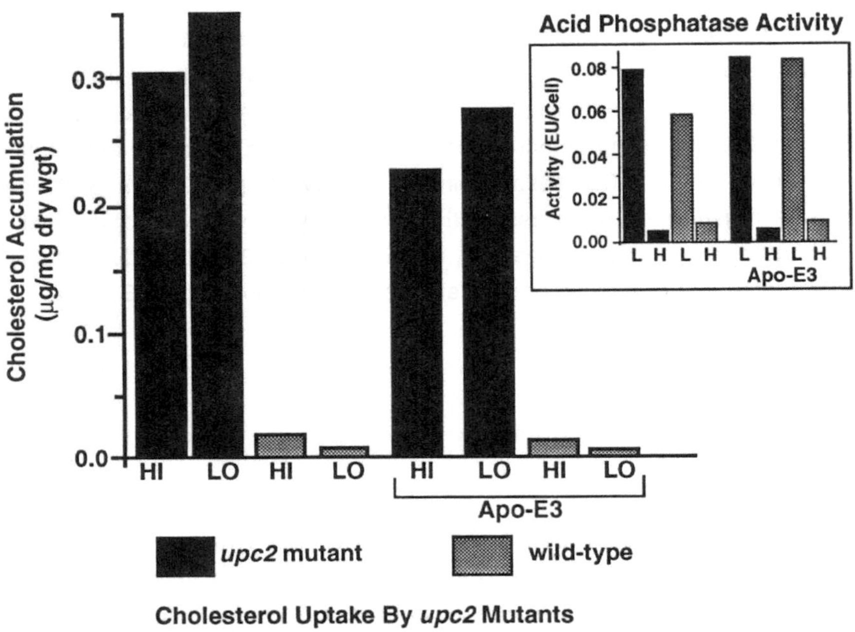

Figure 4. Cholesterol uptake and acid phosphatase activity by yeast in defined media. Triplicate samples of *upc2* and control yeast strains were grown to late exponential phase in the presence of [^{14}C]-cholesterol. H and L denote repressing and derepressing culture conditions, respectively. Strains harboring the apo-E expression plasmid are indicated (YEpE3). Reproduced with permission from Sturley et al. 1991.

Transformation of *upc2* strains with the apo-E expression plasmids resulted in an interesting phenotype. Approximately 40 ng/ml of apo-E was secreted by *upc2* mutants in cholesterol-free media (Figure 5). Furthermore, apo-E secretion was increased 2-3-fold upon the inclusion of cholesterol in the growth media. This response to exogenous cholesterol was not mediated at the transcriptional level, since apo-E mRNA levels were constant under all culture conditions. Concomitant with the increase in secretion following cholesterol uptake by *upc2* strains, approximately 5% of secreted apo-E was associated with lipid; both polar and non-polar lipids were detected in this lipoprotein fraction. Reconstitution of apo-E3 secreted by yeast into DMPC liposomes resulted in the demonstration of LDL receptor binding activity in a 4°C competition assay (not shown).

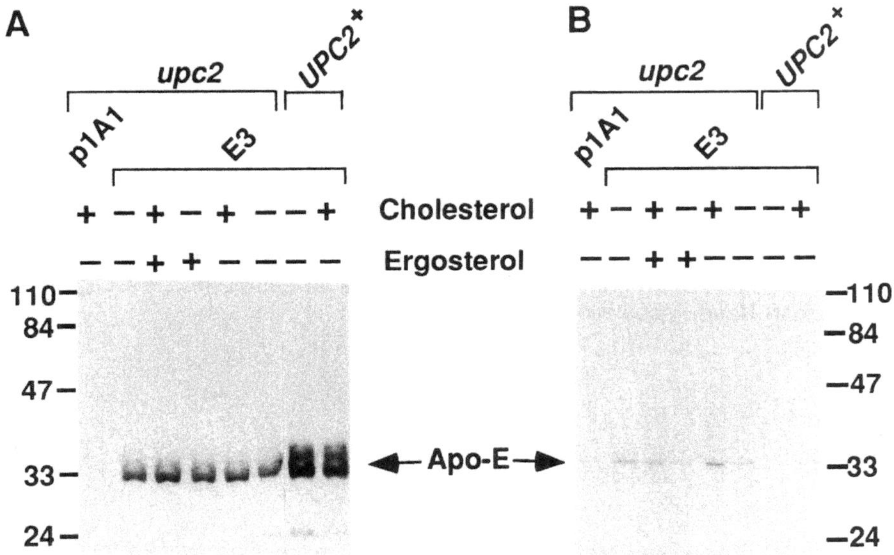

Figure 5. **A.** Expression of apo-E in intracellular extracts from *upc2* mutant and *UPC2* + hosts. Triton X-100 soluble proteins from derepressed cells were analysed by SDS-12%-PAGE and immunoblotting with monoclonal antibodies to human apo-E. **B.** Secretion of apo-E by *upc2* mutant and control (*UPC2* +) yeast strains. Cell-free culture supernatants were concentrated by TCA-Sodium Deoxycholate precipitation and analyzed by immunoblotting. The presence of cholesterol or ergosterol in the culture medium is indicated. p1A1 and E3 denote strains containing control and apo-E expression plasmids, respectively. Reproduced with permission from Sturley et al. 1991.

Conclusions.

The development of the expression systems described here represents a novel alternative to apolipoprotein production in bacterial and mammalian hosts. In terms of yield, the insect systems are several orders of magnitude more productive than any other system yet described for expression of these proteins. The utilization of yeast genetics represents a potentially powerful incursion into the analysis of the host's role in apolipoprotein secretion. We are currently extending these studies to include expression of other apolipoproteins such as apo B and apo (a).

Acknowledgements.

This work was supported by grants from the NHLBI, and the Council for Tobacco Research to A.D.A., the NIH to M.R.C., the NIAID (to P.D.F.), and the NSF (to N.E.B.). D.G.G. was supported by a Biotechnology Training Program, National Research Service Award from the NIGMS. A.D.A is an Established Investigator of the American Heart Association.

References.

Blackhart BD, Yao Z and McCarthy BJ (1990). An expression system for human apolipoprotein B100 in a rat hepatoma cell line. J. Biol Chem 265: 8358-8360.

Gretch DG, Sturley SL, Friesen PD, Beckage NE and Attie AD (1991). Baculovirus-mediated expression of human apolipoprotein-E in *Manduca sexta* larvae generates particles that bind to the LDL receptor. Proc. Natl. Acad. Sci. 88: 8530-8533.

Henry SA (1982). Membrane lipids of yeast: Biochemical and genetic studies. In: The molecular biology of the yeast Saccharomyces: Metabolism and gene expression. New York, Cold Spring Harbor Laboratory. 101-158.

Lalazar A, Weisgraber KH, Rall SC, Giladi H, Innerarity TL, Levanon AZ, et al. (1988). Site-specific mutagenesis of human apolipoprotein E. Receptor binding activity of variants with single amino acid substitutions. J. Biol. Chem. 263: 3542-3545.

Lewis TA, Taylor FR and Parks LW (1985). Involvement of heme biosynthesis in control of sterol uptake by Saccharomyces cerevisiae. J. Bacteriol 163: 199-207.

Lewis TL, Keesler GL, Fenner GP and Parks LW (1988). Pleiotropic mutations in *Saccharomyces cerevisiae* affecting sterol uptake and metabolism. Yeast 4: 93-106.

Luckow VA (1991). Cloning and expression of heterologous genes in insect cell with baculovirus vectors. In: Recombinant DNA Technology and Applications. (eds. R. Prokaop, K.Bajpai and New York, McGraw-Hill. pp. 97-153.

Luckow VA and Summers MD (1988). Trends in the development of Baculovirus expression vectors. Bio/technol. 6: 47-55.
Mallory JB, Kushner PJ, Protter AA, Cofer CL, Appleby VL, Lau K, et al. (1987). Expression and characterization of human apolipoprotein A-I in Chinese hamster ovary cells. J. Biol. Chem. 262: 4241-4247.
Matsuura Y, Possee RD, Overton HA and Bishop DHL (1987). Baculovirus expression vectors: the requirements for high level expression of proteins, including glycoproteins. J. Gen. Virol. 68: 1233-1250.
Nevalainen LT and Makarow M (1988). Intracellular transport in interphase and mitotic yeast cells. Eur. J. Biochem. 178: 39-46.
Reardon CA, Lau YF, Paik YK, Weisgraber KH, Mahley RW and Taylor JM (1986). Expression of the human apolipoprotein E gene in cultured mammalian cells. J. Biol. Chem. 261: 9858-9864.
Ryan RO (1990). Dynamics of insect lipophorin metabolism. J. Lipid Res. 31: 1725-1740.
Shapiro JP, Law JH and Wells MA (1988). Lipid transport in insects. Annu. Rev. Entomol. 33: 297-318.
Simonet WS, Bucay N, Pitas RE, Lauer SJ and Taylor JM (1991). Multiple tissue-specific elements control the apolipoprotein E/C-I gene locus in transgenic mice. J. Biol. Chem. 266: 8651-8654.
Smith GE, Ju G, Ericson BL, Moschera J, Lahm H-W, Chizzonite R, et al. (1985). Modification and secretion of human interleukin 2 produced in insect cells by a baculovirus expression vector. Proc. Natl. Acad. Sci. USA. 82: 8404-8408.
Sturley SL, Culbertson MR and Attie AD (1991). Secretion and lipid association of human apolipoprotein E in *Saccharomyces cerevisiae* requires a host mutation in sterol esterification and uptake. J. Biol. Chem. 266: 16273-16276.
Summers MD and Smith GE (1987). A Manual of Methods for Baculovirus Vectors and Insect Cell Culture Procedures. Tex. Agric. Exp. Stn. Bull.
Vogel T, Weisgraber KH, Zeevi MI, Ben-Artzi H, Levanon AZ, Rall SC, et al. (1985). Human apolipoprotein E expression in *Escherichia coli*: Structural and functional identity of the bacterially produced protein with plasma apolipoprotein E. Proc. Natl. Acad. Sci. USA 82: 8696-8700.
Yao Z, Blackhart BD, Linton MF, Taylor SM, Young SG and McCarthy BJ (1991). Expression of carboxyl-terminally truncated forms of human apolipoprotein B in rat hepatoma cells. Evidence that the length of apolipoprotein B has a major effect on the buoyant density of the secreted lipoproteins. J. Biol. Chem. 266: 3300-3308.

RECOMBINANT HUMAN PROAPOLIPOPROTEIN A-I: EXPERIMENTAL STRATEGIES FOR THE PRODUCTION OF AN AUTHENTIC MOLECULE.

Moguilevsky,N., Varsalona,F., Guillaume,J.-P., Roobol,C.[1] and Bollen,A.
Applied Genetics, U.L.B.,
rue de l'Industrie 24,
B-1400 Nivelles (Belgium)

Summary

A clone coding for human proapolipoprotein A-I (proapo A-I) was isolated from a human liver cDNA library. Expression of recombinant proapo A-I in the cytoplasm of *E.coli* or baculovirus-infected insect cells led to the production of a polypeptide which was characterized in terms of apparent molecular mass, isoelectric point, chemical and enzymatic peptide mapping. In addition, the recombinant protein was assayed *in vitro* for the stimulation of the enzyme lecithin:cholesterol acyltransferase. The data showed that rec proapo A-I was produced at high level in *E.coli* and in insect cells as a stable protein having properties identical to those of the natural product. However, none of these expression systems was able to remove the N-terminal methionine from the recombinant protein. Several methods were thus compared for the production of authentic, N-terminal methionine-free proapo A-I in engineered *Escherichia coli* bacteria. Secretion, mediated by the *Omp*A signal peptide, appeared the best approach, yielding authentic proapo A-I from engineered bacteria. At last, in view of the potential therapeutical application of rec proapo A-I, assessing of DNA contamination in purified batches of the protein was performed using a quantitative polymerase chain reaction assay.

---

[1] UCB-Pharma, chemin du Foriest, B-1420 Braine-l'Alleud (Belgium)

NATO ASI Series, Vol. H 73
Human Apolipoprotein Mutants III
Edited by C. R. Sirtori, G. Franceschini, B.H. Brewer Jr.

Introduction

Epidemiology studies have identified a strong inverse relationship between the risk for atherosclerosis in humans and the level of circulating high density lipoprotein (HDL) and its major protein constituent, apolipoprotein A-I (apo A-I). The biological function of HDL particles is the reverse transport of cholesterol from peripheral tissues to the liver (Glomset, 1968). It has been recently shown that the induction of a high concentration of apo A-I and HDL inhibits the process of atherosclerosis since transgenic mice with high plasma apo A-I and HDL levels were significantly protected from the development of fatty streak lesions (Rubin *et al*, 1991). Apo A-I is synthesized in the liver and intestine as a precursor molecule, proapo A-I, which contains an additional N-terminal hexapeptide (Zannis *et al*,1983). Proapo A-I is secreted into the circulation as a protein phosphatidylcholine complex, a so-called nascent HDL particle (Tall and Small,1980). During its lifetime, this particle will gather cholesterol and proapo A-I will concomitantly be converted to mature apo A-I (Gordon *et al*,1983). Finally, the "adult" HDL particle will be captured by the liver and metabolised.

In this paper, we review the data related to the production and purification of human proapo A-I in *E.coli* and baculovirus-infected insect cells. We stress the difficulties encountered in obtaining N-terminal methionine-free proapo A-I and present several strategies to solve the problem. We selected the secretion approach and show that it leads to the correct processing of preproapo A-I into authentic proapo A-I. At last, we present data pertaining to the monitoring of contaminating DNA in large scale preparations of the recombinant protein.

1. Intracellular production of recproapo A-I

a) in a bacterial system

A cDNA library derived from human liver was used to isolate a clone, pNIV1602, coding for preproapolipoprotein A-I. A coding cassette for the proapo A-I molecule was reconstructed by fusing synthetic sequences, chosen as to optimize expression and specifying the N-terminal methionine and amino acids -6 to +14, to a large DNA fragment derived from pNIV1602 and coding for the C-terminal part of the protein (amino acids 15 to 243). The module was then cloned into an *E.coli* expression vector, pOTS-Nco, (Devare *et al*,1984) carrying the regulatable λ P_L promoter, leading to the recombinant plasmid pNIV1617 (Fig. 5a).
Cells carrying the pNIV1617 plasmid were grown at 30°C in minimal medium, for thermal induction, or at 37°C in rich medium, for chemical induction. Derepression of the P_L promoter was achieved either by shifting the temperature to 42°C or by adding nalidixic acid. Induction was performed for various periods of time. Pelleted bacteria were lysed and the resulting extracts were analyzed by $NaDodSO_4$-polyacrylamide gel electrophoresis, Western blotting, and immunodetection using antibodies raised against plasma apo A-I. Figure 1 shows that a single product, specifically recognized by anti-apo A-I antibodies, is expressed in the bacterial strains carrying plasmid pNIV1617, irrespective of the mode of induction. The protein had a molecular mass of about 28,000 daltons which, within experimental limits, fitted the calculated molecular weight of proapo A-I.
Levels of proapo A-I produced in bacteria were estimated by single radialimmunodiffusion. Under thermal induction conditions in M33 minimal medium, proapo A-I was expressed at about 10% of total soluble proteins versus 3.5% in LB medium. Chemical induction appeared to be less effi-

cient, yielding expression levels on the order of 2% in LB medium and no expression in M33 medium.

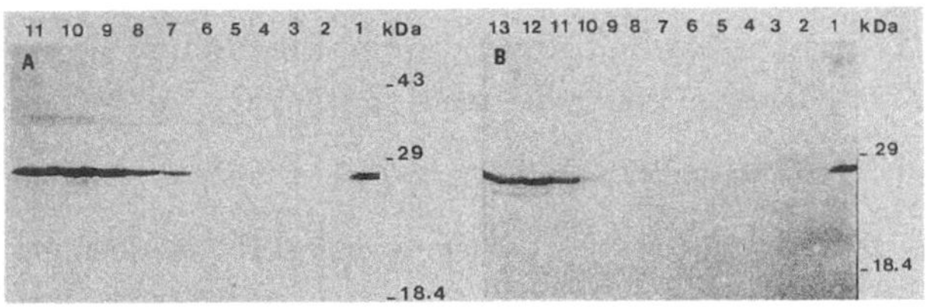

Fig.1. Immunodetection of proapo A-I on Western blots. A. Time course of thermal induction of bacteria carrying pNIV1617 and grown in M33 minimal medium. Lane 1, Plasma derived apo A-I, 500 ng; lanes 2-6, samples taken at 1, 20, 40, 60, and 120 min in noninduction conditions (30°C); lanes 7-11, samples taken at 1, 20, 40, 60, and 120 min after induction (42°C). B. Time course of chemical induction (nalidixic acid) of bacteria carrying pNIV1617 and grown in rich LB medium. Lane 1, Plasma derived apo A-I, 500 ng; lanes 2-7, samples taken at 1, 30, 60, 120, 180 and 240 min in noninduction conditions (37°C); lanes 8-13, samples taken at 1, 30, 60, 120, 180, and 240 min after induction (37°C and nalidixic acid, 60 μg/ml).

b) in baculovirus-infected insect cells

The baculovirus-insect cell expression system utilizes the promoter of the highly expressed polyhedrin gene from *Autographa californica* nuclear polyhedrosis virus (AcNPV) modified for the insertion of foreign genes (Matsuura *et al*, 1987). The cDNA containing the entire coding region of proapo A-I was inserted in proper orientation into the *BamH*I site of transfer vector pAcYM1. The recombinant transfer vector, pNIV1619, was used in conjunction with AcNPV DNA to co-transfect *Spodoptera frugiperda* Sf9 cells. Twenty recombinant viruses were purified by two cycles of plaque assay on Sf9 cells; one of them, vNIV1629, was chosen for expression assays (Fig. 2). The concentration of proapo A-I, produced in insect cells and measured by single radial immunodiffusion, was estimated at 100 mg per litre of culture.

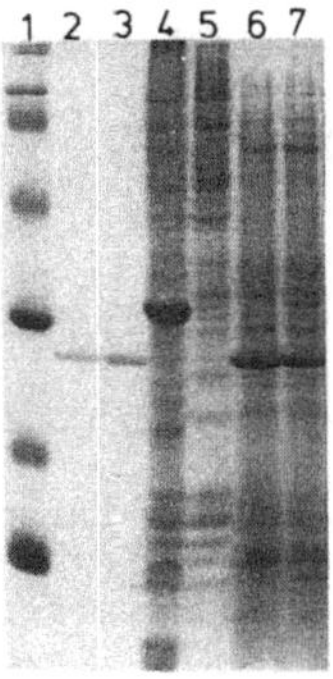

Fig.2. Expression of proapo A-I in baculovirus-infected insect cells. Analysis by SDS-PAGE (Coomassie blue staining). Lane 1, molecular weight standard 500 ng; lanes 2 and 3, apo A-I standards (750 ng); lane 4, wild type baculovirus-infected cells; lane 5, uninfected cells, lanes 6 and 7, recombinant baculovirus-infected cells expressing proapo A-I.

2. Purification and characterization of recproapo A-I

Chromatofocusing on a PBE-74 column was used to purify the recombinant proapo A-I. As shown in Fig. 3, immunoreactive material was resolved from contaminants and was identified by isoelectric focusing mainly as proapo A-I. Moreover, chromatofocusing resolved most apo A-I related isoforms.

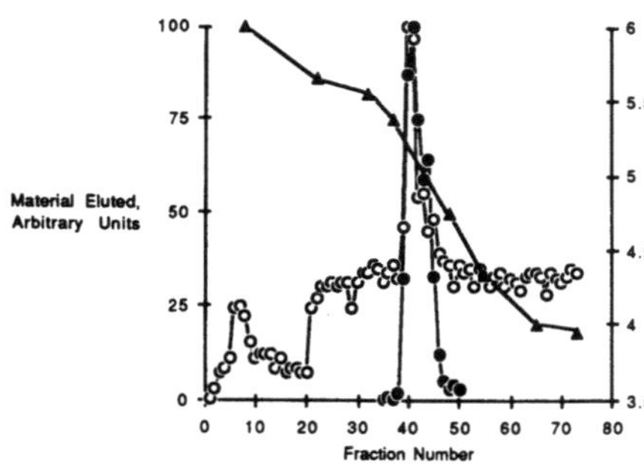

Fig.3. PBE74 chromatofocusing. The major peak of the phenyl sepharose eluate was subjected to chromatofocusing. The eluate was assayed for total protein (o), apo A-I-related immunoreactivity (●), and pH (▲). Fractions 37-48 contain the purified recombinant product.

Pure recombinant proapo A-I was analyzed by $NaDodSO_4$-polyacrylamide gel electrophoresis and isoelectric focusing. Recombinant proapo A-I behaved as a single polypeptide chain with an apparent molecular mass of (27.3±1.1)kD. Mature apo A-I was slightly smaller, having an apparent molecular weight of (26.8±1.1)kD. In isoelectric focusing (data not shown), the recombinant protein showed one major isoform and one minor isoform, having isoelectric points of 5.20 and 4.95, respectively. The relative migration distances of these isoforms coincided exactly with those of plasma proapo A-I and apo A-I, respectively. Partial chemical cleavage at tryptophan residues, was performed on recombinant proapo A-I (major isoform) and on plasma-derived apo A-I. Figure 4 shows the large peptides of expected molecular weight produced in both cases by cleavage at positions 8, 50, 72 and 108 in the proteins.

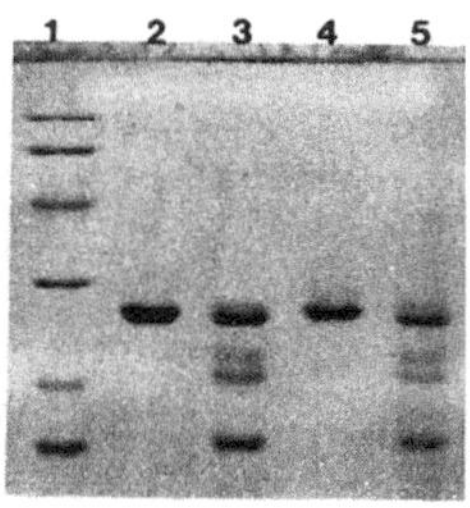

Fig.4. Chemical characterization of proapo A-I. Purified recombinant proapo A-I and human plasma apo A-I were subjected to partial chemical peptide mapping. Lane 1, Molecular mass standards; lane 2, human plasma apo A-I, untreated; lane 3, human plasma apo A-I, cleaved; lane 4, recombinant proapo A-I, untreated; lane 5, recombinant proapo A-I, cleaved.

In addition, digestions of recombinant proapo A-I and of plasma-derived apo A-I with trypsin in very mild conditions yielded similar patterns of fragments (data not shown). Thus, both types of experiments suggest that the recombinant molecule is structurally identical to its natural counterpart.

Recombinant proapo A-I (pI 5.20) and its minor isoform (pI 4.95) were both able to transform synthetic lecithin-cholesterol liposomes into substrates for the enzyme lecithin:cholesterol acyltransferase. As seen in Table 1, the activation capacity of recombinant apoproteins was similar to the one observed with authentic human plasma apo A-I.

3. Strategies for removal of the N-terminal methionine

Although human recombinant proapo A-I, produced in engineered *Escherichia coli* and Baculovirus-infected insect cells, displayed physi-

Source	LCAT (μl)	Apolipoprotein (μM)	LCAT activity (nmoles/hr × liters)
Plasma apo A-I	25	0	2.5
	25	3	250
Recombinant proapo A-I	100	0	100
Major isoform (pI 5.20)	100	3	850
	100	6	1,600
Minor isoform (pI 4.95)	100	0	100
	100	3	700
	100	6	700

Table 1. Stimulation of Lecithin:Cholesterol Acyltransferase Activity by Recombinant Proapo A-I.

cal and functional properties that were indistinguishable from those of its natural counterpart, none of the expression systems used was able to remove the N-terminal methionine from the recombinant protein. In view of the potential therapeutic application of rec proapo A-I, it became essential to design systems for removing this additional amino acid. Although methionylated proapo A-I could be cleaved in plasma to yield authentic apo A-I, this processing would release a methionylated hexapeptide which, by definition, constitutes a foreign entity and thus carries the potential to evoke an undesired immune response.

Several methods were compared with respect to the production of authentic, N-terminal methionine-free proapolipoprotein A-I in engineered *Escherichia coli* bacteria. A first approach consisted of treating the purified methionylated recombinant protein with an aminopeptidase, purified from *Aeromonas* proteolytica (Prescott *et al*, 1971). This approach however was not satisfactory since the processing to authentic proapo A-I was only partial and was accompanied by considerable degradation.

In a second approach, we decided to design recombinant proapo A-I molecules carrying built-in recognition sites for specific enzymes. Enterokinase and kallikrein, whose recognition sites in proteins are well known (Maroux *et al*,1971; Fiedler and Leysath,1979), were selected

since i) these cleavage sites were not present in the sequence of rec proapo A-I and, ii) the specificity of cleavage was very high. In addition, it was anticipated that the addition of several charged amino acids to proapo A-I would confer a significant charge difference to the fused protein. This would facilitate the subsequent separation of the cleaved proapo A-I from uncleaved material. Two plasmids were constructed, pNIV1624 and pNIV1625, which carried the proapo A-I DNA sequence fused downstream to the sequences encoding the enterokinase or kallikrein recognition and cleavage sites respectively (Fig.5b and 5c).
These plasmids were introduced by transformation into *E.coli* strain AR120. Appropriate clones were selected, characterized and then tested for expression after chemical induction of the P_L promoter present on the expression plasmids. Bacteria carrying pNIV1624 and pNIV1625 expressed the fusion proteins up to 5% of total soluble proteins. The products were purified according to previously published procedures (Moguilevsky *et al*,1989) and then incubated with enterokinase or kallikrein at various enzyme/product ratios and for various periods of time. Digestion products were analyzed on SDS-polyacrylamide gels in the case of enterokinase and by isoelectrofocusing in the case of kallikrein.
Cleavage of proapo A-I from the fusion protein encoded by pNIV1624 by enterokinase was accompanied by significant proteolysis to smaller peptides, and cleavage of proapo A-I from the fusion protein of pNIV1625 by kallikrein was more selective and gave better yield. The charge difference between authentic proapo A-I and its fusion precursor allowed a clear distinction between substrate and product on isoelectrofocusing gels. Both types of digestion, however, required relatively high enzyme/product ratios, concomitantly resulting in a considerable nonspecific proteolytic degradation.

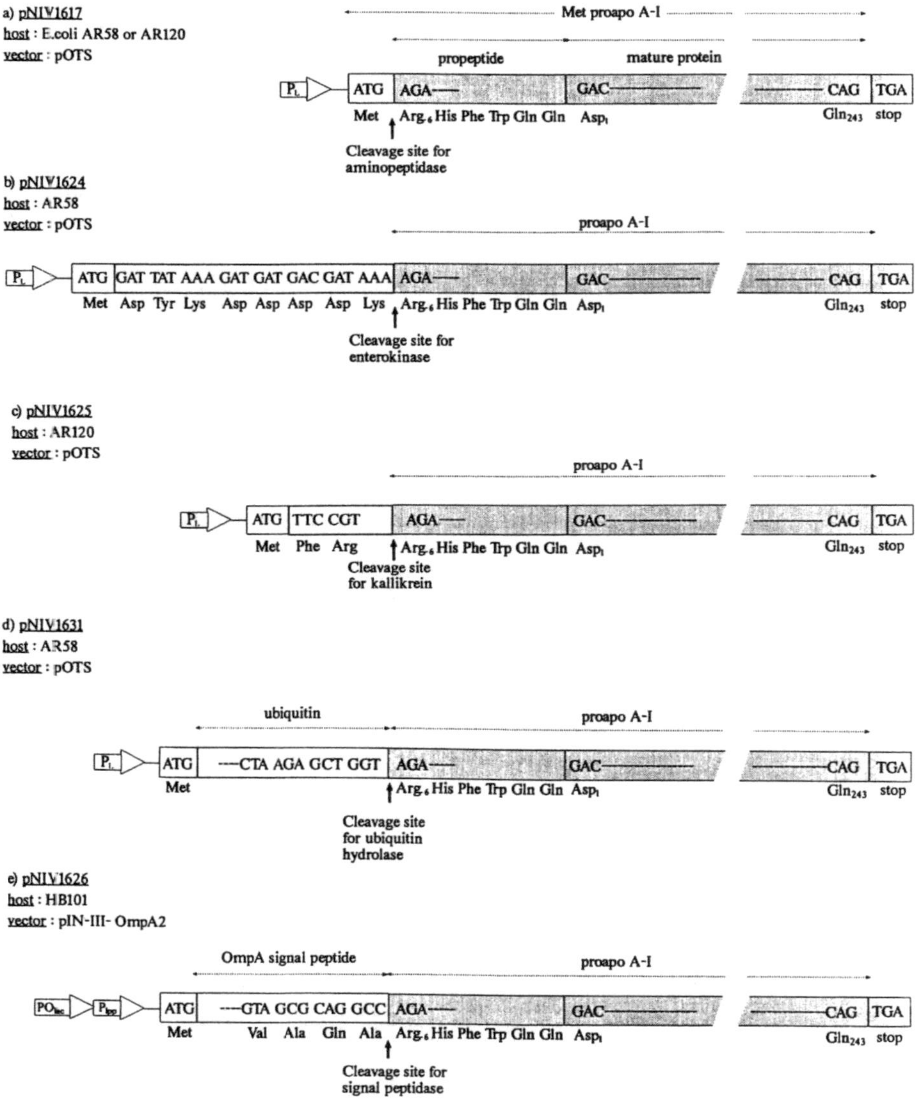

Fig.5. Schematic outlines of the different proapo A-I coding modules.

In another *in vitro* cleavage strategy, we designed a DNA module coding for a fusion between the sequence specifying yeast ubiquitin, a highly conserved 76 amino acid protein (Ecker *et al*, 1987) and that encoding proapo A-I (Fig.5d). The plasmid carrying this fusion cassette, pNIV1631, was introduced into *E.coli* AR58 cells and, upon thermal induction, bacteria carrying pNIV1631 produced the ubiquitin-proapo A-I fusion at about 13% of total protein. Liu *et al* (1989) have demonstrated that ubiquitin hydrolase as present in crude yeast extracts is able to completely cleave *in vitro* ubiquitin fused to foreign polypeptides, provided that the essential Gly76 is present upstream from the foreign protein (Jonnala-Gadda *et al*,1989). Therefore, we incubated crude extracts of bacteria carrying pNIV1631 with various amounts of crude extract obtained from the yeast strain TCDY1 and for different periods of time. Digestion products were analyzed on SDS-polyacrylamide gels followed by immunoblotting using sheep anti apo A-I polyclonal antibodies. The cleavage of the fusion did occur, but was far from efficient and required relatively high amounts of the yeast extract. The cleaved immunoreactive product had the molecular weight expected for authentic proapo A-I (28 kDa) and no significant unspecific degradation was observed.

4. Secretion of human rec proapo A-I in E.coli

In the *Escherichia coli* system, it is possible to guide a cloned gene product to three different locations in the cell; the cytoplasm, outside the cytoplasmic membrane, or into the outer membrane. The secretion of a cloned gene product across the cytoplasmic membrane can be achieved by fusing an appropriate signal peptide to the amino-terminal end of the gene product. Such a secretion cloning system has significant advantages over the other cloning systems. Firstly, the N-terminal amino acid residue of the secreted product can be identical to the natural gene product as a result of the cleavage of the signal peptide by a specific signal peptidase. Secondly, protease activity is considered to be much less in the periplasmic space than in the cytoplasm. Therefore, the gene product is more stable in the former than the latter.

Aiming at the *in situ* conversion of a precursor into authentic proapo A-I, we constructed plasmid pNIV1626 (Fig.5e), a derivative of the secretion expression vector pIN-III-*Omp*A2 (Ghrayeb *et al*,1984). It carried, under the control of the isopropyl-1-thio-β-D-galactopyranoside (IPTG)-inducible PO_{lac}plpp tandem promoter, a fused DNA sequence corresponding to the signal peptide of the bacterial outer membrane protein (*Omp*A) and to proapo A-I (Fig. 5e).

Expression of the cloned DNA is ensured jointly by the IPTG-inducible *lac* promoter operator DNA fragment and the strong *lpp* promoter.

Bacteria transformed with pNIV1626 were cultured in LB medium, induced with IPTG (100 μM) for 2 hours at 37°C and analyzed.

If the gene products induced by IPTG are secreted across the cytoplasmic membrane, they should be found in the periplasmic cell fraction. In the case of pNIV1626, the periplasmic fraction was prepared from the cells by a conventional osmotic shock procedure (Koshland and Botstein, 1980). The proapo A-I is a major component of this fraction and comprises about 7.5% of the total cell protein. A larger protein was also observed upon induction of pNIV1626 in the membrane fraction. The localization of the larger induced protein in the membrane fraction suggests that this protein is the unprocessed hybrid preprotein having a hydrophobic uncleaved OmpA signal peptide.

Fractions corresponding to cytoplasm and periplasm were separated onto 2-dimensional polyacrylamide gels. In the periplasmic fraction (Fig.6), the recombinant protein behaved as authentic proapo A-I, indicating that the processing of the signal peptide has occurred efficiently. N-terminal amino acid analysis of the purified processed protein yielded the sequence Arg His Phe Trp Gln Gln Asp Glu Pro Pro which corresponded to the sequence of natural proapo A-I. In addition, C-terminal analysis of the processed material showed that it ended with a Gln residue, as in the natural protein, and thus that proteolytic degradation has not occurred.

Pure recombinant proapo A-I was analyzed by 12%-SDS-polyacrylamide gel electrophoresis and isoelectric focusing. Recombinant proapo A-I behaved as a single polypeptide chain with an apparent molecular mass of 27.3±1.1 KDa. In isoelectric focusing, the recombinant protein showed one major isoform (pI=5.20) and one minor isoform (pI=4.95) corresponding exactly to the isoforms of plasma proapo A-I and apo A-I. The chemical and enzymatic characterization of proapo A-I in the conditions described before (Moguilevsky *et al*, 1989) suggest that the authentic recombinant proapo A-I is structurally identical to the natural molecule.

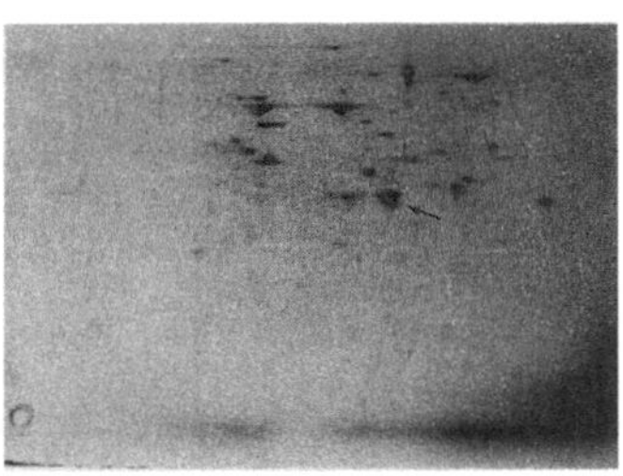

Fig.6. Analysis by 2-dimensional PAGE of the periplasmic fraction of the secretion strain. Fraction corresponding to the periplasm was separated onto 2-dimensional polyacrylamide gels. The arrow indicates the position of the authentic proapo A-I. The identity of this spot was verified by immunoblotting.

5. Assessing DNA contamination in secreted human recombinant proapo A-I produced in *E.coli*

Fermentation of engineered bacteria in a large scale allows the purification of gram quantities of recombinant proteins. In view of its potential therapeutical application, recombinant proapo A-I will have to conform to Health Authorities regulations, namely as regards the level of contaminating DNA in the preparation. Eventual treatment of atherosclerotic patients with the recombinant protein will probably require large amounts of material. Assuming a dosis in the order of the gram, contaminating DNA should not exceed 10 pg, that is 10 fg DNA per mg of protein. Currently used detection methods based on DNA hybridization are not sensitive enough to achieve that threshold of sensitivity. We therefore relied on the polymerase chain reaction to evaluate batches of purified recombinant proapo A-I for the presence of contaminating DNA.

As a first approach, we decided to detect residual recombinant plasmid DNA in the purified protein conservatively assuming that at least 2% of total bacterial DNA would consist of the plasmidic species. This seems reasonable since the recombinant proapo A-I expression plasmid is still present in multicopies in the host at the end of a fermentation process. Biotinylated amplification primers, specifying and flanking a 141 bp DNA fragment of the proapo A-I sequence, were used in the assay and amplified products were detected by electrophoresis in agarose gel.

The general procedure consisted of establishing first a standard curve using natural proapo A-I, isolated from human plasma and purified in identical conditions as the recombinant product. Aliquots (1mg) of the purified natural protein were mixed with various amounts of linearized plasmid DNA and digested with proteinase K. Digestion products were eliminated by phenol extraction and the residual DNA was precipitated by ethanol, dried and resuspended into the PCR reaction buffer.

The polymerase chain reaction was performed for 35 cycles, which constituted the optimal number of cycles to obtain the required detection threshold. In these conditions, the system detected 0.2 pg of plasmidic DNA per gram of protein (Fig.7).

Batches of recombinant proapo A-I were analyzed by PCR using the same protocol. One milligram samples, usually 5, of each batch were treated as described above, submitted to the amplification reaction and contaminating DNA was evaluated. Depending on the batch, contamination was inferior to 10 pg of total DNA/gram of proapo A-I.

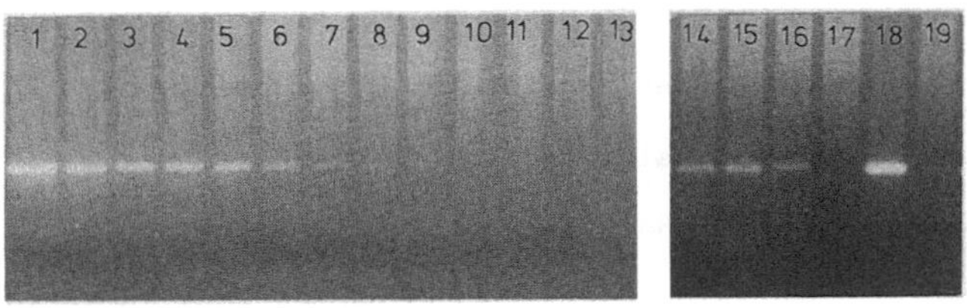

Fig.7. Quantitative estimation by PCR of contaminating DNA in purified proapo A-I. Lanes 1-10, per lane, products of PCR of serial 1/10 dilution from 1 ng to 1 atg of recombinant plasmid, corresponding to 20 μg to 0.02 pg of plasmidic DNA per g of recombinant proapo A-I. Lanes 11-13, products of PCR of purified proapo A-I, batch 1 containing less than 0.2 pg plasmidic DNA/g protein. Lanes 14-16, idem 11-13, for batch 2 containing more than 200 pg plasmidic DNA/g protein. Lane 17, distilled water. Lane 18, positive control: recombinant secretion plasmid linearized by StuI, 20 ng. Lane 19, negative control, secretion plasmid without insert. A 4% agarose gel for the separation of PCR products was run at 150 V for 90 min and then stained with ethidium bromide. Amplification conditions: denaturation 94°C for 2 min, annealing 65°C for 1 min, polymerization 70°C for 2 min.

Discussion

Techniques to produce apo A-I in large quantities have been developed using plasma as starting material (Ross and Carson, 1985; Brewer *et al*, 1986). Attempts to express mature apo A-I in recombinant bacteria proved difficult due to the instability of the protein (Lorenzetti *et al*, 1986; Mallory *et al*, 1987; Moguilevsky *et al*, 1989). To circumvent this problem, we inserted the sequence coding for the propeptide upstream to the apo A-I cDNA. This resulted in a dramatic increase in the efficiency of expression, as was observed also by Isacchi *et al* (1989). The introduction of the propeptide thus allowed the expression of proapo A-I at high level in *E.coli*, as a stable and undegraded protein having physical and functional properties indistinguishable from those of the natural product. However, proapo A-I expressed in the cytoplasm of *E.coli* or in insect cells retained the N-terminal methionine. Our attempts to express proapo A-I in another eukaryotic expression system, the yeast, was unsuccessful, the expression level being very low.

Apo A-I has been expressed also in various eukaryotic cells such as Chinese hamster (Mallory *et al*,1987; Forte *et al*,1990; Brissette *et al*,1991), in 3T3 fibroblasts (Lamon-Fava *et al*,1987), in rat myogenic cells (Ruiz-Opazo and Zannis, 1987), in murine pituitary cell line (Fennewald *et al*, 1988) and in monkey kidney COS cells (Dixon *et al*, 1989). However, all these expression systems allowed a low yield of production as compared with the level of expression obtained in *E.coli*. Moreover, in most of the cases, apo A-I was secreted into the medium associated with lipids. The baculovirus-infected insect cells system allowed a high production level but presented two inconvenients: the presence of a N-terminal methionine and the difficulty of scaling-up the insect cell expression system. The bacterial expression system thus appeared the most appropriate to produce proapo A-I at a high scale.

To solve the problem of the excendentary N-terminal methionine in proapo A-I, produced in the cytoplasm of *E.coli.*, five systems were tested; advantages and inconveniences of the different strategies are listed in Table 2. It can be concluded that all *in vitro* systems required very pure enzymes to minimize non specific proteolytic degradation. The ubiquitin hydrolase enzyme, appeared to be an exception in this respect, since even as a crude preparation, it generated mainly the expected cleavage product. The use of purified ubiquitin hydrolase would *a fortiori* constitute an efficient way to recover an authentic protein from fusions produced in *E.coli*. Alternatively, concomitant *in vivo*

Table 2 : Comparison of different strategies to produce authentic recombinant proapo A-I from bacteria

Enzyme involved in the process	Advantages	Inconveniences	Possible improvements
1. Aminopeptidase (from Aeromonas proteolytica)	removal of N-ter Methionine	- non specific unless Glu, Asp or X-Pro as second amino acid - requires high degree of purity - requires post-cleavage purification of the product	purer enzyme use of methionine amino-peptidase (37)
2. Enterokinase (porcine or bovine)	- specificity (Asp,Lys) - charge difference between precursor and final product allows easy purification	- partial cleavage - proteolytic degradation - requires high degree of purity of both substrate and enzyme - expensive	purer enzyme
3. Kallikrein (human)	- specificity (Phe Arg) - fairly efficient	charge difference between precursor and final product too small for easy purification after treatment	Increase length and charge of the N-terminal extension
4. Ubiquitin hydrolase (yeast extract)	- cleavage possible on crude bacterial extracts - efficient cleavage - no degradation products - increased expression level due to the fusion	- requires high amounts of yeast extracts	- use purified ubiquitin hydrolase - Clone ubiquitin hydrolase gene on exxpression plasmid in order to obtain in vivo conversion of the precursor (38)
5. Signal peptidase (bacterial)	- in vivo conversion - fairly efficient - low level of unspecific degradation - ease of purification from membrane-associated fractions - no inclusion bodies despite high level of expression	- variability in the conversion rate depending on culture conditions - difficulty to perform reproducible osmotic shocks on large scale preparations	- perform excretion of the authentic product in the culture medium via the permeabilization process (35,36)

Strategies 1,2 and 3: all require post- cleavage purification, pure enzymes and are expected to be forbiddingly expensive production of ubiquitin hydrolase by recombinant bacteria carrying the corresponding gene on the expression plasmid might even be more appropriate. The ubiquitin fusion approach conferred also an additional advantage since it appeared to significantly increase the level of expression of the desired protein (this work; Butt *et al*,1989). The *in vivo* approach used here, i.e. the secretion of proapo A-I via a bacterial signal peptide, appeared rather efficient and specific. The system, however, required fine tuning of culture conditions of recombinant bacteria to assure reproducibility in maturation levels. In addition, extrapolation of osmotic shock procedures to large batches of bacteria might be difficult to achieve. This inconvenience could be overcome by the use of *E.coli* cells harboring either an excretion vector (Kato *et al*,1987) or a system leading to membrane permeability (Hsiung *et al*,1989) and thus to actual excretion of the product in the culture medium.

Retrospectively, the ubiquitin fusion and the signal peptide strategies thus appeared the most appropriate for implementation in a protocol for the production of authentic proapo A-I from *Escherichia coli*.

Acknowledgements

This work was supported by U.C.B. Pharmaceutical Division (Braine-l'Alleud, Belgium) and the Ministry of the Walloon Region, Belgium (Service des Technologies Nouvelles).
The authors are indebted to Drs. Lars A. Carlson (Karolinska Institute, Stockholm, Sweden), Ernst Wülfert and Jean Gobert (UCB-Pharma, Braine-l'Alleud, Belgium) for many helpful discussions.

References

Brewer HB, Roman R, Meng M, and Bishop C(1986). Isolation and characterization of apolipoproteins A-I, A-II and A-IV. Methods Enzymol. 128:223-246.

Brissette L, Cahuzac-Bec N, Desforges M, Bec JL, Marcel YL and Rassart E (1991). Expression of recombinant human apolipoprotein A-I in chinese hamster ovary cells and *Escherichia coli*. Protein expression and purification 2:296-303.

Butt TR, Jonnala-Gadda S, Monia BP, Sternberg EJ, Marsh JA, Stadel JM, Ecker DJ and Crooke ST (1989). Ubiquitin fusion augments the yield of cloned gene products in *Escherichia coli*. Proc. Natl. Acad. Sci. USA 86: 2540-2544.

Devare SG, Shatzman A, Robbins KC, Rosenberg M. and Aaronson S (1984). Expression of PDGF-related transforming protein of simian sarcoma virus in *E.coli*. Cell 36: 43-49.

Dixon JL, Battini R, Ferrari S, Redman CM, and Banerjee D (1989). Expression and secrétion of chicken apolipoprotein A-I in transfected COS cells. Biochim. Biophys. Acta 1009: 47-53.

Ecker DJ, Khan MI, Marsh J, Butt TR and Crooke ST (1987). Chemical synthesis and expression of a cassette adapted ubiquitin gene. J. Biol. Chem. 262: 3524-3527.

Fennewald SM, Hamilton Jr RL, Gordon JI (1988). Expression of human preproapo A-I and pre(Δ pro)apo A-I in murine pituitary cell line (AtT-20). J. Biol. Chem. 263: 15568-15577.

Fiedler F and Leysath G (1979). Substrate specificity of porcine pancreatic kallikrein. Adv. Exp. Med. Biol. 120: 261-271.

Forte TM, McCall MR, Amacher S, Nordhausen RW, Vigne JL, and Mallory JB (1990). Physical and chemical characteristics of apolipoprotein A-I-lipid complexes produced by Chinese hamster ovary cells transfected with the human apolipoprotein A-I gene. Biochim. Biophys. Acta 1047: 11-18.

Ghrayeb J, Kimura H, Takahara M, Hsiung H, Masui Y and Inouye M (1984). Secretion cloning vectors in *Escherichia coli*. EMBO J. 3: 2437-2442.

Glomset JA (1968). The plasma lecithin: Cholesterol acyltransferase reaction. J. Lipid Res. 9: 155-167.

Gordon JI, Sims H, Lentz SR, Edelstein C, Scanu AM and Strauss AW (1983). Proteolytic processing of human preproapolipoprotein A-I. J. Biol. Chem. 258: 4037-4044.

Hopp TP, Prickett KS, Price VL, Libby RT, March CJ, Cerretti DP, Urdal DL, and Conlon PJ (1988). A short polypeptide marker sequence useful for recombinant protein identification and purification. Biotechnology 6: 1204-1210.

Hsiung HM, Cantrell A, Luirink J, Oudega B, Veros AJ and Becker GW (1989). Use of bacteriocin release protein in *E.coli* for excretion of human growth hormone into the culture medium. Biotechnology 7: 267-271.

Isacchi A, Sarmientos P, Lorenzetti R and Soria M (1989) Mature apolipoprotein A-I and its precursor proapo A-I: influence of the sequence at the 5' end of the gene on the efficiency of expression in *Escherichia coli*. Gene 81 :129-137.

Jonnala-Gadda S, Butt TR, Monia BP, Mirabelli CK, Gotlieb L, Ecker DJ and Crooke ST (1989). Multiple (α-NH-ubiquitin) protein endoproteases in cells. J. Biol. Chem. 264: 10637-10642.

Kato C, Kobayashi T, Kudo T, Furusato T, Murakami Y, Tanaka T, Baba H, Oishi T, Ohtsuka E, Ikehara M, Yanagida T, Kato H, Moriyama S, Horikoshi K (1987). Construction of an excretion vector and extracellular production of human growth hormone from *Escherichia coli*. Gene 54: 197-202.

Koshland D and Botstein D (1980). Secretion of beta-lactamase requires the carboxy end of the protein. Cell 20: 749-760.

Lamon-Fava S, Ordovas JM, Mandel G, Forte M, Goodman, Schaefer EJ (1987) Secretion of human apolipoprotein particles following transfection of the human apolipoprotein A-I genes into 3T3 cells. J. Biol. Chem. 262: 8944-8947.

Liu CC, Miller HI, Kohn WJ and Silber JI (1989). Purification of a ubiquitin protein peptidase from yeast with efficient *in vitro* assays. J. Biol. Chem. 264: 20331-20338.

Lorenzetti R, Sidoli R, Malomba R, Monaco L, Martineau D, Lappi DA and Soria M (1986) Expression of the human apolipoprotein A-I gene fused to the *E.coli* gene for β-galactosidase. FEBS Lett. 194:343-346.

Mallory JB, Kushner PJ, Protter AA, Cofer CL, Appleby VL, Lau K, Shilling JW and Vigne JL (1987) Expression and characterization of human apolipoprotein A-I in chinese hamster ovary cells. J. Biol. Chem. 262: 4241-4247.

Maroux S, Baratti J, and Desnuelle P (1971). Purification and specificity of porcine enterokinase. J. Biol. Chem. 246: 5031-5039.

Matsuura Y, Possee RD, Overton HA and Bishop DHL (1987). Baculovirus expression vectors: the requirements for high level expression of proteins, including glycoproteins. J. Gen. Virol. 68: 1233-1250.

Moguilevsky N, Roobol C, Loriau R, Guillaume JP, Jacobs P, Cravador A, Herzog A, Brouwers L, Scarso A, Gilles P, Holmquist L, Carlson LA and Bollen A (1989) Production of human recombinant proapolipoprotein A-I in *Escherichia coli*: purification and biochemical characterization. DNA 8:429-436.

Monaco L, Bond HM, Howell KE and Cortese R (1987) A recombinant apo A-I protein A hybrid reproduces the binding parameters of HDL to its receptor. EMBO J. 6:3253-3260.

Prescott JM, Wilkes SH, Wagner FW and Wilson KJ (1971). Aeromonas Aminopeptidase: Improved isolation and some physical properties. J. Biol. Chem. 246: 1756-1764.

Ross SE and Carson SD (1985) Rapid chromatographic purification of apolipoproteins A-I and A-II from human plasma. Anal. Biochem. 149:166-168.

Rubin ME, Krauss RM, Spangler EA, Verstuyft JG and Clift SM (1991) Inhibition of early atherogenesis in transgenic mice by human apolipoprotein A-I. Nature 353:265-267.

Ruiz-Opazo N, and Zannis VI (1988). Expression of the human apolipoprotein A-I gene in rat myogenic L6E9 cells. J. Biol. Chem. 263: 1739-1744.

Tall AR and Small DM (1980). Body cholesterol removal: role of plasma high-density lipoproteins. Adv. Lip. Res. 17: 1-51.

Zannis VI, Karthanasis SK, Keutmann T, Goldberger G, and Breslow JL (1983). Intracellular and extracellular processing of human apolipoprotein A-I: Secreted apolipoprotein A-I isoprotein 2 is a propeptide. Proc. Natl. Acad. Sci. USA 80: 2574-2578.

THE ROLE OF HDL APOLIPOPROTEINS IN MEDIATING HDL RECEPTOR BINDING AND INTRACELLULAR CHOLESTEROL EFFLUX

Armando J. Mendez and John F. Oram
Division of Metabolism, Endocrinology & Nutrition
University of Washington
Seattle, Washington, USA

INTRODUCTION

The role of high density lipoproteins (HDL) in reverse cholesterol transport has been postulated to be in the removal of excess cholesterol from peripheral tissues for delivery to the liver for excretion or reutilization (Glomset). While *in vivo* evidence for this role of HDL is scant, many studies utilizing a variety of cultured cell types have demonstrated that HDL is able to promote the net efflux of cellular cholesterol (reviewed in Phillips et al). More recently, Badimon and coworkers (Badimon et al 1989 and Badimon et al 1990) were able to demonstrate that intravenous injection of purified HDL into cholesterol fed rabbits resulted in significantly decreased aortic atherosclerotic involvement in treated versus control animals. Thus, providing *in vivo* evidence for the role of HDL in preventing cholesterol accumulation.

The mechanisms by which HDL promote cellular cholesterol efflux remain poorly understood. Several studies have suggested that HDL lipids are responsible for mediating efflux by passive desorption/diffusion (of cholesterol from the plasma membrane,) events driven by the equilibration of cholesterol between donor (cells) and acceptors (e.g. HDL) (Phillips et al, Fielding and Fielding). In contrast, studies from our laboratory have demonstrated that binding of HDL to receptor sites on the cell surface is responsible for the translocation and efflux of intracellular cholesterol and that intact apolipoproteins (apos) are required for binding and efflux to occur (Slotte et al, Oram et al 1991). The present studies were performed to further delineate the role of HDL apos in mediating HDL receptor binding and promoting cellular cholesterol efflux. Using trypsin digestion of HDL apos, it was possible to obtain lipoprotein particles which differed only in their content of intact HDL associated apos but were identical in terms of lipid composition. These particles were then used to assess the role of HDL apos in high affinity cell surface binding and promoting intracellular cholesterol efflux from cultured cells.

NATO ASI Series, Vol. H 73
Human Apolipoprotein Mutants III
Edited by C. R. Sirtori, G. Franceschini, B.H. Brewer Jr.

METHODS

HDL3 was isolated by ultracentrifugation in the density range 1.12 to 1.21 g/ml and subjected to heparin-Sepharose chromatography to remove apo E containing particles. HDL was incubated with trypsin at a protein ratio of 40:1 (w/w respectively) at 37°C for various times and the reaction was stopped by the addition of phynylmethylsulfonylfloride to a final concentration of 1 mM. HDL was separated from trypsin and digested peptides by Sephadex G-75 chromatography. Reconstitution of trypsin-modified HDL3 was achieved by incubation of particles treated with trypsin for 5 or 60 minutes with purified apos A-I, A-II, or E for 2 to 4 hours at room temperature. Free apos were separated from the lipoproteins by ultracentrifugation at 1.21 g/ml. Control HDL3 and trypsin-HDL were treated similarly but without added apos.

Human skin fibroblasts were cultured and maintained as previously described (Brinton, et al). Cholesterol efflux was determined by either selective labeling of intracellular sterol pools in cholesterol-loaded cells using ^{3}H-mevalonolactone or by measuring the changes in the acyl CoA cholesterol: acyl transferase (ACAT) substrate pool by the incorporation of ^{14}C-oleate into cholestryl esters (Oram et al, 1991). Plasma membrane pools of cholesterol were labeled by short pulse incubations with ^{3}H-cholesterol (Oram et al, 1991). Cholesterol mass was measured by the cholesterol oxidase method. Cell surface binding of ^{125}I-HDL was done as described (Oram) or with the indicated modifications.

RESULTS

Trypsin treatment of HDL. HDL3 was subjected to trypsin treatment for various times from 5 to 120 minutes. The resulting particles had nearly identical lipid composition to control HDL and retained alpha mobility by agarose gel electrophoresis. Particle size, determined by nondenaturing gradient gel electropheresis was unaltered with up to 20 minutes of trypsin treatment, however by 30 minutes a band with larger apparent size (similar to HDL_2) was observed, and by 120 minutes nearly all of the particles had shifted to the larger band. Sodium dodecyl sulfate polyacylamide gel electrophoresis of the trypsin treated HDL3 particles showed that after 5 minutes of trypsin modification 85% of the apo A-I and 60% of the apo A-II remained intact. These values progressively decreased and by 60 minutes less than 30% of the apo A-I and 20% of the apo A-II remained undigested. The relative loss of HDL3 associated protein after trypsin treatment could also be quantitated by the protein to phospholipid ratio, since no changes in

TREATMENT	μg Protein/μmole Phospholipid	% of Control
Control HDL	1.232	------
Trypsin 5 min	1.047	85
Trypsin 10 min	1.017	83
Trypsin 20 min	0.978	79
Trypsin 30 min	0.938	76
Trypsin 60 min	0.850	68
Trypsin-HDL (5min) + apo A-I	1.175	95
Trypsin HDL (60 min) + apo A-I	1.078	88
Trypsin HDL (60 min) + apo A-II	1.320	107
Trypsin HDL (60 min) + apo E	1.112	90

Table 1. Changes in HDL_3 protein content after trypsin treatment and reconstitution with apos. Protein to phospholipid ratios of control and modified HDL_3 were calculated to show the relative change in HDL_3 protein content following modifications. Results are representative of at least three different preparations.

HDL3 lipids were observed. Table 1 shows there was a 15% to 30% decrease in total HDL3 associated protein relative to phospholipid after modification with trypsin for 5 to 60 minutes. In all studies described, control and modified HDL3 particles were normalized to the same phospholipid concentration. In order to assess the role of different HDL apos in mediating cellular binding and cholesterol efflux, various apos were reconstituted back into the trypsin-modified particles. Incubation of trypsin-modified HDL3 with the purified apos increased the protein content of the particles (Table 1).

Reconstitution with purified apos had no effect on particle size distribution relative to the parent trypsin-modified HDL3. Reconstitution with apo A-II resulted in particles which contained almost exclusively apo A-II, similar to particles produced by exchange of A-II for A-I previously reported (Edelstein, et al). Trypsin-modified particles and apo reconstituted particles were used in cell binding and cholesterol efflux experiments.

Cell binding studies. Preliminary experiments demonstrated that trypsin-modified and control HDL3 were similarly effective at competing with ^{125}I-HDL3 for high affinity biding sites on cholesterol-loaded fibroblasts using direct competition binding assays. However, direct competition binding studies are confounded by the possibility of apo exchange between labeled ligand and cold competitor. Direct binding studies with ^{125}I-trypsin HDL showed an increase in total cell associated binding relative to control HDL due to increased non-specific or low affinity binding. Thus making it difficult to quantitate high affinity binding for the modified particles.

To better compare high affinity binding of control and modified particles under conditions where apo exchange is minimized a sequential competitive binding assay was developed. In this method, cells were first incubated with the competing ligand for 3 hours at 4°C, and after washing to remove excess competitor, labeled ligand was added and the incubation continued for 1 hour. Since cold competitor and labeled ligand were added during sequential incubations, the possibility of artifacts due to interparticle apo exchange was minimized. Dissociation of HDL from high affinity sites is slow at 0°C (Biesbrock et al) so that the majority of the binding sites remain occupied during the second incubation blocking the potential interaction of labeled lipoprotein with the same site.

Using the sequential competitive binding assay, cells were first incubated with increasing concentrations of cold HDL_3 followed by 5 µg/ml ^{125}I-HDL_3 during the second incubation. HDL_3 at concentrations greater than 20 µg/ml (approximately 30µM phospholipid) blocked 45% of the total ^{125}I-HDL_3 in a saturable manner. Half maximal competition occurred at $\approx$ 10 µg/ml HDL_3. It was estimated that the Kd for these binding sites was in the order of 10^{-8}M, similar to values from direct binding studies (Brisbrock et al, Oram et al 1983). These data show that the sequential competitive binding assay readily detects high affinity binding sites for HDL_3. When HDL_3 modified with trypsin for 60 minutes was used as the competing ligand, there was no inhibition of subsequent binding with ^{125}I-HDL_3, suggesting that these particles bind to distinct cell surface sites from those recognized by native HDL_3. Thus, it appears that trypsin digestion destroys a labile fraction of apoproteins necessary for high affinity cell surface binding.

The extent of trypsin treatment necessary to abolish the ability of HDL_3 to block high affinity cell surface binding was examined. Table 2 shows that at saturating concentrations of HDL_3 (40µM phospholipid), 45% of ^{125}I-HDL_3 binding is blocked. However after only 5 minutes of trypsin treatment, resulting in the loss of less than 20% of the intact HDL associated apos, the ability to block

Competing Ligand	% Inhibition of ^{125}I-HDL Binding (mean ± S.D.)
Control HDL	45.4 ± 3.3
Trypsin-HDL (5 min)	8.2 ± 4.9
Trypsin-HDL (60 min)	3.3 ± 2.2
Trypsin-HDL (5 min)+ apo A-I	38.6 ± 2.2
Trypsin-HDL (60 min)+ apo A-I	26.1 ± 2.7

Table 2. Effect of trypsin-modification on HDL binding to high affinity cell surface sites on cholesterol-loaded fibroblasts. Binding of ^{125}I-HDL_3 to cells was determined after pre-incubation with 40μM phospholipid of the indicated competitor using the sequential competitive binding assay. HDL was treated with trypsin for 5 or 60 minutes, and aliquots of the resulting particles were reconstituted with apo A-I. Binding in the absence of competitor was 17.5 ± 0.6 ng ^{125}I-HDL/mg cell protein.

binding is lost. Trypsin modification for up to 60 minutes had no further effect. These data show that only a subfraction of the total HDL associated apos, revealed by sensitivity to trypsin, are involved in high affinity HDL binding. Reconstitution of trypsin-modified HDL with apo A-I restored the ability of the particles to compete with ^{125}I-HDL for high affinity binding sites. (Table 2). These results demonstrate that HDL apoproteins, and not lipids, are responsible for high affinity HDL binding.

Cholesterol efflux studies. The effect of trypsin treatment on HDL_3-mediated intracellular cholesterol efflux was examined. Confluent cultures of cholesterol-loaded human skin fibroblasts were incubated with ^{3}H-mevalonolactone to enrich intracellular pools with labeled sterol. Previous studies have demonstrated that in cholesterol-loaded cells newly synthesized sterols accumulate within intracellular pools, and are available for esterification by ACAT. (Oram et al 1991). Efflux of labeled sterols is dependent on HDL particles which can bind to high affinity sites, promoting the translocation of cholesterol from intracellular pools to the plasma membrane (Slotte et al; Oram et al 1991). Once present in the plasma membrane, cholesterol is available for efflux to appropriate acceptors by disruption/diffusion events. When cholesterol-loaded cells labeled with ^{3}H-mevalonolactone were incubated with 20μM phospholipid of HDL treated with trypsin for only 5 minutes the ability to promote ^{3}H-sterol efflux was only 50% of that obtained with control HDL_3 at the same concentration. Increasing trypsin treatment for up to 60 minutes had no further effect. These data suggest that the most trypsin labile apos are responsible for intracellular cholesterol translocation and efflux. Efflux observed by these particles most likely represents uptake and exchange of labeled sterols present in the plasma membrane prior to the addition of lipoproteins.

Intracellular cholesterol efflux could also be estimated by the ability of HDL_3 to deplete the ACAT cholesterol substrate pool. Cholesterol-loaded cells are first incubated with test media, then after washing, were pulse labeled for one hour with media containg ^{14}C-oleate. Incorporation of radiolabel into cellular cholesteryl esters by the ACAT reaction was quantitated. Fibroblasts incubated with HDL_3 for 6 hours decreased cholesterol esterification from 1809 to 984 pmol/mg cell protein compared to cells incubated with medium alone. HDL_3 modified by incubation with trypsin for only 5 minutes completely abolished the ability of the lipoprotein to affect cholesterol esterification (no significant change from control). Protease treatment beyond 5 minutes had no additional affect. These data, in agreement with results examining high affinity cell surface binding and intracellular cholesterol efflux, demonstrate that a trypsin labile subfraction of HDL apos, accounting for less than 20% of the total HDL associated apos, are responsible for depleting cells of intracellular cholesterol available for esterification by ACAT.

When HDL modified with trypsin for 5 or 60 minutes was reconstituted with apo A-I or apo A-II the ability to promote efflux of intracellular ^3H-sterols and inhibit cholesterol esterification was restored. The reconstituted particles were as effective as control HDL. Reconstitution of modified particles with apo E also improved the ability of the particles to inhibit esterification, but to a lesser degree than apos A-I or A-II. These data demonstrate that apos A-I or A-II reassociated with the trypsin-modified HDL in an active confirmation able to restore the ability of the particles to promote intracellular efflux.

The effect of trypsin modification or the ability of HDL to promote the net mass efflux of cholesterol was examined. Cholesterol-loaded fibroblasts were incubated with various lipoproteins for 24 hours and the decrease in cell cholesterol, relative to control incubations was determined. Control HDL (40μM Phospholipid) caused a 10.1 ± 2.8 μg/mg cell protein decrease in cellular cholesterol. However, HDL treated with trypsin for 60 minutes decreased cholesterol by 2.8 ± 3.3 μg/mg which was not significantly different from control cells. Reconstitution of trypsin-HDL with either apo A-I or apo A-II, restored the ability to promote net cholesterol efflux, depleting cells of 8.8 ± 2.4 and 12.1 ± 0.6 μg cholesterol/mg cell protein, respectively. These data, in agreement with the radioisotope studies show that HDL apolipoproteins and not lipids are necessary to promote net cholesterol efflux.

The ability of trypsin-modified HDL to act as cholesterol acceptors, was tested, since a defect in this function of HDL could partially explain the above results. Non-cholesterol-loaded cells were labeled with ^{3}H-cholesterol to constant specific activity. Under these conditions at least 90% of fibroblast cellular cholesterol is localized to the plasma membrane (Lange et al). Cells were then incubated with medium alone or containing control HDL_3, HDL_3 treated with trypsin for 5 or 60 minutes or trypsin modified HDL reconstituted with apo A-I. In each cases, the rate of cholesterol efflux to the lipoproteins was identical over a 6 hour period. Thus trypsin modification had no effect on the ability of the lipoproteins to promote cholesterol efflux by passive desorption/diffusion mediated mechanisms.

CONCLUSIONS

These data demonstrate, in agreement with our previous observations (Oram, et al 1991), that a subtraction of HDL apos are responsible for HDL receptor binding and HDL-mediated intracellular cholesterol efflux. Based on trypsin sensitivity it was demonstrated that destruction of less than 20% of intact HDL associated apos resulted in a loss of HDL function. These data also demonstrate a correlation between the ability of HDL particles to bind to high affinity cell surface sites and the ability to promote intracellular cholesterol efflux. In contrast, the exchange and uptake of a plasma membrane cholesterol appears to be independent of HDL apos and the ability of particles to bind to high affinity cell surface sites. Thus this function of HDL appears to depend on the lipid composition of the particles (Phillips et al).

The loss of HDL function after proteolysis of only a fraction of the total apos suggests that these apos may exist in a particularly trypsin-labile confirmation. One possibility (depicted in Figure 1) is that there is a freely exchangeable pool of HDL associated apos, that once exposed from a lipid (and presumably protected) environment are readily degraded by the protease. The freely exchangeable apos may also be suitable ligands for HDL receptors. Once an appropriate ligand binds to the receptor site, translocation of cholesterol from intracellular pools to the plasma membrane is stimulated, possibly by protein kinase C mediated events (Mendez, et al). This hypothesis would explain the current data, and also those of

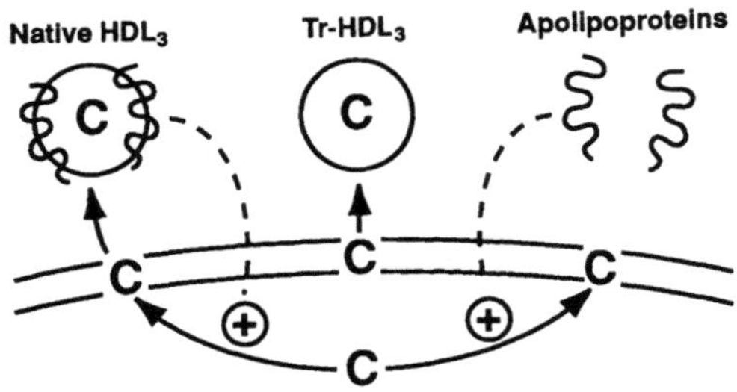

Figure 1. Scheme illustrating how intact HDL_3, trypsin-modified HDL_3 (Tr-HDL_3) and "free" apo A-I affect cellular cholesterol (c) translocation and efflux.

Hara and Yokayama who demonstrated that lipid free apos are active promoters of cholesterol efflux from cultured macrophages. In contrast, trypsinization destroys the "active" apos, and these particles can no longer stimulate intracellular cholesterol efflux, but can act as acceptors of membrane cholesterol. Reconstitution of trypsin-modified HDL with purified apos would probably replenish and enrich the hypothetical exchangeable pool since the method depends on exchange of free apos into the modified particles. This would also explain why these particles, which contain less total associated apos than control HDL, are effective at promoting intracellular cholesterol efflux and competing for high affinity binding.

In summary, a trypsin-labile pool of HDL3 apos, accounting for less than 20% of the total HDL3 associated apos, are required for high affinity binding of particles to HDL-receptors and to promote the efflux of intracellular cholesterol. The trypsin lability of this small pool of apos suggests that these proteins exist in a unique confirmation which may confer biological activity. Further studies to identify this pool of apos will increase our understanding of HDL in reverse cholesterol transport.

Literature references

Badimon, J., L. Badimon, et al. (1990). Regression of atherosclerotic lesions by high density lipoprotein plasma fraction in the cholesterol-fed rabbit J. Clin. Invest. 85:1234-41

Badimon, J., L. Badimon, et al. (1989). High density lipoprotein plasma fractions inhibit aortic fatty streaks in cholesterol-fed rabbits Lab Invest 60:455-61

Biesbroeck, R., J. Oram, et al. (1983). Specific high affinity binding of high density lipoproteins to cultured human skin fibroblasts and arterial smooth muscle cells J. Clin Invest 71:525-539

Brinton, E., J. Oram, et al. (1986). Binding of high density lipoproteins to cultured fibroblasts after chemical alteration of apoprotein amino acid residues J. Biol Chem 261:495-503

Edelstein, C., M. Haluri, et al. (1982). On the mechanism of the deplacement of apolipoprotein A-I by apolipoprotein A-II from the high density lipoprotein surface J. Biol Chem 257:7189-7195

Fielding, C. and P. Fielding (1982). Cholesterol transport between cells and body fluids Med Clinics N. Amer 66:363-373

Glomset, J. (1968). The plasma lecithin: cholesterol acyltransferase reaction J. Lipid Res 9:155-167

Hara, H. and S. Yokoyamea (1991). Interaction of free apolipoproteins with macrophages J. Biol Chem 266:3080-3086

Lange, Y., M. Swarsgood, et al. (1989). Plasma membrane contains half the phospholipid and 90% of the cholesterol and sphingomyelin cultured human fibroblasts J. Biol Chem 264:3786-3793

Mendez, A., J. Oram, et al. (1991). Protein kinase C as a mediator of high density lipoprotein receptor dependent efflux of intracellular cholesterol J. Biol. Chem. 266:10104-10111

Oram, J. (1986). Receptor mediated transport of cholesterol between cultured cells and high density lipoproteins Methods Enzymol 129:645-659

Oram, J., E. Brinton, et al. (1983). Regulation of high density lipoprotein receptor activity in cultured human skin fibroblasts and human arterial smooth muscle cells J. Clin Invest 72:1611-1621

Oram, J., A. Mendez, et al. (1991). High density lipoprotein apolipoproteins mediate removal of sterol from intracellular pools but not from plasma membranes of cholesterol-loaded fibroblasts Arterio and Throm 11:403-414

Phillips, M., W. Johnson, et al. (1987). Mechanisms and consequences of cellular cholesterol exchange and transfer Biochim Biophys Acta 906:223-276

Slotte, J., J. Oram, et al. (1989). Binding of high density of lipoprotein to cell receptors promotes translocation of cholesterol from intracellular membranes to the cell surface J. Biol Chem 262:12904-12907

THE METABOLISM OF TRUNCATED FORMS OF APOLIPOPROTEIN B IN HYPOBETALIPOPROTEINEMIA

Gustav Schonfeld, M.D.
Elaine S. Krul, Ph.D.
Washington University School of Medicine
4566 Scott Ave., Box 8046
St. Louis, MO 63110 USA

INTRODUCTION

In recent years explanations have been found for the very low levels of total- and LDL-cholesterol seen in some humans (Young, 1990). In addition to patients with "classical" abetalipoproteinemia, homozygous hypobetalipoproteinemia and chylomicron retention disease whose total plasma cholesterol levels approximate 40 mg/dl (~ 1mM) and who may suffer from the consequences of severe fat malabsorption (Herbert et al., 1985; Huang et al., 1991; Pessah et al., 1991), there are persons whose total plasma cholesterols are of the order of 100-120 mg/dl (~ 2.5-3.0mM, <5th percentile for age and sex) and who appear to be in good health. Indeed, some affected individuals may attain very long lifespans. We now know that many of these cases of uncomplicated hypobetalipoproteinemia are associated with genetically determined truncations of apoB-100, (the major apolipoprotein of LDL), and that the low LDL levels are due to alterations in the metabolism of the truncated apoB-containing lipoprotein particles.

Our aim is to review the studies performed in our own laboratory that support the above conclusions.

Hypobeta Kindreds

During the last 3 years we have screened by immunoblotting plasmas of individuals with <10th percentile concentrations of total cholesterol, for the presence of apoB-100 truncations. We have found six kindreds with six different truncations (Krul et al., 1989; Talmud, et al., 1989; Wagner, et al., 1991; Krul, et al., 1992, and unpublished observations) (Figure 1). Each proband had <5th percentile levels of total- and LDL-cholesterol and apoB in plasma. The mean lipid and apoB values of the other affected members also were very low, and significantly less than the values of unaffected members (Table 1). HDL-cholesterol levels were low in some but not all affected subjects. Based on immunoblotting analyses with monoclonal antibodies directed to various regions of apoB-100, all of the truncations appeared to be deletions of carboxyl end of the molecule.

NATO ASI Series, Vol. H 73
Human Apolipoprotein Mutants III
Edited by C. R. Sirtori, G. Franceschini, B.H. Brewer Jr.

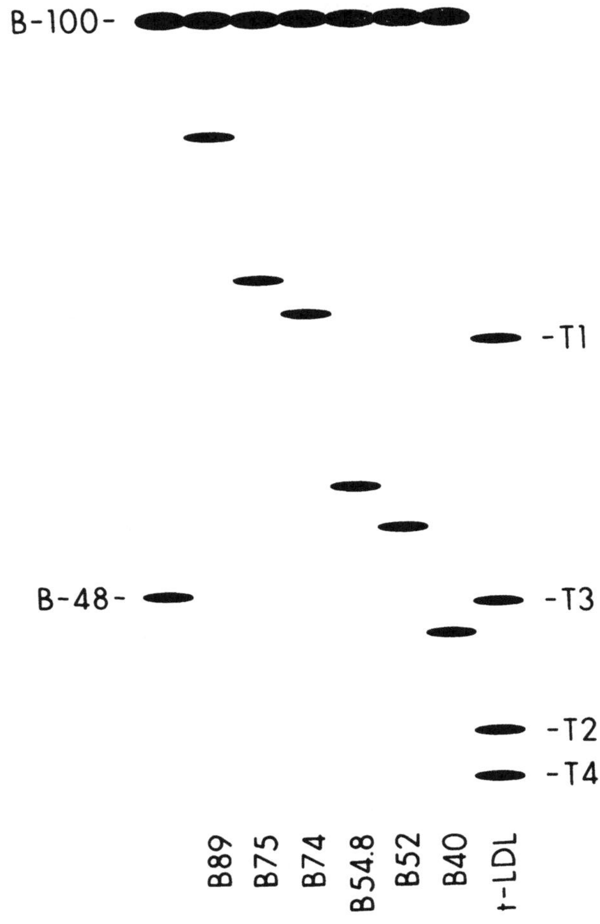

FIGURE 1 - Schematic of Western blot analyses performed on the plasmas of individuals heterozygous for truncated variants of apoB. ApoB is immunoprecipitated from plasma and applied onto 3-6% gradient SDS-PAGE gels. The protein bands are electrotransferred to Immobilon-P membranes and are immunoblotted with a monoclonal antibody (C1.4) directed towards the amino terminus of apoB-100. The heterozygotes are indicated along the bottom of the schematic gel. ApoB-100 and apoB-48 as well as thrombin-generated fragments (T1-T4) of apoB-100 are shown in this Figure for comparison.

TABLE 1
SUMMARY OF CLINICAL CHARACTERISTICS OF HYPOBETALIPOPROTEINEMIC KINDREDS HAVING TRUNCATED FORMS OF APOB

	AGE	BMI ($kg \cdot m^{-2}$)	TCHOL ($mg \cdot dl^{-1}$)	TG ($mg \cdot dl^{-1}$)	LDL-C ($mg \cdot dl^{-1}$)	HDL-C ($mg \cdot dl^{-1}$)	ApoB ($mg \cdot dl^{-1}$)
B89/B40 (3)[a]	49 ± 3	30.5 ± 5.3	70 ± 4	84 ± 37	15 ± 5	40 ± 10	18 ± 7
B100/B40 (1)	36	26.5	108	72	49	45	29
B100/B89 (2)	32 ± 12	27.1 ± 2.0	139 ± 26	102 ± 12	78 ± 19	40 ± 9	81[b]
B100/B100 (2)	31 ± 10	24.9 ± 1.4	197 ± 27	71 ± 3	120 ± 18	63 ± 10	98 ± 17
B40/B100 (2)[a]	44 ± 13	23.5 ± 1.4	103 ± 15	118 ± 4	36 ± 18	44 ± 4	43 ± 10
B100/B100 (1)	30	21.6	175	226	83	44	91
B52/B100 (11)	36 ± 23	21.1 ± 3.0	114 ± 28***	59 ± 38**	46 ± 21***	54 ± 13	47 ± 25***
B100/B100 (15)	39 ± 27	21.4 ± 4.0	146 ± 82	146 ± 82	108 ± 37	47 ± 13	106 ± 25
B54.8/B100 (11)	48 ± 19	29.5 ± 2.4	123 ± 20***	69 ± 24**	49 ± 14***	60 ± 15	26 ± 8***
B100/B100 (15)	33 ± 17	28.6 ± 5.6	186 ± 42	166 ± 111	104 ± 33	49 ± 14	72 ± 23
B74/B100 (2)[a]	53 ± 24	24.3 ± 0.5	119 ± 16	98 ± 17	46 ± 8	54 ± 12	41 ± 1
B100/B100 (2)	53 ± 15	27.5 ± 1.3	180 ± 16	148 ± 47	96 ± 8	54 ± 2	87 ± 13
B75/B100 (4)	52 ± 20	25.5 ± 3.5	115 ± 14***	92 ± 50	48 ± 11**	49 ± 4	28 ± 9*
B100/B100 (9)	44 ± 22	25.7 ± 7.4	192 ± 31	143 ± 76	109 ± 30	54 ± 14	90 ± 3

[a], Statistics not possible due to too few individuals sampled in each group; [b], Value for 1 individual only; *, **, and *** correspond to $p < 0.01$, $p < 0.005$ and $p < 0.001$ by unpaired Student's T-test or Mann-Whitney Two-Sample test compared to unaffected members in the same kindred

<u>ApoB Gene Defects</u>

Knowledge of the approximate molecular weight of the truncation gained from Western blot analyses (Figure 1) and the availability of the DNA sequence of the apoB-gene permitted us to amplify, by polymerase chain reaction (PCR), those regions of the apoB gene suspected of harboring mutant sequences. The PCR products were then ligated into a sequencing vector for subsequent cloning, amplification and sequencing (Krul et al., 1989; Talmud, et al., 1989; Wagner, et al., 1991; Krul, et al., 1992). To date the relevant regions of 5 of the six truncations have been sequenced (Table 2). Four of the defects consist of single to quadruple base deletions resulting in frame-shifts and the downstream formation of premature stop codons. One defect is a C→T transition also resulting in the formation of a stop codon. The predicted sizes of the proteins matched the sizes obtained on the SDS-gels. Furthermore, within each kindred, the protein defects cosegregated with the putative gene mutations. The latter were identified in affected members of the various kindreds either by hybridization with allele specific oligonucleotides, unique endonuclease polymorphisms, or by haplotyping of apoB using several genetic markers of apoB (RFLP's, VNTR's, insertion-deletion polymorphism). Thus, although the mutations have not been expressed in cultured cells, we are reasonably confident that we have identified the genetic mutations responsible for the apoB truncations. In all but 3 cases the individuals were simple heterozygotes for the defect. Three sisters turned out to be compound heterozygotes for 2 different defects (apoB-89/40) (Table 1).

TABLE 2
SUMMARY OF MUTATIONS GIVING RISE TO TRUNCATED APOLIPOPROTEIN B IDENTIFIED IN ST. LOUIS AREA

	Mutation	**Plasma lipoproteins containing truncation**
B40	Deletion of nucleotides 5693-5694	VLDL, LDL, HDL
B52	Deletion of nucleotides 7275-7279	VLDL, LDL
B54.8	C→T transition at nucleotide 7655	VLDL, LDL
B74	Not yet identified	VLDL, LDL
B75	Deletion of nucleotide 10366	VLDL, LDL
B89	Deletion of nucleotide 12309	VLDL, LDL

Lipoprotein Associations of Short apoBs

We sought to discover which major density classes of lipoproteins contained apoB truncations by subjecting ultracentrifugally isolated VLDL, LDL and HDL to SDS-gel electrophoresis and immunoblotting. All of the truncations were found in VLDL and LDL. In each case, the truncated proteins were present in much smaller amounts than the full length apoB (10-30% of apoB-100). In addition, apoB-40 also was found in the HDL density range. Next we examined whether the truncated apoB and normal sized apoB-100 were associated with the same or different particles. This was done by performing one or more of 3 procedures: a) density gradient ultracentrifugation of VLDL (d<1.006) fractions, b) size fractionation of whole plasma on Sepharose 6B columns in the FPLC System (Figure 2), and c) size fractionation by non-denaturing polyacrylamide gel electrophoresis (GGE). ApoBs were identified and/or quantified in the density and the column fractions, and in the GGE gels by immunoblotting. We concluded that the full sized apoB and the truncated apoB were not on the same particles. Indeed, it was possible clearly to separate two subpopulations of LDL in the cases of the apoB-89 and apoB-75 mutations (Figure 2). The larger LDL populations contained only apoB-100, while the smaller populations contained the truncations.

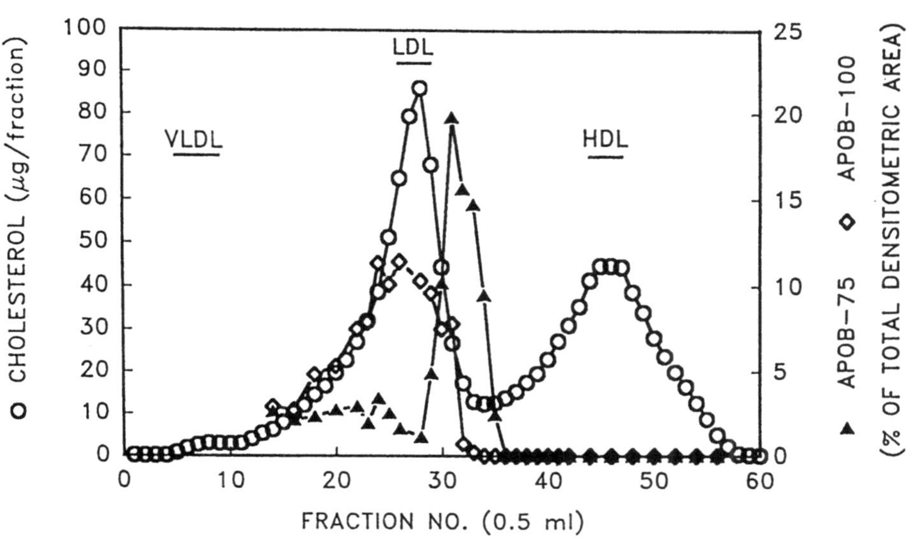

FIGURE 2 - Separation of plasma apoB-100 from apoB-75. Plasma (1.5 ml) was chromatographed on 2 Superose 6 columns and fractions were analyzed for cholesterol or apoB-100 and apoB-75 by immunoblotting. (The initial 12 ml was diverted to waste). Bands corresponding to apoB-100 and apoB-75 were scanned by densitometry and the areas under the protein peaks were quantified. Areas corresponding to apoB-100 or apoB-75 were summed and the values represent % of total densitometric area fraction. (Reproduced with permission from *The Journal of Lipid Research*).

Physiologic Studies

In order to identify the physiologic bases for the low apoB levels in plasma, we performed kinetic studies in affected subjects. The presence of two lipoprotein populations one with normal and the other with mutant apoB suggested that the two populations probably were metabolized independently of each other and that the metabolism of each set of particles could be studied separately from the other.

The kinetic studies were performed by intravenously infusing ^{13}C-leucine or ^{15}N-glycine as endogenous precursors over 8 hours. Approximately 25 timed plasma samples were collected between 0 and 100 hours. VLDL, IDL, and LDL were isolated from each sample by ultracentrifugation. The two forms of apoB in each lipoprotein fraction were separated by SDS-gel electrophoresis. The appropriate bands were cut out of the gels and proteins were hydrolyzed with 6N HCl. The resulting amino acids were recovered on cation exchange columns, derivatized and analyzed by gas-chromatography-mass spectrometry. The calculated enrichment data were subjected to multicompartmental kinetic analysis using the SAAM program (Parhofer et al., 1991). The computer drawn curves in Figure 3 represent best fit mean values of the data for 1-3 subjects with the same mutation.

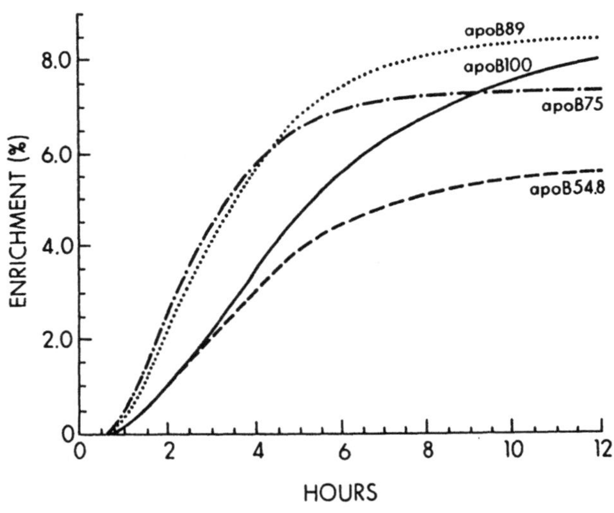

FIGURE 3 - Production of VLDL-apoB as measured by the % enrichment with either ^{13}C-leucine or ^{15}N-glycine in 4 groups of subjects heterozygous for the apoB truncations indicated. Curves represent the means of data obtained from 1 to 3 subjects per group.

Clearly the physiologic bases for the hypobetalipoproteinemias found in the subjects with various truncations of apoB are determined by the sizes of the truncations. Because most of the subjects studied are simple heterozygotes, the behaviors of apoB-100 and the truncated apoB can be directly compared in the same person. For example, concentrations of apoB-89-containing lipoproteins are low relative to the apoB-100 lipoproteins of the same subjects because the apoB-89 particles are cleared by LDL-receptor mediated mechanisms more rapidly than the apoB-100 particles. Both VLDL populations are secreted at comparable rates and become readily converted to IDL. However, a larger proportion of apoB-89 IDL than normal apoB-100 IDL is removed from plasma directly. The remaining apoB-89 IDL are converted to apoB-89 LDL, and the latter are removed more rapidly than apoB-100 LDL (Krul, et al., 1989; ,Parhofer et al., 1990, Parhofer et al., 1992). In sum, the relatively low levels of apoB-89 lipoproteins in plasma are due to the more rapid removal of apoB-89 than apoB-100 lipoproteins at every stage of apoB-89 catabolism. By contrast, apoB-75 VLDL are secreted at ~30% of the rates of apoB-100 VLDL in the same subjects, and they are also removed more rapidly than apoB-100-containing lipoproteins, i.e., two different physiologic mechanisms work cooperatively to achieve the relatively low plasma concentrations of apoB-75 (Krul et al., 1992). ApoB-54.8 are secreted much more slowly than the respective apoB-100 lipoproteins, while their removal rates (FCR's) are similar to those of apoB-100 (unpublished). Since the putative LDL-receptor recognition site is absent in apoB-54.8, a non-LDL-receptor mechanism must be involved in their clearance from plasma. While these "internal" comparisons of apoB truncations with apoB-100 lipoproteins in any given heterozygote do explain differences in the plasma levels of apoB-100 and the truncation in the given subject, these comparisons do not explain why apoB-100 levels are only ~50% of those expected based on the simple heterozygosity of these individuals. A preliminary comparison of secretory rates in the heterozygotes to apoB-100 secretory rates in normolipidemic controls, reveals that heterozygotes produce apoB-100 at ~30% of control rates. If confirmed, this could explain on a physiologic level the lower than expected plasma levels of apoB-100 lipoproteins found in the heterozygotes. The molecular and cell biologic bases for the altered rates of secretion and/or removal are not known for any of the truncations. However, from data we have collected we can conclude that the length of the apoB molecule that associates with a lipoprotein particle in liver or intestine greatly affects its intracellular assembly, secretion and its subsequent interaction with lipoprotein receptors.

REFERENCES

Herbert PN, Hyams JS, Bernier DN, Berman MM, Saritelli AL, Lynch KM, Nichols AV, and Forte TM (1985) Apolipoprotein B-100 deficiency: Intestinal steatosis despite apolipoprotein B-48 synthesis. J Clin Invest 76:403-412

Huang LS, Kayden H, Sokol RJ, and Breslow JL. (1991) ApoB gene nonsense and splicing mutations in a compound heterozygote for familial hypobetalipoproteinemia. J Lipid Res 32:1341-1348

Krul ES, Kinoshita M, Talmud P, Humphries SE, Turner S, Goldberg AC, Cook K, Boerwinkle E, and Schonfeld G (1989) Two distinct truncated apolipoprotein B species in a kindred with hypobetalipoproteinemia. Arteriosclerosis 9:856-868

Krul ES, Parhofer KG, Barrett PHR, Wagner RD, and Schonfeld G (1992) ApoB-75, a truncation of apolipoprotein B associated with familial hypobetalipoproteinemia: Genetic and kinetic studies. J Lipid Res 33:1037-1050

Parhofer KG, Daugherty A, Kinoshita M, and Schonfeld G (1990) Enhanced clearance from plasma of low density lipoproteins containing a truncated apolipoprotein, apoB-89. J Lipid Res 31:2001-2007

Parhofer KG, Barrett PHR, Bier DM, and Schonfeld G (1991) Determination of kinetic parameters of apolipoprotein B metabolism using amino acids labeled with stable isotopes. J Lipid Res 32:1311-1323

Parhofer KG, Barrett PHR, Bier DM and Schonfeld G (1992) Lipoproteins containing the truncated apolipoprotein, apoB-89, are cleared from human plasma more rapidly than apoB-100 containing lipoproteins in vivo. J Clin Invest 89:1931-1937

Pessah M, Benlian P, Beucler I, Loux N, Schmitz J, Junien C, and Infante R. (1991) Anderson's Disease - genetic exclusion of the apolipoprotein-B gene in two families. J Clin Invest 87:367-370

Scriver CR, Baudet AL, Sly WS, and Valle D (1989) Metabolic Basis of Inherited Disease. McGraw-Hill, New York.

Talmud P, King-Underwood L, Krul ES, Schonfeld G, and Humphries S (1989) The molecular basis of truncated forms of apolipoprotein B in a kindred with compound heterozygous hypobetalipoproteinemia. J Lipid Res 30:1773-1779

Wagner RD, Krul ES, Tang JJ, Parhofer KG, Garlock K, Talmud P, and Schonfeld G (1991) ApoB-54.8, a truncated apolipoprotein found primarily in VLDL, is associated with a nonsense mutation in the apoB gene and hypobetalipoproteinemia. J Lipid Res 32:1001-1011

Young SG (1990) Recent progress in understanding apolipoprotein B. Circulation 82:1574-1594

APOLIPOPROTEIN B: GENETIC VARIANTS PROVIDE INSIGHT INTO STRUCTURE AND FUNCTION

Philippa Talmud, Alison Dunning, Steve Humphries
Dept of Medicine
Rayne Institute
University College Middlesex School of Medicine
London, UK

In order to fulfil its function as a secretory protein, lipid transporter and ligand for the LDL-receptor, apolipoprotein B (apo B) has a number of functional domains; a signal peptide that directs the translocation of the protein across the endoplasmic reticulum to the Golgi apparatus, a region that binds to the LDL-receptor and a domain for lipid binding to maintain the integrity of the lipoprotein particle and stability during lipolysis. Much of our understanding of the structure-function aspects of apo B have come from studies of rare genetic variants. Several genetic disorders of lipoprotein metabolism are directly linked to the apo B gene or associated with the biosynthesis, assembly or secretion of apo B containing lipoproteins.

RARE MUTATIONS

Familial Hypobetalipoproteinaemia (HBL): HBL is an autosomal codominant disorder with heterozygotes having total cholesterol, LDL cholesterol and apo B levels below the 5th centile. Cosegregation studies have shown that the mutation is in the apo B gene (Leppert et al., 1988). All the mutations to date result in truncated proteins and most have been initially identified at the protein level using polyacrylamide gel electrophoresis of plasma or LDL to get an estimate of protein size. In a recent screening carried out on blood donors in St Louis, Missouri, the frequency of HBL was estimated to be less than 0.01% (Wagner et al., 1991). In the homozygous form, patients have trace levels of apo B-containing

NATO ASI Series, Vol. H 73
Human Apolipoprotein Mutants III
Edited by C. R. Sirtori, G. Franceschini, B.H. Brewer Jr.

lipoproteins often associated with neuropathies and retinopathies, the result of fat-soluble vitamin deficiency. The extent of the severity appears to be related to the size of the truncated apo B protein and whether it can associate with lipid.

These truncated apo B species help define the functional domains of apo B, namely the lipid-binding and receptor-binding regions. There is now accumulated evidence about the lipid binding capabilities of these truncated proteins from the density range in which they are identified, that provides insight into their lipid binding potential.
More than 15 truncated apo B mutations causing HBL have now been characterised at the molecular level. A map of the apo B gene showing the position of the mutations that lead to truncated proteins appear in Figure 1. Except in the case of the apo B25, the result of a deletion of the entire exon 21, all the truncated forms reported to date are due to C→T transitions or base deletions. In addition all the reported mutations have been unique to the kindreds they have been identified in and no two unrelated families have the same mutation.

Definition of a lipid-binding domain: Where the mutation creates a protein greatly reduced in size, for example the truncated apo B species apo B25 (Huang et al., 1989) and apo B29 (Collins et al., 1988) no plasma apo B was detected. It-has been suggested that these small truncated species lack a domain necessary for lipoprotein association and stability. The amino acid residues in the amino terminus of the protein are hydrophilic and would not bind well to lipid. The truncated species apo B31 (Young et al., 1990) and apo B37 (Young et al, 1987) are associated with small amounts of triglyceride-rich lipoproteins and occur primarily in the HDL density range; this reflects the small amount of associated lipid and not an HDL-type particle.

In vitro expression studies have further clarified the lipid binding domain. Expression of cDNA constructs in either transfected rat hepatoma cell lines or HepG2 cells have clarified that constructs smaller than apo B28 associate to a minor extent with lipid but are found primarily in the infranatant (Yao et al, 1991; Graham et al, 1989). Thus there appears to be a critical size which the protein must exceed in order to associate with lipid. In the region of apo B31 and apo B37 there is an increased amount of hydrophobic amino acids which can bind to lipid. However in view of the fact that naturally occurring apo B25 and apo B29 are not detected in the plasma it would seem that truncated apo B proteins smaller than apo B31, although detected in vitro, are probably very unstable and in vivo degrade after secretion. The apo B46 (Young et al, 1989) apo B50 (Hardman et al., 1989) and apo B54.8 (Wagner et al., 1991) are found primarily in the VLDL density range although apo B46 is seen in LDL and HDL ranges and apo B54.8 in LDL density range. The larger apo B species are found in a similar density range as full length apo B100. Thus, the truncated apo B proteins help define the region in the amino-terminal end of the protein important for the association with lipids.

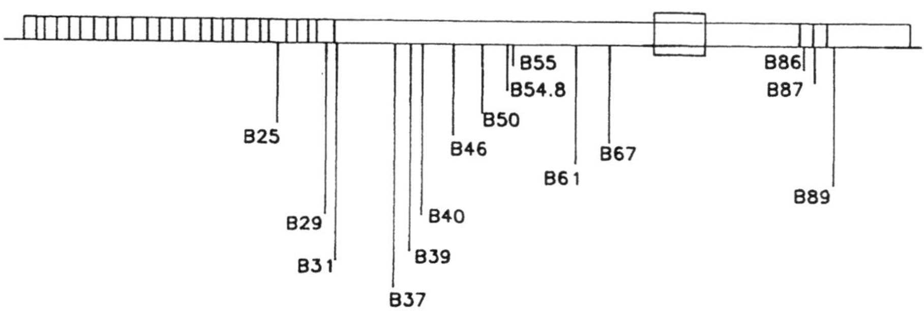

FIGURE 1. Schematic map of the apo B cDNA identifying reported mutations that lead to HBL. Vetical lines on the cDNA map represent the introns. The open box depicts the proposed LDL-receptor binding domain

Definition of additional LDL-R binding domains: The mechanism causing low levels of these truncated forms in the plasma is unclear. Most of the truncated apo B proteins, identified to date, terminate before the putative receptor-binding domain that binds to the LDL-R. Thus an increase in catabolism is unlikely to be an explanation for the low concentrations of apo B and some other mechanism/s must explain this. Instability of the mutant lipoproteins may contribute to their low levels. The larger truncated proteins apo B87 and apo B89, which do encode the primary LDL-R binding domain provide evidence of a secondary binding domain on apo B for the LDL-R. In the case of the apo B89, degradation and binding were increased on cultured fibroblasts when compared to apo B100 (Krul et al., 1989) and an enhanced clearance of apo B89 in LDL-turnover studies performed in rabbits, suggests that apo B89 has an increased affinity for the LDL receptor (Parhofer et al., 1990). The absence of the carboxy-terminal 500 amino acids in apo B89, which are hydrophobic and firmly embedded in lipid, may result in the release of the constraint put on the primary receptor binding domain or make more available a secondary binding region. The functional domains defined by the truncated proteins are shown in Figure 2

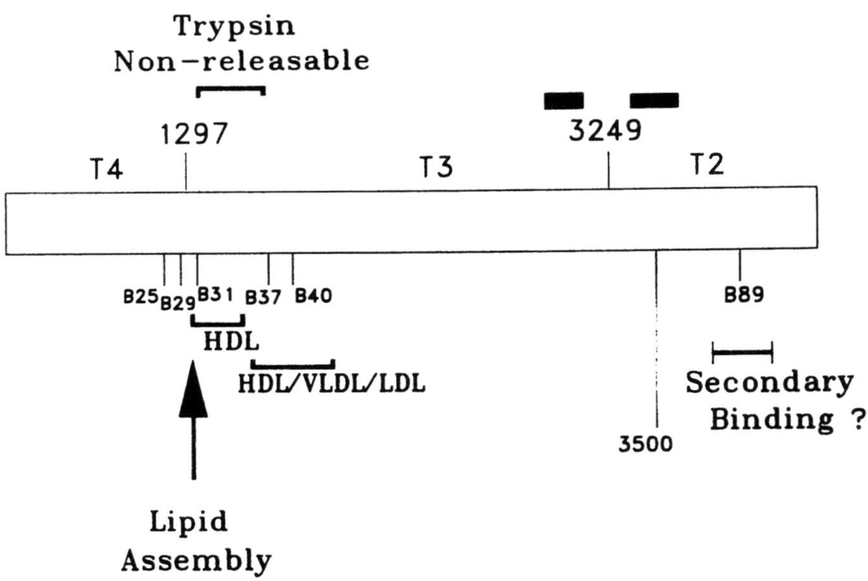

Figure 2. Schematic map of the apo B protein showing the thrombolytic fragments. The filled blocks represent the putative receptor binding domain. Lipid binding and receptor binding domains defined further by the truncated proteins are shown.

Why are plasma apo B levels so low in heterozygous HBL patients? It is still unclear , however, some heterozygous HBL patients have LDL levels below the 5th centile and not at the 50th centile, despite the presence of one normal apo B100 allele. Some insight comes from LDL-turnover studies in a patient with apo B55 (Talmud et al., 1992) which revealed a reduced LDL synthetic rate (5.5mg/dl/day) compared to the normal range of 10-15mg/dl/day (Converse et al., 1990). Apo B synthesis is constitutive in Hep G2 cells and the control of apo B output is thought to be either due to an increase in the mRNA translational efficiency or the decreased degradation of intracellular apo B in the presence of oleate

(Pullinger et al,. 1989). Thus it is possible that in the hepatocytes of the heterozygous HBL patients, while mRNAs coding for both the truncated and normal apo B proteins are translated, the truncated species interfere with the assembly of apo B100 into normal lipoproteins, leading to an overall reduction in plasma apo B-containing lipoproteins.

Familial Defective apo B100 (FDB): It is well known that defects in the LDL-R that destroy function lead to FH. Similarly, defects in the apo B gene that would reduced binding to the LDL-R could result in raised cholesterol levels. The disorder Familial Defective apo B100 (FDB) has been proposed for such mutations. A number of laboratories have carried out systematic searches, using molecular biology techniques, to identify patients who have mutations in the apo B gene that would alter receptor binding. To date there is only one form of FDB and that is due to a mutation that substitutes codon Arg_{3500} by Gln, hence called the apo B3500 mutation. Originally observed in a patient who showed reduced clearance of autologous LDL, when compared to clearance of LDL from a normal donor the mutation was identified as a G→A substitution altering codon CGG to CAG (Soria et al., 1989). This single amino acid change reduces binding of the LDL containing apo B-Gln_{3500} resulting in the accumulation of such LDL while LDL-Arg_{3500} is cleared with normal efficiency. The frequency of FDB has been estimated to be 1/500-1/700. Although residue 3500 is 120 amino acids to the COOH-terminal side of the initially defined receptor binding domain, the Arg→Gln substitution clearly affects the interaction with the receptor. It was observed that monoclonal antibody MB47 binds with a greater affinity to the defective LDL-Gln_{3500} suggesting that the Arg_{3500}→Gln substitution has a large effect on the conformation of that region of the protein (Weisgraber et al., 1988). Furthermore Lund-Katz et al (1989), using ^{13}C NMR studies have shown that the six lysine residues within the region of amino acid 3500 have altered pKs in the presence of Gln_{3500}. Lysine residues are known to be involved in the binding of apo B100 to the LDL-R and the substitution of Arg_{3500}→Gln by altering the pKs of the lysines, alters the microenvironment of the receptor binding domain.

Thus mutations in patients with HBL or FDB help define both the lipid-binding and receptor-binding domain of apo B and provide evidence that sequence changes outside the primary binding domain may affect the affinity of LDL for the receptor. Furthermore the molecular defects leading to HBL illustrate the genetic heterogeneity that underlies this syndrome.

COMMON POLYMORPHISMS

Association studies with common polymorphisms of the apo B gene have identified those polymorphisms which show a consistant association with differences in lipid levels. The association studies themselves do not provide information as to whether the polymorphisms are functional or whether the effects seen are due to linkage disequlibrium with functional sequence changes elsewhere in the gene. We have been particularly interested in identifying the functional sequence changes that underlie these effects.

XbaI polymorphism: Several Restriction Fragment Length Polymorphisms (RFLPs) of the apo B gene have been identified and using these RFLPs, population studies have shown that variation at a polymorphic XbaI site within the apo B gene is associated with differences in serum cholesterol levels, with the X+ allele (presence of the XbaI cutting site) being associated with raised levels (reviewed by Humphries, 1988). Although this association has not been seen in every study, it has been observed in many different laboratories, using samples from different countries, and the consistent nature of this finding makes it highly unlikely that the observation is due to chance alone. The size of the effect associated with this RFLP is modest explaining 3 - 8% of the sample variance in cholesterol levels, and is thus of the same order of magnitude as is seen with variation at the apo E gene (Davignon et al., 1988). The base change that creates the XbaI site does not alter an amino acid, as it is in the third (wobble) position of a codon for threonine 2488, and so cannot be the direct cause of the effects seen. We thus postulate a second sequence variation, elsewhere in the gene and in linkage disequilibrium with the XbaI RFLP, that leads to altered serum lipid levels. This sequence difference could occur in the promoter region of the apo B gene where it may alter transcription of the gene and thus affect production of the protein from the intestine or liver. Alternatively, as with apo E2, it may change an amino acid in the protein and alter the affinity of LDL for its receptor. Evidence in favour of the latter mechanism has come from two in vivo studies that have shown that compared with the X- allele, the X+ allele is associated with a reduced fractional catabolic rate (FCR) of autologous LDL (Demant et al., 1988; Houlston et al., 1988). These data thus predict that LDL from individuals with the genotype X-X- would bind to the LDL-receptor with increased affinity compared to LDL from individuals with the genotype X+X+. Support for this prediction has been obtained using a competative binding and internalisation assay of radiolabelled LDL to normal fibroblasts (Series et al., 1989), where "X+" LDL competed poorly compared to "X-" LDL, as expected from the turnover data. This is compatible with the hypothesis that a common amino acid change in the apo B protein, may affect the affinity for the LDL receptor.

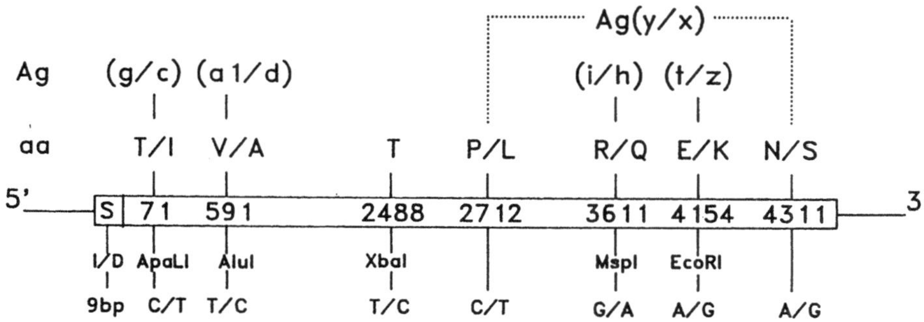

Figure 3. Map of the apo B gene showing the relative position of the identified apo B Ag polymorphisms, the signal peptide polymorphism, and the XbaI RFLP variable site.

Ag polymorphisms: Several common protein polymorphisms of apo B are known (Fig 3). These were first identified in 1961 as Antigenic determinants (Ag) using antibodies found in multiple transfused individuals (Allison and Blumberg, 1961). Five pairs of epitopes were identified by these studies, with each pair representing alleles at closely linked loci (Breuget et al., 1990). With the cloning and sequencing of the apo B gene, the molecular basis of four of these polymorphisms has been deduced from association studies. The c/g and a1/d, are located in the N-terminal region of the protein which is trypsin-accessable and to be the result of single base changes in the gene that cause respectively the substitution of Ile for Thr (Ma et al., 1989) and Val for Ala (Wang et al., 1990) amino acid. The c/g epitopes have been associated with differences in plasma cholesterol levels (both LDL-C and HDL-C) in a large study of children from Finland (Tikkanen et al., 1988), but this association was not observed in an earlier smaller study (Young et al., 1987). No association between the a1/d epitopes and lipid levels have been reported. The Ag epitope pairs h/i and t/z occur in the COOH-terminal part of the protein, in peptides that are not released by digestion of LDL by trypsin. The h/i polymorphism at residue 3611 is of interest because, like the apo B 3500 mutation (Innerarity, 1990), it is the result of a change from Arg to Gln. This amino acid region is likely to be located on the surface of the LDL-particle since it alters an epitope recognised by antibodies and is close to the region of apo B that interacts with the LDL-receptor. However, unlike apo B 3500, Arg_{3611}-Gln does not appear to alter the affinity of the particle for the LDL-receptor, and has not been consistently associated with a significant effect on lipid levels (Xu et al., 1989; Rajput-Williams et al., 1988). The t/z polymorphism is unique in that it substitutes a positively charged amino acid Lys for a negatively charged Glu (Ma et al., 1987). Associations have been noted between this polymorphism and differences in VLDL-triglyceride levels and plasma lipid levels although this association has not been confirmed in other studies, (see Humphries, 1988). Until recently, the position of the remaining Ag epitope pair x/y had not been determined. This polymorphism has been associated with differences in plasma cholesterol and triglyceride levels (Berg et al., 1976), but the effect is small, and only reached statistical significance after combining data from a large number of individuals (n = 1087) from different studies. On average, individuals carrying one or more "x" allele had cholesterol levels roughly 3% lower and triglyceride levels 16% lower than those with only the "y" allele.

To identify the sequence creating the x/y epitope a search was made for sequence differences between a group of 10 individuals (Dunning et al., 1991) using PCR amplification of different exons of the gene followed by chemical mismatch cleavage. An A to G base change was identified in exon 29 of the apoB gene that results in the substitution of serine for asparagine at residue 4311 of mature apo B100 (Dunning et al., 1992). In a recent publication, Huang et al (1990) have reported a C to T base change in exon 26 that causes the substitution of leucine for proline at residue 2712 of apo B. We have found complete allelic association between the alleles at both these sites and Ag (x/y) in a sample of 118 Finnish individuals (37). This implies that either one, or both of these substitutions combined, could be the molecular basis of the Ag(x/y) antigenic determinants - the allele encoding Leu_{2712} plus Ser_{4311} representing the Ag(x) epitope and that encoding Pro_{2712} plus Asn_{4311} the Ag(y). It is also possible that the epitope is created by an additional sequence change yet undetected in the gene that is in complete association with the changes at residues 2712 and 4310. One approach to distinguish these three possibilities would be by expressing in HepG2 cells, constructs of the apo B gene containing the different amino acid substitutions.

The Pro-Asn / Leu-Ser polymorphism shows strong allelic association with the XbaI polymorphism with the Leu-Ser (Ag x) being found exclusively on chromosomes lacking the XbaI site. In a sample of 90 healthy, Swedish individuals the Leu_{2712}/Ser_{4311} allele is associated with reduced serum levels of LDL cholesterol and apo B and with raised levels of High Density Lipoprotein, thus confirming the finding of Berg et al., (1976). However these differences are of smaller effect than those associated with the XbaI Restriction Fragment Length Polymorphism (RFLP) of the apo B gene in this sample (Dunning et al., 1992). It thus appears that although the two amino acid changes are associated with some effect on LDL-function, they do not explain all the variation seen with the XbaI RFLP and further functional genetic polymorphism at this locus remain to be discovered.

Signal peptide polymorphism: Another example of a common polymorphism with a possible functional role is the insertion/deletion polymorphism of the apo B signal peptide which has been shown to be associated with differences in triglyceride levels (Xu et al, 1990) and baseline atherosclerosis score (Peacock et al, 1992). Whether the presence or absence of the three amino acids seen in the insertion and deletion alleles respectively have a direct effect and affect the rate of translocation across the endoplasmic reticulum or whether the association is to an indirect effect due to linkage disequlibrium with a second sequence change. We will test the hypothesis by expressing these signal peptide alleles, linked to a yeast secretory protein in a yeast expression system.

Thus both the rare mutations and the common polymorphisms of the apo B gene help define the those regions of the apo B protein with functional properties. The consistency of the association between the polymorphism at the apo B locus and differences in plasma levels of apoprotein and lipoprotein suggest that in addition, when the functional sequence changes are identified these genetic markers will be useful in assessing an individuals risk of developing hyperlipidaemia and thus CAD.

ACKNOWLEDGEMENTS

Support for this work was provided by the British Heart Foundation (grant RG16), and the Charing Cross Sunley Research Trust. The authors would like to thank Ms Gina Deeley for help in preparing the manuscript.

REFERENCES

Allison AC, Blumberg BS (1961) An isoprecipitation reaction distinguishing human serum protein types. Lancet 1 1:634-637

Berg K, Hames C, Dahlen G, Frick MH, Krishan I (1976) Genetic variation in serum low density lipoproteins and lipid levels in man. Proc Natl Acad Sci USA 73:937-940

Breguet G, Butler R, Butler-Brenner E, Sanchez-Mazas A (1990) A worldwide population study of the Ag-system haplotypes, a genetic polymorphism of human low-density lipoprotein. Am J Hum Genet 46:502-517

Collins DR, Knott TJ, Pease RJ, Powell LM, Wallis SC, Robertson S, Pullinger CR, Milne RW, Marcel YL, Humphries SE, Talmud PJ, Lloyd JK, Miller NE, Muller D, Scott J (1988) Truncated variants of apolipoprotein B cause hypobetalipoproteinemia. Nucl Acid Res 16:8361

Converse CA, Keegan WA, Huq L, Series J, Caslake M, McLachlan J, Packard CJ, Shepherd J (1990) Further Epidemiological Studies in Lipid Metabolism in Retinitis Pigmentosa, *in:* Inherited and Enviromentally Induced Retinal Degeneration, M.M.Lavail, J.G.Hollyfield and R.E.Anderson (Eds), Eds, Alan M Liss, New York

Davignon J, Gregg RE, Sing CF (1988) Apolipoprotein E polymorphism and atherosclerosis. Arteriosclerosis 8:1-21

Demant T, Houlston RS, Caslake MJ et al (1988) Catabolic rate of low density lipoprotein is influenced by variation in the apolipoprotein B gene. J Clin Invest 82:797-802

Dunning AM, Houlston R, Frostegard J, Revill J, Nilsson J, Hamsten A, Talmud P, Humphries S (1991) Genetic evidence that the putative receptor binding domain of apolipoprotein B (residues 3130 to 3630) is not the only region of the protein involved in interaction with the low- density lipoprotein receptor. Biochem Biophys Acta 1096:231-237

Dunning AM, Tikkanen MJ, Enholm C, Butler R, Humphries SE (1988) Relationships between DNA and protein polymorphisms of apolipoprotein B. Hum Genet 78:325-329

Dunning AM, Renges H-H, Xu C-F, Peacock R, Brasseur R, Laxer G, Tikkanen MJ, Butler R, Saha N, Hamsten A, Rossenue M, Talmud P and Humphries SE (1991) Complete Allelic association between two amino acid substitutions in apolipoprotein B and the x/y antigen group polymorphism. Am J Hum Genet (In Press).

Graham LD, Pullinger CR, Jones TC, Knott TJ, Scott J (1989) Expression and secretion of apolipoprotein (apo)-B constructs by HepG2 cells. Circulation 80:II:1854

Hardman DA, Pullinger CR, kane JP, Malloy MJ (1989) Molecular defect in normotriglyceridaemic abetalipoproteinaemia, Circulation, 80:(II) 466

Houlston RS, Turner PR, Revill J, Lewis B, Humphries SE (1988) The fractional catabolic rate of low density lipoprotein in normal individuals is influenced by variation in the apolipoprotein B gene. Atherosclerosis 71:81-85

Huang LS, Gavish D, Breslow JL (1990) Sequence polymorphism in the human apo B gene

at position 8344. Nucleic Acids Res 18:5922.
Huang L-S, Ripps ME, Korman SH, Deckelbaum RJ, Breslow, JL (1989) Hypobetalipoproteinemia due to an apolipoprotein B gene exon 21 deletion derived by Alu-Alu recombination. J Biol Chem 264:11394
Humphries SE (1988) DNA polymorphisms of the apolipoprotein genes -their use in the investigation of the genetic component of hyperlipidaemia and atherosclerosis. Atherosclerosis 72:89-108.
Krul ES, Kinoshita M, Talmud P, Humphries SE (1989) Turner, S., Goldberg, A. C., Cook, K., Boerwinkle, E., and Schonfeld, G., Two distinct truncated apolipoprotein B species in a kindred with hypobetalipoproteinemia. Arteriosclerosis 9:856
Leppert M, Breslow JL, Wu L, Hasstedt S, O'Connell P, Lathrop M, Williams RR, White R, Lalouel J-M (1988) Inference of a molecular defect of apolipoprotein B in hypobetalipoproteinemia by linkage analysis in a large kindred. Clin Invest 8266:847
Lund-Katz S, Innerarity TL, Curtiss LK, Krauss RM, Phillips MC (1989) C-NMR evidence that substitution of Gln for Arg 3500 in familial defective Apo B-100 disrupts the conformation of the receptor binding domain. Arteriosclerosis 9:715a
Ma Y, Wang X, Butler R, Schumaker VN (1989) Bsp 12861 restriction fragment length polymorphism detects Ag(c/g) locus of human apolipoprotein B in all 17 persons studied Arteriosclerosis 9:242-246.
Ma Y, Schumaker V, Butler R, Sparkes RS (1987) Two DNA restriction fragment length polymorphism (RFLPs) associated with Ag (t/z) and Ag (c/g) antigenic sites of human apolipoprotein B. Arteriosclerosis 3:301-305.
Parhofer KG, Daugherty A, Kinoshita M, chonfeld G (1990) Enhanced clearance from plasma of low density lipoproteins containing a truncated apolipoprotein, Apo B-89. J Lipid Res 31:2001
Peacock R, Dunning A, Hamsten A, Tornvall P, Humphries S, Talmud P (1992) Apolipoprotein B gene polymorphisms, lipoproteins and coronary atherosclerosis: A study of young myocardial infarction survivors and healthy population-based individuals. Atherosclerosis 92:151-164.
Pullinger CR, North JD, Teng BB, Rifici VA, Ronhild de Brito AE, Scott J (1989) The apolipoprotein B gene is constitutively expressed in HepG2 cells: regulation of secretion by oleic acid, albumin, and insulin, and measurement of the mRNA half-life. J Lipid Res 30:1065
Rajput-Williams J, Knott TJ, Wallis SC et al (1988) Variation of apolipoprotein B-gene is associated with obesity, high blood cholesterol levels , and increased risk of coronary heart disease. Lancet 1442-1446
Series J, Cameron I, Caslake M, Gaffney D, Packard CJ, Shepherd J (1989) The XbaI polymorphism of the apolipoprotein B gene influences the degradation of low density lipoprotein in vivo. Biochim Biophys Acta 1003:183-188
Soria LF, Ludwig EH, Clarke HR, Vega GL, Grundy SM, McCarthy BJ (1989) Association between a specific apolipoprotein B mutation and familial defective apolipoprotein B-100. Proc Natl Acad Sci USA 86:587
Talmud PJ, Converse C, Krul E, Haq L, McIlwaine GG, Series JJ, Boyd P, Schonfeld G, Dunning A, Humphries SE (1992) A novel truncated apolipoprotein B (apo b55) in a patient with hypobetalipopriteinaemia and autosomal dominant retinitis pigmentosa. Clinical Genetics (Submitted)
Talmud P, King Underwood L, Krul E, Schonfeld G, Humphries S (1989) The molecular basis of truncated forms of apolipoprotein B in a kindred with compound heterozygous hypobetalipoproteinemia. J Lipid Res 30:1773-1779

Tikkanen MJ, Viikari J, Kakerblom HK, Pesonen E (1988) Apolipoprotein B polymorphism and altered apolipoprotein B and low density lipoprotein cholesterol concentrations in Finnish children. Brit Med J 296:169-170

Wagner RD, Krul ES, Tang J, Parhofer KG, Garlock K, Talmud P, Schonfeld G (1991) Apo B 54.8, a truncated apolipoprotein found primarily in VLDL is associated with a nonsense mutation in the apo B gene and hypobetalipoproteinemia. J Lipid Res 32:1001

Wang X, Schlapfer P, Ma Y, Butler R, Elovson J, Schumaker VN (1990) Apolipoprotein B: The Ag($_{1}$/d) immunogenetic polymorphism coincides with a T-to-C substitution at nucleotide 1981, creating an alu 1 restriction site. Arteriosclerosis 8:429-435

Weisgraber KH, Innerarity TL, Newhouse YM, Young SG, Arnold KS, Krauss RM, Vega G L (1988) Familial defective apolipoprotein B-100: enhanced binding of monoclonal antibody MB47 to abnormal low density lipoproteins. Proc Natl Acad Sci USA 85:9758

Xu C-F, Nanjee N, Tikkanen MJ et al (1989) Apolipoprotein B amino acid 3611 substitution from arginine to glutamine creates the Ag (h/i) epitope: the polymorphism is not associated with differences in serum cholesterol and apolipoprotein B levels. Hum Genet 82:322-326

Xu C-F, Tikkanen MJ, Huttunen JK, Pietinen P, Butler R, Humphries SE, Talmud P (1990) Apolipoprotein B signal peptide insertion/deletion polymorphism is associated with AG epitopes and involved in the determination of serum triglyceride levels J.Lipid Res. 31: 1255-1261.

Yao Z, Blackhart BD, Linton MF, Taylor SM, Young SG, McCarthy BJ (1991) Expression of carboxyl-terminally truncated forms of human apolipoprotein B in rat hepatoma cells : evidence that the length of apolipoprotein B has a major effect on the buoyant density of the secreted lipoproteins. J Biol Chem 266:3300

Young SG, Hubl ST, Smith RS, Snyder SM, Terdiman JF, (1990) Familial hypobetalipoproteinemia caused by a mutation in the apolipoprotein B gene that results in a truncated specied of apolipoprotein B (B-31). A unique mutation that helps to define the portion of the apolipoprotein B molecule required for the formation of byoyant,triglyceride-rich lipoproteins. J Clin Invest 85:933

Young SG, Bertics SJ, Scott TM et al (1987) Apolipoprotein B allotypes $MB19_1$ and MB_2 in subjects with coronary artery disease and hypercholesterolemia. Arteriosclerosis 7:61-65

Young SG, Bertics SJ, Artiss LK, Witztum JL (1987) Characterization of an abnormal species of apolipoprotein B, apolipoprotein B-37, associated with familial hypobetalipoproteinemia. J Clin Invest 79:1831

Young SG, Hubl ST, Chappel DA, Smith RS, Claiborne F, Snyder SM Terdiman JF, (1989) Familial hypobetalipoproteinemia associated with a mutant species of apolipoprotein B (B-46). N Engl J Med 32:1604

THE MUTATIONS OF APOLIPOPROTEIN B GENE CAUSING HYPOBETALIPOPROTEINEMIA: MOLECULAR AND FUNCTIONAL ABNORMALITIES IN APO B-87$_{Padova}$

C. Gabelli, C.Bilato, S.Martini,*G.E. Tennyson, *H.B. Brewer Jr, G. Crepaldi and G. Baggio.

Department of Internal Medicine, University of Padova, Italy and *Molecular Disease Branch, NHLBI, NIH, Bethesda,Md, USA.

INTRODUCTION

Familial Hypobetalipoproteinemia and ApoB Gene Mutations

Human familial hypobetalipoproteinemia (FHBL) is a genetic disorder, inherited as an autosomal dominant trait, characterized by decreased levels of low density lipoprotein (LDL) cholesterol and apoB. The classical clinical picture of the homozygous patient includes profound hypocholesterolemia with virtual absence of LDL cholesterol and apoB, hypotriglyceridemia, fat malabsorption, clotting abnormalities due to vitamin K deficiency and red cell acanthocytosis [Kane J.P, 1989]. In addition, severe neurologic disorders may became evident after the first decade of life that seem related to vitamin E malabsorption [Muller D.P.R., 1983]. Heterozygous individuals are usually free from symptoms but their LDL cholesterol and apoB concentrations are half the normal value or less.

Over the last five years the analysis of the apoB gene in a number of kindreds with FHBL has revealed a variety of mutations which interfere with the full-length traslation of the apoB molecule [Young S.G., 1990; Gabelli C., 1992]. Most of them are due to point mutations with a single nucleotide substitution (frequently G to C), or a deletion followed by a frameshift, with the formation of a premature stop codon. A list of the various mutations associated with hypobetalipoproteinemia is given in table I.

NATO ASI Series, Vol. H 73
Human Apolipoprotein Mutants III
Edited by C. R. Sirtori, G. Franceschini, B.H. Brewer Jr.

Table 1. Mutations of apoB Gene Associated with FHBL.

	Number of * amino acids	Mutation	ApoB in ** plasma	Reference
Null allele	-	G→T at the 5' end of intron 5	Absent	Huang LS 1989
$ApoB_{9}$	411	649 bp deletion	Absent	
$ApoB_{25}$	1025(73)	649 bp deletion	Absent	Huang LS 1989a
$ApoB_{29}$	1305	C→T at nucleotide 4125	Absent	Collins DR 1988
$ApoB_{31}$	1425(2)	Deletion of nucleotide 4480	HDL, >1.21 g/ml	Young SG 1990
$ApoB_{37}$	1728(1)	Deletion of nucleotides 5391-5394	HDL,VLDL, LDL	Young SG 1987
$ApoB_{39}$	1799(5)	Deletion of nucleotide 5591	VLDL,LDL	Collins DR 1988
$ApoB_{40}$	1829(1)	Deletion of nucleotides 5693-5694	HDL,VLDL, LDL	Krul ES 1989
$ApoB_{46}$	2057	C→T at nuclotide 6381	VLDL,LDL, HDL	Young SG 1989
$ApoB_{50}$	2251	C→T at nuclotide 6963	VLDL	Hardman DA 1991
$ApoB_{54.8}$	2485	C→T at nucleotide 7665	VLDL,LDL	Wagner RD 1991
$ApoB_{61}$	2784(12)	Deletion of nucleotides 8525-8561	VLDL,LDL	Pullinger CR 1990
$ApoB_{67}$	3040(1)	Deletion of nucleotide 9327	VLDL,LDL	Welty FK 1991
$ApoB_{75}$	3386	C deletion at nucleotide 10366	VLDL,LDL	Krul ES 1991
$ApoB_{83}$	3749	C→T at nucleotide 11458	VLDL,LDL	Farese RV 1991
$ApoB_{86}$	3896(20)	Deletion of nucleotide 11840	VLDL,LDL	Linton MF 1990
$ApoB_{87}$	3978(37)	Deletion of nucleotide 12032	VLDL,LDL	Gabelli C 1989
$ApoB_{89}$	4039(6)	Deletion of nucleotide 12309	VLDL,LDL	Talmud P 1989

*The number of amino acids out-of-frame in the COOH-terminal portion are indicated in parenthesis. **Lipoprotein classes in which the truncated form is found in circulation.

The heterogeneity of the molecular defect is associated to a variable severity of the clinical manifestations in homozygotes. The data obtained after the definition of the disease at the molecular level indicate that the symptoms are severe (malabsorption, cerebellar ataxia, acanthocytosis, blood clotting abnormalities and retinitis pigmentosa) when the synthesis of apoB is abolished or only very short apoB forms, as apoB-9 and apoB-25, are encoded. Recently a mutation causing a splicing defect has been described [Huang L.S., 1991]. Examples in other apolipoprotein genes [Fojo S.S., 1988] have shown that such mutations in|erfere with the formation of normal mRNA and lead to a great reduction or total lack of the synthesis of a normal-sized protein. Fat malabsorption is still present in the patient homozygous for apoB-39 [Collins D.R., 1988]. In the apoB-50 homozygous individual, where the B-48 pathway is preserved, the classical clinical picture of the disease almost disappears [Hardman D.A., 1991].

In general it has been observed that longer truncated apoB species are associated with milder symptoms. The observed differences are probably due to a variable ability of the apoB mutant to deliver lipids and fat soluble vitamins from the intestine and the liver to the periferal tissues.
The metabolic abnormalities causing the hypocholesterolemia in these patients are still poorly understood. Recently, Hardman and coworkers have shown that apoB-50 is mainly found in VLDL, in an in vivo kinetic study these lipoproteins have an half-life of 50 min and virtually no apoB-50 is converted to LDL. Therefore apoB-50, and probably forms with similar size, are removed from circulation quite efficiently and can deliver, together with apoB-48, exogenous lipids to the specific tissues.

The metabolism of longer truncated forms of apoB have been recently evaluated in our laboratory and by others investigators. We present here a brief resume of the clinical, molecular and functional abnormalities in the apoB-87_{Padova} family and discuss the structural and functional implications of the results.

METHODS

Subjects

Plasma was obtained from each subject from fasting blood samples collected in a mixture of protease inhibitors.

ApoB molecular weight was analyzed in lipoprotein fractions after separation of apolipoproteins by 3-10% density gradient polyacrylamide gel electrophoresis containing 3% Sodium-Dodecyl-Sulphate (SDS-PAGE). The electrophoresis was performed in a 160x140x1.5 mm chamber using a Protean I Bio-Rad apparatus (Bio-Rad, Richmond, CA) according to the manufacturer's instruction. After separation the proteins were stained with 0.1 % Coomassie brilliant blue R-250 or transferred to nitrocellulose paper for immunoblot analysis. Western blot was performed as described by Towbin et al.[1979] using polyclonal and monoclonal first antibodies and ^{125}iodine second antibodies followed by autoradiography. Specific murine monoclonal antibodies againts different epitopes of apoB-100 were kindly provided by dr. Y. Marcel.

Genomic DNA Amplification and Sequence Analysis

Genomic DNA was prepared from leukocytes as described [Sambrook J, 1989]. The polymerase chain reaction was used to amplify a region of the apoB gene encompassing exon 27 and 28, from 1 μg of genomoc DNA of the proband. PCR products of the predicted size were isolated from 2% NuSieve/1% agarose gels (FMC BioProducts) and either cloned into the BamHI-HindIII sites of pGEM3Z (Promega) and sequenced [Sanger F, 1977] or sequenced directly [Mitchell LG, 1989] and resolved trough 6% polyacrylamide/7M urea gels.

In vivo LDL turnover

LDL kinetic was evaluated in a apoB-87 homozygous individual and in two normal volunteers. All subjects were free of renal, hepatic hematologic and tyroid abnormalities. No medication was allowed before and during the turnover study, potassium iodide was given (200 mg t.i.d.) to prevent tyroid uptake of the radioactive isotope. Written informed consent was asked to each individual. LDL from the apoB-87 proband were isolated starting from 1.220 ml of plasma separated by apheresis. Normal LDL were obtained from one normal volunteer with E3/3 phenotype. LDL preparations were performed by sequential ultracentrifugation (d=1.019-1.063 g/ml), concentrated and dialyzed to a final protein concentration of 2.5 g/dl. One ml of each isolated LDL fraction was filtered though a 0.22 μm filter (Millipore) and radioiodinated using a modification of the iodine monochloride method [Bilheimer DW, 1972] utilizing ^{125}I for apoB-87-LDL and ^{131}I for normal LDL. Unbound iodine was removed by gel filtration chromatography and dialysis. Each subject received a bolus injection of both preparations at 8 am. Blood samples were collected after 10 min, 30 min, 1 hr, 3 hrs, 6 hrs, 12 hrs, 24 hrs, and thereafter daily for one week. Radioactivity disappearance from the LDL compartment was followed using a Packard 5230 autogamma spectrometer. The data were analyzed by computer curve fitting using a SAAM 27 program [Berman M, 1977].

RESULTS

A 42-years old woman was referred to the Lipid Clinic of the University of Padova for persistent low plasma cholesterol levels. Subsequent evaluation revealed that this individual and a 49-years old sister had a LDL cholesterol concentration below 8 mg/dl while the mother and all the first degree relatives had a LDL cholesterol lower than the 50th percentile for age and sex. The two sister were found homozygous for the defect. Molecular weight estimation in SDS-PAGE and epitope mapping using a panel of monoclonal antibodies indicated that the mutant apoB molecule had a molecular weight corresponding to 87% of normal B-100 (apoB-87$_{Padova}$) and the truncation occurred in the carboxyl-terminal sequence after amino acid 3926.

Enzymatic amplification of enomic DNA obtained from the proband followed by subcloning and sequence analysis showed a single nucleotide G deletion at the immediate 5' end of exon 28 of the apoB gene, resulting in a shifted reading frame which adds 37 nonhomologous amino acids and prematurely terminates the translation after amino acid 3978 (Fig.1).

```
                     INTRON 27 |      EXON 28

Normal            5'...TGTTTCAG GGAATGGGAAGGAAAAGCG.......CTG-3'
ApoB                   -------- nGluTrpGluGlyLysAla.......Leu
                                                          aa 4536

ApoB-87           5'...TGTTTCAG -GAATGGGAAGGAAAAGCGC...CGC-3'
                       --------  nAsnGlyLysGluLysArg...Pro
                                                         aa 3878
```

Fig. 1: Schematic view of the mutation (deletion of the first nuclotide of exon 28) of the ApoB gene in the ApoB-87$_{Padova}$.

To evaluate the metabolic basis of the hypocholesterolemia in this family an <u>in vivo</u> turnover study was performed. After simultaneous injection of normal LDL and B-87-LDL we observed a similar behavior of all the study subjects (one B-87$_{Padova}$ homozygote and two normal controls). LDL containing ApoB-87 were catabolized to a faster rate compared to normal LDL, both in the hypobeta and in the normal individuals, with an half life of bout one third of the normal value. Catabolism of normal LDL in the homozygous patient was only slightly faster compared to normal. Determination of pool size, residence time, and production rate of autologous LDL gave additional information regarding the metabolism of apoB in the subjects under investigation (Table 2). ApoB pool size in LDL of the patient was severely reduced compared to normal and the residence time was 1.01 days compared to a mean value of 2.45 days in the controls. Therefore the calculated production rate of apoB in LDL in the patient with FHBL due to the presence of apoB-87 Padova is severely reduced, indicating both decreased production and increased LDL catabolism cause hypocholesterolemia in these patients.

Table 2: Parameters of autologous LDL metabolism in the apoB-87$_{padova}$ proband and two normal control subjects.

	LDL apoB (mg/dl)	RT (days)	FCR (day^{-1})	PR (mg/kg/day)
Proband	1.6	1.01	0.99	0.7
Normal 1	64.0	2.10	0.47	13.5
Normal 2	55.0	2.80	0.35	8.7

RT= residence time; FCR= fractional catabolic rate; PR=production rate.

DISCUSSION

The variable degree of severity of the disease in the homozygous patients from separate kindreds rises the question whether the hypocholesterolemia may be caused by distinct abnormalities of intracellular or plasma apoB metabolism.

Very short apoB forms (apoB-9) most probably lack the ability to form stable particles with lipids and are not properly secreted from the cells. According to recent in vitro data truncated apoB species as short as 17-18% of the normal-sized apoB-100 can be secreted from the transfected cells, but without a significant amount of lipids and the protein is found in the 1.23 g/ml bottom fraction after centrifugation [Herscovitz H, 1991; Yao Z, 1991]. However both apoB-25 and apoB-29 are not found in plasma suggesting that in vivo the rate of secretion does not parallel what is found in vitro or that these molecules are rapidly removed after secretion.

ApoB-31 is the shortest apoB mutant found in plasma where is mainly associated with particles in the density range kf the high density lipoprotein (HDL). The results of the in vitro expression of apoB-31 and apoB-37 from rat hepatoma cells indicate that these species are efficiently assembled with lipids and are secreted from cells as lipoproteins [Yao Z, 1991]. ApoB-37 does not bind to the LDL of cells [Young SG, 1987a]. All the in vitro results do not offer a positive indication about the in vivo metabolism and further studies are required to clarify whether the apoB forms containing the amino-terminal 30-40% of apoB-100 are secreted at a low rate or undergo rapid clearance via an LDL receptor independent pathway.

The results of the present series of studies together with recent data from other investigators on $apoB_{89}$ and $apoB_{75}$ establish that LDL containind longer truncated apoB molecules are removed from circulation with an efficiency higher than normal and explain, at least in part, the low plasma cholesterol in these patients. In vitro studies with $apoB_{87}$-LDL indicate that the fast catabolism is due to an incrased recognition of the particles by the LDL receptor [Gabelli C, 1992a]. These data suggest that the carboxyl-terminal of apoB-100 might actually have effect of decreasing the affinity of apoB-100 for the LDL receptor. In keeping with such hypothesis Fantappie' and coworkers [1992] recently observed that the interaction of different murine monoclonal antibodies with human low density lipoproteins can change the binding to the LDL receptor in opposite directions. Using a monoclonal antibody specific for the carboxyl-terminal region of apoB (aa 4082-4306) they found that the LDL binding was increased up to threefold.

Even if the hypothesis that the removal of the terminal part of the apoB may increase the recognition of the molecule by the receptor is a very interesting one, it needs to be considered with caution. Other factors may contribute to the increased binding of LDL containing short apoB species for the LDL receptor:

a) an increased apoE/apoB molar ratio in the particle and,

b) a decreased diameter of the LDL particles which can be taken up more efficiently by the receptor as proposed by Chappell [1991].

ACKNOWLEDGMENTS:

The present research was partially supported by C.N.R.: Progetto Finalizzato Invecchiamento, Sottoprogetto 3 - Invecchiamento e Malattie Endocrino-Metaboliche, Meccanismi Patogenetici. Pub. Number: 92.3.202

REFERENCES

Berman M, Weiss M. (1977) SAAM Manual. (DHEW Publ. 78-180) NIH Bethesda MD.
Bilheimer DW, Eisenberg S, Levy RI. (1972) The metabolism of very low density lipoprotein. Preliminary in vitro and in vivo observations. Biochim Biophys Acta 260:212-221.
Chappell DA, Fry GL, Waknitz MA, Berns JJ (1991) Ligand size as a determinant for catabolism by the low density lipoprotein (LDL) receptor pathway. A lattice moldel for LDL binding. J Biol Chem 266:19296-19302.
Collins DR, Knott TJ, Pease RJ, Powell LM, Wallis SC, Robertson S, Pullinger CR, Milne YL, Humphries SE, Talmud PJ, Lloyd JK, Miller NE, Muller D, Scott J (1988) Truncated Variants of Apolipoprotein B Cause Hypobetalipoproteinemia. Nucleic Acids Res 16:8361-8375.
Fantappie' S, Corsini A, Sidoli A, Uboldi P, Granata A, Zanelli T, Rossi P, Marcovina S, Fumagalli R, Catapano AL (1992) Monoclonal antibodies to human low density lipoprotein identify distict areas on apolipoprotein B-100 relevant to low density lipoprotein-receptor interaction J Lipid Res 33:1111-1121.
Farese RV, Garg A, Pierotti VR, Vega GL, Young SG, Grundy SM (1991) A Truncated Species of ApolipoproteinB, ApoB-83, Associated with Hypobetalipoproteinemia. Clin Res 39:276A.
Gabelli C, Baggio G, Bilato C, Martini S, Previato L, Marcel Y, Corsini A, Crepaldi G (1989) Identification of a New Apo B Variant Associated with Hypobetalipoproteinemia. Circulation 80-II:276.
Gabelli C (1992). The lipoprotein metabolism of apolipoprotein B mutants. Curr. Opin. Lipidol. 3:208-214.
Gabelli C, Martini M, Bilato C, Bertolini S, Corti MC, Crepaldi G, Baggio G (1992a) Apolipoprotein B Mutants: Impact on Low Density Lipoprotein Levels and Metabolism. In:Atherosclerosis IX. Y Stein,editor, R and L Creative Communications, Tel Aviv.
Hardman DA, CR Pullinger, RL Hamilton, JP Kane, MJ Malloy (1991) Molecular and metabolic basis for the metabolic disorder normotriglyceridemic abetalipoproteinemia. J. Clin. Invest. 88:1722-1729.
Herscovitz H, Hadzopoulou-Cladaras M, Walsh MT, Cladaras C, Zannis VI, Small DM. (1991) Expression, Secretion, and Lipid-binding Characterization of the N-terminal 17% of Apolipoprotein B. Proc Natl Acad Sci USA 88:7313-7317.
Huang LS, ME Ripps, SH Korman, RJ Deckelbaum, JL Breslow (1989) Hypobetalipoproteinemia due to an apolipoprotein B gene exon 21 deletion derived by Alu-Alu recombination. J.Biol. Chem. 264:11394-1140.
Huang LS, Kayden H, Sokol RJ, Breslow JL (1991) Apo B Gene Nonsense and Splicing Mutations in a Compound Heterozygote for Familial Hypobetalipoproteinemia. J Lipid Res 32:1341-1348.
Kane JP, Havel RJ (1989) Disorder of biogenesis and secretion of lipoproteins containing the B apolipoprotein. In The metabolic basis of inherited disease. C.R. Scriver, A.L. Beaudet, W.S. Sly, D. Valle, editors. McGraw-Hill, New-York, 1139-1164.

Krul ES, Kinoshita M, Talmud P, Humphries SE, Turner S, Goldberg AC, Cook K, Boerwinkle E, Schonfeld G (1989) Two Distinct Truncated Apolipoprotein B Species in a Kindred with Hypobetalipoproteinemia. Arteriosclerosis 9:856-868.
Krul ES, Wagner RD, Parhofer KG, Barrett H, Schonfeld G (1991) Apolipoprotein B-75: A Truncated Form of Apolipoprotein B with Increased Affinity for the LDL Receptor. Circulation 84-II:340.
Linton MF, Pierotti VR, Hubl ST, Young SG (1990) An Apo-B Gene Mutation Causing Familial Hypobetalipoproteinemia Analyzed By Examining the Apo-B cDNA Amplified from the Fibroblast RNA of an Affected Subject. Clin Res 38:286A.
Mitchell LG, Merril CR (1889) Affinity generation of single-stranded DNA for dideoxy sequencing following the polymerase chain reaction. Anal Biochem 178:239-242.
Muller DPR, Lloyd JK, Wolf OH (1983) Vitamin E and neurologic function. Lancet 1:225-228.
Pullinger CR, Hillas E, Hadman DA, Chi Chen G, Naya-Vigne JM, Laluel JM, Williams RR, Kane JP (1990) A Mutation in the Apolipoprotein B Gene Results in a Truncated Variant, ApoB-61, in VLDL and LDL in a Kindred with Hypobetalipoproteinemia. Circulation 82-II:424.
Sambrook J, Fritsch EF, Maniatis T, (1989) In Molecular Cloning: A Laboratory Manual. Cold Spring Harbour Press.
Sanger F, Nicklen S, Coulsen AR. (1977) DNA sequencing with chain terminating inhibitors. Proc Natl Acad USA 74:5463-5467.
Talmud P, King-Underwood L, Krul E, Schonfeld G, Humphries S (1989) The Molecular Basis of Truncated Forms of Apolipoprotein B in a Kindred with Compound Heterozygous Hypobetalipoproteinemia. J Lipid Res 30:1773-1779.
Tennyson GE, Gabelli C, Baggio G, Bilato C, Brewer HB jr (1990) molecular Defect in the Apolipoprotein B Gene in a Patient with Hypobetalipoproteinemia and Three Distinct ApoB Species. Clin Res 38:482A.
Towbin H, Staehlin T, Gordon J (1979) Electrophoretic transfer of proteins from polyacrylamide gels to nitrocellulose sheet. Proc. Natl. Acad. Sci. USA. 76:4350-4354.
Yao Z, Blackhart BD, Linton MF, Taylor SM, Young SG, McCarthy BJ (1991) Expression of Carboxyl-terminally Truncated Forms of Human Apolipoprotein B in Rat Hepatoma Cells. Evidence That the Length of Apolipoprotein B Has a Major Effect on the Buoyant Density of the Secreted Lipoproteins. J Biol Chem 266:3300-3308.
Young SG, Bertics SJ, Curtiss LK, Witztum JL (1987) Characterization of an Abnormal Species of Apolipoprotein B, Apolipoprotein B-37, Associated with Familial Hypobetalipoproteinemia. J Clin Invest 79:1831-1841.
Young SG, Peralta FP, Dubois BW, Curtiss LK, Boyles JK, Witztum JL (1987a) Lipoprotein B37, a Naturally Occurring Lipoprotein Containing the Amino-terminal Portion of Apolipoprotein B100, Does Not Bind to the Apolipoprotein B,E (Low Density Lipoprotein) Receptor. J Biol Chem 262:16604-16611.
Young SG, Hubl ST, Chappell DA, Smith RS, Claiborne F, Snyder SM, Terdiman JF (1989) Familial Hypobetalipoproteinemia Associated with a Mutant Species of Apolipoprotein B (B-46). New Engl J Med 320:1604-1610.

Young SG (1990) Recent progress in understanding apolipoprotein B . Circulation 82:1574-1594.
Young SG, Hubl ST, Smith RS, Snyder SM, Terdiman JF (1990a) Familial Hypobetalipoproteinemia Caused by a Mutation in the Apolipoprotein B Gene That Results in a Truncated Species of Apolipoprotein B (B-31). J Clin Invest 85:933-942.
Wagner RD, Krul ES, Tang J, Parhofer KG, Garlock K, Talmud P, Schonfeld G (1991) ApoB-54.8, a Truncated Apolipoprotein Found Primarily in VLDL, is Associated with a Nonsense Mutation in the ApoB Gene and Hypobetalipoproteinemia. J Lipid Res 32:1001-1011.
Welty FK, Hubl ST, Pierotti VR, Young SG (1991) A Truncated Species of Apolipoprotein B (B-67) in a Kindred with Familial Hypercholesterolemia. J Clin Invest 87:1748-1754.

CAUSES UNDERLYING THE REDUCED RESPONSE TO SIMVASTATIN TREATMENT IN HYPERCHOLESTEROLEMIC PATIENTS.

Alberto Corsini, Ph.D., Maria Mazzotti, Ph.D., Tiziano Zanelli, Ph.D., Patrizia Uboldi, B.S., Domenico Sommariva, M.D.°, Franco Pazzucconi, M.D.+, Cesare R. Sirtori, M.D.+, Antonio Gaddi, M.D.*, Giancarlo Descovich, M.D.*, Agnese Granata, B.S., Remo Fumagalli, M.D., and Alberico L. Catapano, Ph.D..

Institute of Pharmacological Sciences, University of Milan, Milan, °Bollate Hospital, Milan, +Center E. Grossi Paoletti, University of Milan, *Atherosclerosis Center and Department of Internal Medicine S. Orsola Hospital, University of Bologna, Italy.

ABSTRACT

Simvastatin, a competitive inhibitor of HMG-CoA reductase, effectively reduces elevated plasma cholesterol levels by up regulating the LDL receptor expression. Therefore in patients whose hypercholesterolemia (HC) is due to a defective LDL-receptor interaction vastatins may not be as effective as expected. To verify this hypothesis we studied the possible causes for the poor response (<15% decrease of LDL cholesterol) to simvastatin (40 mg/die) in 11 HC patients. Biochemical defects were identified in 5 patients. Three patients presented with binding-defective LDL, without the 3500 mutation, and the remaining two had normal LDL but their serum contained factors interfering with the LDL-receptor interaction. From these results we conclude that distinct biochemical defects might contribute to the poor response to simvastatin in hypercholesterolemic patients.

Key words: LDL receptor, binding defective LDL, hypolipidaemic drugs, apolipoprotein B, probucol.

NATO ASI Series, Vol. H 73
Human Apolipoprotein Mutants III
Edited by C. R. Sirtori, G. Franceschini, B.H. Brewer Jr.

INTRODUCTION

Hypercholesterolemia or, more accurately, an elevated low density lipoprotein (LDL) cholesterol level, is a major risk factor for the development of coronary heart disease (CHD) (Stamler, 1979; Lowering Blood Cholesterol to Prevent Heart Disease, Consensus Conference 1985)). Lowering of total and LDL cholesterol can reduce the incidence of ischemic heart disease and retard the development of coronary atherosclerosis (Lipid Reaserch, Clinics Program, 1984; Frick et al., 1987; Brown et al., 1990; Kane et al., 1990).

LDL transport about two-thirds of plasma cholesterol (Brown and Goldstein, 1986) and their fate is largely determined by the apo B,E (LDL) receptors, that bind apolipoprotein (Apo) B-100, the main protein constituent of LDL (Goldstein and Brown, 1989; Brown and Goldstein, 1987). The importance of the apo B-100/LDL receptor interaction in maintaining cholesterol homeostasis is best illustrated by familial hypercholesterolemia (FH), a genetic disease in which a defective LDL receptor expression triggers an increase of plasma LDL cholesterol levels and premature CHD (Brown and Goldstein, 1986; Goldstein and Brown, 1989; Brown and Goldstein, 1987). Patients with heterozygous FH express about 50% of LDL receptors, and have a two to three fold increase of plasma LDL levels (Brown and Goldstein, 1987).

Given the importance of the LDL receptor activity in determining plasma cholesterol, a pharmacological approach aimed at raising the LDL receptor expression should result in a significant lowering of LDL cholesterol (Brown and Goldstein, 1981). Recently lovastatin and other vastatins, such as simvastatin and pravastatin, have become available; these drugs are highly effective, at relatively low doses, in lowering LDL cholesterol (Grundy, 1988). Their primary action is to competitively inhibit 3-hydroxy-3-methylglutaryl Coenzyme A (HMG-CoA) reductase, the rate limiting enzyme of cholesterol synthesis (Grundy, 1988).

Shortage of cholesterol at the cellular level results in an up regulation of LDL receptor expression. Recently it was shown that vastatin therapy is associated with a significant increase of LDL-receptor activity in circulating monocytes and in liver (Hagemenas et al., 1990; Reihner et al., 1990; Raveh et al., 1990), thus accounting for the lower plasma total and LDL cholesterol levels. According to this postulated mechanism, lovastatin will not reduce LDL cholesterol levels in homozygous FH (receptor negative) patients (Grundy, 1988).

Most individuals with primary hypercholesterolemia, however possess a normal number of LDL receptors. Therefore, in most cases of primary hypercholesterolemia other causes must contribute to the increase of plasma LDL, such as a defective clearance or increased production of LDL (Grundy and Vega, 1990).

In principle, elevated LDL concentrations that results from an inefficient clearance of LDL may derive from receptor defects (classical FH) (Goldstein and Brown, 1989), ligand defects (familial defective apo B-100) (Vega and Grundy, 1986; Innerarity et al., 1987; Soria et al., 1989; Innerarity et al., 1990; Corsini et al., 1991) and from the presence of factors (i.e. antibodies) interfering with the LDL-receptor interaction (Riesen and Noseda, 1975; Lewis et al., 1975; Beaumont and Beaumont, 1977; Corsini et al., 1986). If primary hypercholesterolemia is characterized by a defect in the LDL-receptor interaction (presence of binding-defective LDL or of autoantibodies impairing the ligand-receptor interaction), vastatins might not be as effective as expected in their hypercholesterolemic activity.

In this report we focus on the possible causes for the poor response of 11 hypercholesterolemic patients to simvastatin treatment (simvastatin non responders).

PATIENTS AND METHODS

Patients. A total of 11 hypercholesterolemic patients

(WHO type IIa or IIb) were referred to us from different Lipid Clinics because of poor response (<15% reduction of LDL cholesterol) to simvastatin treatment. All but one were given the maximal therapeutic dose of simvastatin (40 mg/die) for 4-6 weeks. The remaining patient received 20 mg/die. All routine laboratory tests were normal and known secondary causes of hypercholesterolemia were excluded. Baseline plasma lipids and lipoprotein lipids of the patients and their response to treatment are presented in Table 1. Blood samples for lipid and lipoprotein analysis were obtained after a 12 hour fast, at the beginning and end of therapy.

DNA samples, prepared by standard procedures from patients, were analyzed by the polymerase chain reaction (PRC) and allele specific oligonucleotide hybridization (Soria et al., 1989).

Materials. Minimum essential medium (MEM), fetal calf serum, trypsin-EDTA (X 1), penicillin (10,000 U/ml), streptomycin (10 mg/ml), tricine buffer (1M, pH 7.4), nonessential aminoacids (X 100), were purchased from Gibco (Grand Island, NY); disposable culture flasks and Petri dishes were from Corning Glass Works (Corning, NY), filters from Millipore Corp. (Bedford, MA). Sodium ^{125}I-iodide, carrier free in 0.1 N NaOH was purchased from Amersham (Amersham, United Kingdom).

Sephadex G-25 were obtained from Pharmacia (Uppsala, Sweden). Kits for enzymatic determination of plasma lipids were from Boehringer (Mannheim, Germany). All electrophoresis material and equipment were purchased from Bio-Rad Laboratories (Richmond, CA). All others reagents were analytical grade.

Plasma lipids and lipoproteins. Plasma was obtained in EDTA (0.1%), cholesterol and triglycerides were determined by enzymatic procedures. Lipoproteins were isolated by ultracentrifugation (Havel et al., 1955) using the 1.019-1.063 g/ml density cut at 40,000 rpm in a Beckman 50.2 Ti rotor (Beckman Instruments, Inc., Palo Alto, CA) for 18 h at 12°C. Lipoprotein triglyceride and cholesterol contents were determined by enzymatic procedures. The LDL-apoprotein moiety was characterized by sodium dodecylsulfate electrophoresis (SDS-PAGE, 4%) (Laemmli, 1970).

TABLE 1

BIOCHEMICAL AND CLINICAL CHARACTERIZATION OF PATIENTS AT BASELINE AND PERCENTAGE REDUCTION IN PLASMA AND LDL CHOLESTEROL AFTER SIMVASTATIN TREATMENT (40 mg/day, 4 weeks)

				CHOLESTEROL mg/dl			% DECREASE IN CHOLESTEROL		
	SEX	AGE	CHD	TOTAL	HDL	LDL	TG	TOTAL	LDL
L.A.+	F	42	+	678	52	608	92	7	9
R.F.	M	32	+	220	27	154	197	10	10
N.N.	M	55	+	252	40	176	182	9	9
Z.I.	M	68	-	400	40	336	120	13	15
E.P.	M	42	-	339	41	280	91	12	15
P.S.	F	55	+	365	55	256	269	9	11
P.N.	F	45	-	340	40	270	150	10	10
R.G.	M	44	-	336	39	260	183	9	15
M.G.	M	26	-	380	50	314	81	13	15
N.G.*	F	35	-	273	58	232	199	9	12
P.A.	F	50	-	405	31	309	327	12	10

* The patient was treated for 4 weeks with simvastatin (20 mg/d)
CHD = coronary heart disease
\+ The patient died of CHD in September 1991.

Lipoprotein-deficient serum (LPDS) was obtained as the serum fraction at d< 1.25 g/ml by ultracentrifugation for 72 h at 40,000 rpm at 12°C in a Beckman 60 Ti Rotor (Goldstein et al., 1983). Densities were adjusted by addition of solid KBr to the solutions.

All lipoproteins and LPDS were dialyzed extensively

against NaCl 0.15 M, EDTA 0.3 mM, pH 7.4, stored at 4°C after sterilization through a Millipore 0.45-μm filter and used within 15 days. LPDS was stored at -20°C and thawed just before use. For some experiments LDL were partially delipidated with diethyl ether according to Innerarity and Mahley (1978).

LDL were labeled with ^{125}I according to Bilheimer et al. (1972). Specific activity was 150-230 cpm/ng protein. Free ^{125}I was removed by column cromatography on a Sephadex G-25 column eluted with 0.15 M NaCl, 0.3 EDTA, pH 7.4. Fractions containig lipoproteins were collected and dialyzed against 0.15 M NaCl, 0.3 mM EDTA, pH 7.4, overnight at 4°C. Free iodine accounted for 1% of total radioactivity. Lipid-associated radioactivity was always less then 5%.

Apo B was separated by SDS polyacrylamide (4%) gel electrophoresis (Laemmli, 1970).

Cell culture. Human skin fibroblasts were grown from explants of skin biopsies obtained from patients and from normolipidemic healthy individuals. Cells were grown in monolayers in a medium containing MEM supplemented with 10% fetal calf serum, nonessential aminoacid solution (1%, vol/vol), penicillin (100 U/ml), streptomycin (100 ug/ml), tricine buffer (20 mM, pH 7.4), and $NaHCO_3$ (24 mM) and maintained in 75-cm^2 plastic flasks at 37°C in a humidified atmosphere of 95% air, 5% Co_2.
For all experiments, cells from the stock flasks were dissociated with 0.05% trypsin-0.2% EDTA at confluency (5-15 passages), seeded in 35 mm plastic petri dishes (1.5 x 10^5 cells), and used just before reaching confluency, usually 6 d after plating. The medium was changed every 2-3 d (Corsini et al., 1986). Cell viability, assessed by Trypan blue exclusion, was always greater than 95%. For lipoprotein competition experiments confluent cells were preincubated 24 h at 37°C in a medium containing 10% human LPDS to induce LDL receptors (Goldstein et al., 1983; Corsini et al.,

1986). After preincubation fresh medium containing the appropriate amount of normal human ^{125}I-LDL protein (7.5 μg/ml) was added, along with increasing concentrations of unlabeled LDL either from controls or from patients, to the medium containing 10% LPDS. The cells were incubated for further 5 h at 37°C. For total uptake (binding + internalization) of LDL, cell monolayers were directly digested in 0.1 NaOH. LDL degradation was measured as the accumulation of non-iodide trichloroacetic acid soluble ^{125}I in the incubation medium in excess of that occurring in the absence of cells (Goldstein et al., 1983; Innerarity et al., 1990). Non specific uptake and degradation were determined by adding a 100-fold excess of unlabeled LDL. Each experimental point represents the average value of triplicate incubations, as specified in figures and tables. The amount of cellular uptake and proteolytic degradation of LDL was determined in binding experiments performed at 37°C (Goldstein et al., 1983). The amount of unlabeled LDL required to displace 50% of the ^{125}I-labeled ligand was calculated by linear regression analysis of the logarithm of concentration (μg of protein/ml) vs. probit and read from a probit transformation table (Fisher and Yates, 1953).

The ability of LPDS to interfere with the LDL-receptor interaction was investigated after preincubation of cells for 24 h at 37°C in a medium containing increasing concentration of LPDS from patients and from controls. ^{125}I-LDL (7.5 μg/ml) were then added and the incubation carried out for further 5 h at 37°C. Uptake and degradation of unlabeled LDL were determined as described above (Innerarity et al., 1990).

Oligonucleotides. All oligonucleotides were synthesized on an Applied Biosystems Synthesizer 380B as described by Soria et al. (1989). Probes were end-labeled with T4 polynucleotide kinase and AT^{32}P to 10^6 cpm/ng.

Polymerase Chain Reaction. The PCR was used for local amplification of a section of the human apo B gene. Reactions were performed in an automated Perkin Elmer Cetus Thermal Cycle using thermostable Thermus acquaticus DNA polymerase obtained from Cetus (Saiki et al., 1988).

Plasmids. Plasmid p^{46a}, containing an EcoRI-EcoRI 1549 fragment from exon 26 subcloned from G.R. Lambda ZAP DNA library, represents the wild-type allele. Plasmid P^{43}, containing an EcoRi-EcoRi 1549-bp fragment from exon 25 subcloned from the G.R. Lambda ZAP library, represents the mutant allele, i.e., possessing the GA substitution at the codon for amino acid 3500 (Soria et al., 1989).

RESULTS

Hypercholesterolemic patients were referred to us by different Lipid Clinics because they experienced only a slight reduction (maximum 15%) of total and LDL cholesterol levels after 4-6 weeks of simvastatin treatment even at the maximal therapeutic dose (40 mg/day). These results should be compared with the reductions induced by simvastatin on plasma and LDL cholesterol in heterozygous FH and in other forms of primary hypercholesterolemia (35-44%) (Grundy, 1988; Stossel 1988; Grundy et al., 1990). Table 1 shows the plasma lipids and lipoproteins of the patients and their response to simvastatin treatment.

To investigate the mechanism(s) responsible for resistence to simvastatin in hypercholesterolemic patients, two possible causes were investigated: a) the presence of binding defective LDL, and b) the presence of serum factor(s) interfering with the LDL-receptor interaction.

a) Binding-defective LDL

- Case report: Patient R.F.

This patient, with coronary heart disease, had elevated levels of LDL cholesterol and triglycerides (phenotype II B). Family history was positive for familial combined hyperlipidemia, a disorder characterized by multiple phenotypes of hyperlipoproteinemia. Cholesterol ranged between 220 and 355 mg/dl and TG levels between 166 and 325 mg/dl. Raised apo B levels (> 140 mg/dl) were also detected.

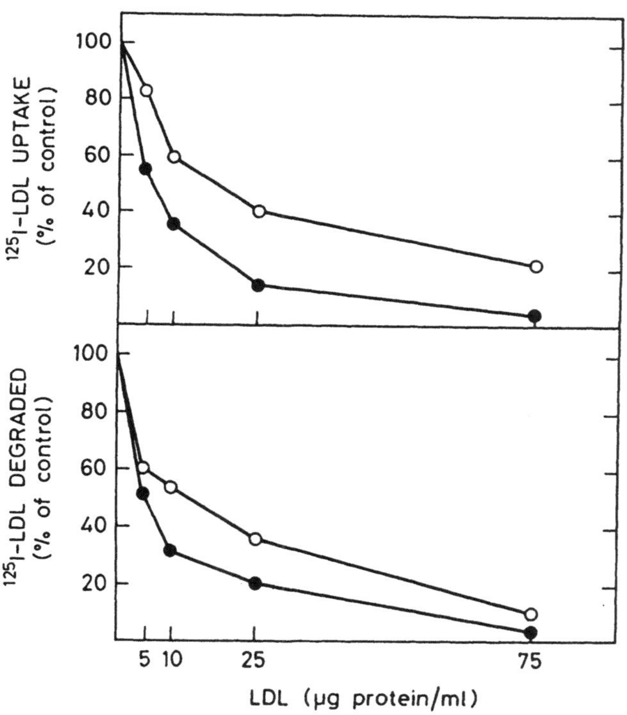

Fig. 1. Ability of LDL from R.F. (o - o) and a control (• - •) to compete with normal ^{125}I-LDL for uptake and degradation by cultured fibroblasts.
Confluent cells were preincubated for 24 h in a medium containing 10% LPDS and then incubated for 5 h with the indicated concentrations of unlabeled LDL in the presence of normal ^{125}I-LDL (7.5 μg/ml). Experimental conditions are as described in Materials and Methods. Data are the average of triplicates that did not differ by more than 10%.

LDL obtained from patient R.F. displayed a reduced affinity (2,5 fold) for the LDL receptor (Fig. 1). The average IC_{50} from patient R.F. vs control were 17.8 vs 5.1 μg/ml and 10.4 vs 4.1 for uptake and degradation, respectively. Binding-defective LDL were also detected in his children but not in his wife (data not shown).

Analysis of the LDL lipid composition showed an increase of the protein/cholesterol ratio (1.4 vs 0.7) and a decrease of the protein/triglyceride ratio (1.12 vs 2.9) that might potentially affect, at least in part, the LDL-receptor interaction (Kleinman et al., 1985). An abnormal (smaller) LDL size was also detected by FPLC (not shown). Analysis of R.F. LDL protein moiety by SDS-PAGE showed that the MW of apo B-100 was normal (data not shown). Genetic analysis of DNA failed to detect a point mutation leading to an Arg to Gln substitution at position 3500 of the mature protein, thus ruling out the diagnosis of FDB (Fig. 2).

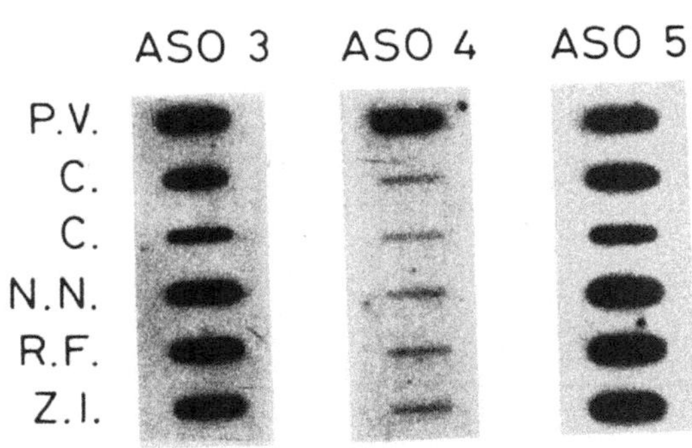

Fig. 2. Presence of the mutation in the codon for aa 3500 of the human apo B gene in binding-defective, simvastatin non responder, patients. Rows A, B, and C, respectively are autoradiograms of slot blots probed with ^{32}P labelled oligonucleotides Aso 3, Aso 4 and Aso 5. Aso 3 and Aso 4 are a pair of 15 mer oligonucleotides representing the normal and mutant alleles. Aso 5 is a 15 mer the amplified region whose sequence was identical in both alleles. P.V. is the DNA from a patient with familial defective apo B-100 (Corsini, 1991).

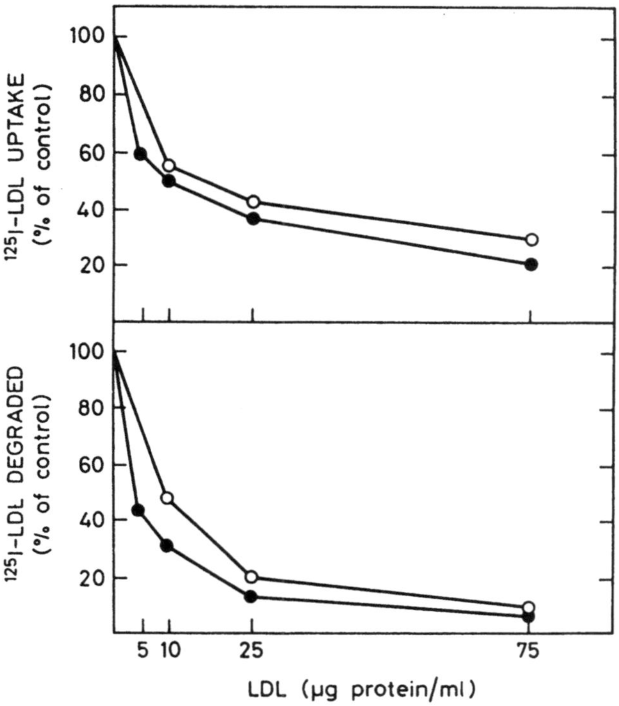

Fig. 3. Ability of LDL from N.N. (o - o) and a control (● - ●) to compete with normal ^{125}I-LDL for uptake and degradation by cultured fibroblasts.
Confluent cells were preincubated for 24 h in a medium containing 10% LPDS and then incubated for 5 h with the indicated concentrations of unlabeled LDL in the presence of normal ^{125}I-LDL (7.5 μg/ml). Experimental conditions are as described in Materials and Methods. Data are the average of triplicates that did not differ by more than 10%.

- Case report: Patient N.N.

He had had a myocardial infarction at age of 50 and coronary angioplasty at age of 53. Family history was positive and hypercholesterolemia was detected in his daughter (309 mg/dl).

"In vitro" LDL from patient N.N. were less effective (1.7 fold) in competing with control ^{125}I-LDL for uptake and degradation by HSF, as compared to LDL from normolipidemic subject (Fig. 3). The average IC_{50} from patient N.N. vs control were 14.2 vs 9.4 μg/ml and 8.1 vs 2.9 μg/ml for uptake and degradation, respectively. Binding-defective LDL were also detected in his daughter but not in his normolipidemic son (data not shown).

LDL lipid composition as well as apoprotein moiety were normal. Analysis of DNA failed to detect the point mutation leading to an Arg to Gln substitution at position 3500 of the apo B-100 mature protein, thus ruling out the diagnosis of FDB (Fig. 2).

- Case report: Patient Z.I.

This patient is affected by primary hypercholesterolemia (phenotype IIa) with family history positive and no clinical signs of CHD.

In an "in vitro" cell binding assay Z.I. LDL were less effective (1.7 fold) than normal LDL in competing with ^{125}I-LDL for uptake and degradation by HSF in the average IC_{50} from patient Z.I. vs control were 25.2 vs 14.5 μg/ml and 25.7 vs 12.7 μg/ml for uptake and degradation, respectively (Fig. 4).

Analysis of DNA failed to detect the point mutation leading to an Arg to Gln substitution at position 3500 of the apo B-100 mature protein, thus ruling out the diagnosis of FDB (Fig.2).

b) Serum factor(s) interfering with the LDL/LDL receptor interaction.

- Case report: Patient L.A.

Patient L.A. is a 42-yr-old woman, with a 15-yr history of rheumatoid arthritis and hypercholesterolemia (in the range of 550-680 mg/dl), with xanthomas at the elbows and thickening of Achilles tendon. Upon screening of the family members, hypercholesterolemia was detected in her

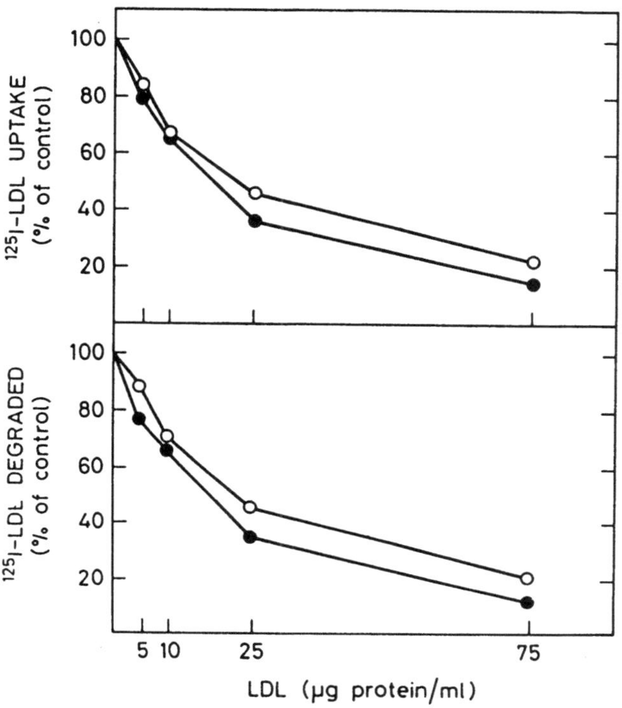

Fig. 4. Ability of LDL from Z.I. (o - o) and a control (●- ●) to compete with normal ^{125}I-LDL for uptake and degradation by cultured fibroblasts.
Confluent cells were preincubated for 24 h in a medium containing 10% LPDS and then incubated for 5 h with the indicated concentrations of unlabeled LDL in the presence of normal ^{125}I-LDL (7.5 μg/ml). Experimental conditions are as described in Materials and Methods. Data are the average of triplicates that did not differ by more than 10%.

mother (total cholesterol 288 mg/dl), in two brothers (252 and 592 mg/dl, respectively) but not in the father (226 mg/dl) and son. Although family history was positive, the absence of hypercholesterolemia in father and son rules out the diagnosis of homozygous FH. At age 42 she suddenly died at home during the night. Table 1 shows plasma lipids and lipoprotein levels; compared to controls, she had a four-fold increase of LDL mass with normal HDL-cholesterol and triglycerides levels. The patient was first treated with simvastatin alone (40 mg/die) for 4 weeks, then in association with cholestyramine (12 g/die). Only slight decreases of total and LDL cholesterol were observed (7-9%). In a first set of experiments we investigated the ability of LDL obtained from the patient and from controls to compete with normal ^{125}I-LDL for uptake and degradation by cultured fibroblasts. As shown in Fig. 5, L.A. LDL were as effective as control lipoproteins in competing with normal ^{125}I-LDL for the receptor (IC_{50} for uptake 14.7 and 16.4 respectively), thus ruling out the presence of binding defective LDL. We investigated the possible interference of the patients lipoprotein deficient serum (LPDS) on the LDL-receptor interaction. LPDS from the patient clearly inhibited, in a dose-dependent manner, uptake degradation of ^{125}-LDL by human skin fibroblasts, while control LPDS had no effect (Fig. 5).

- Case report: Patient P.S.

Patient P.S., with primary hypercholesterolemia and CHD, only showed a slight decrease (9-11%) of total and LDL cholesterol after 4 weeks of simvastatin treatment (40 mg/day) (Tab. 1). Family history was positive. As shown in Fig. 6, P.S. LDL were as effective as control LDL in competing with ^{125}I-LDL for the receptor (IC_{50} is for uptake were 4.4 and 4.7 respectively), thus ruling out the presence of binding-defective LDL.

We investigated the possible effect of patient's and control serum (LPDS) on the uptake and degradation of ^{125}I-LDL by HSF. LPDS from the patient inhibited, in a dose-dependent manner, these processes while control LPDS did not (Fig. 6).

Also the remaining simvastatin non-responders were investigated (Tab. 1). However, neither the presence of binding defective LDL nor serum factors interfering with the LDL-receptor interaction were detected (data not shown). Other mechanism(s) must therefore contribute to the lack of response to simvastatin treatment (Yamamoto et al., 1988; Stossel, 1988; Hagamenas et al., 1990) in these subjects.

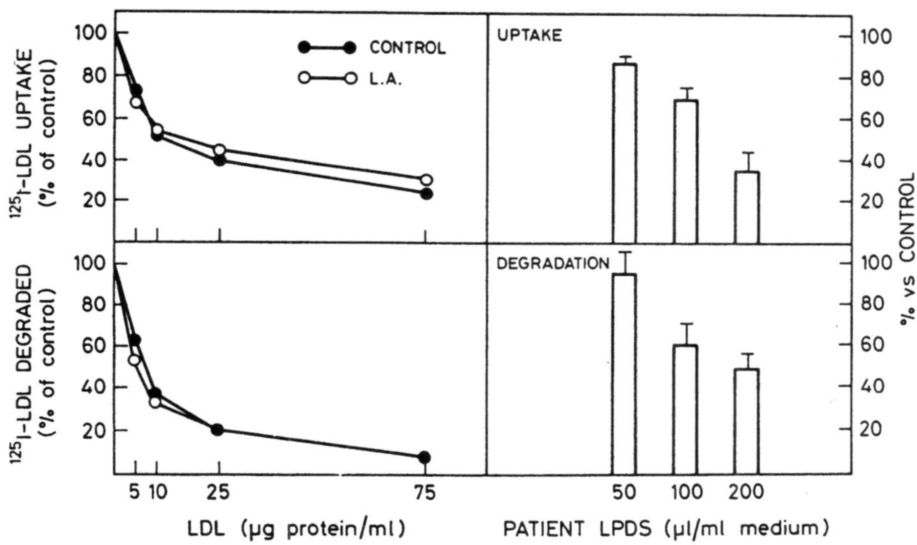

Fig. 5. Effect of LDL and LPDS from patient L.A. and a control on the uptake and degradation of normal ^{125}I-LDL by cultured skin fibroblasts.

Left panels. Confluent cells were preincubated for 24 h in a medium containing 10% control LPDS and then incubated for 5 h with the indicated concentrations of unlabeled LDL in the presence of normal ^{125}I-LDL (7.5 µg/ml). Data are the average of triplicates that did not differ by more than 10%.

Right panels. Confluent cells were preincubated at 37°C with the indicated amounts of control and L.A. LPDS. After 24 h, ^{125}I-LDL (7.5 µg/ml) were added to the medium and the incubation continued for further 5 h. The protein content of control and L.A. LPDS was 25 mg/ml. Specific uptake and degradation were determined as described in Materials and Methods. Data are the average of triplicates that did not differ by more than 10%.

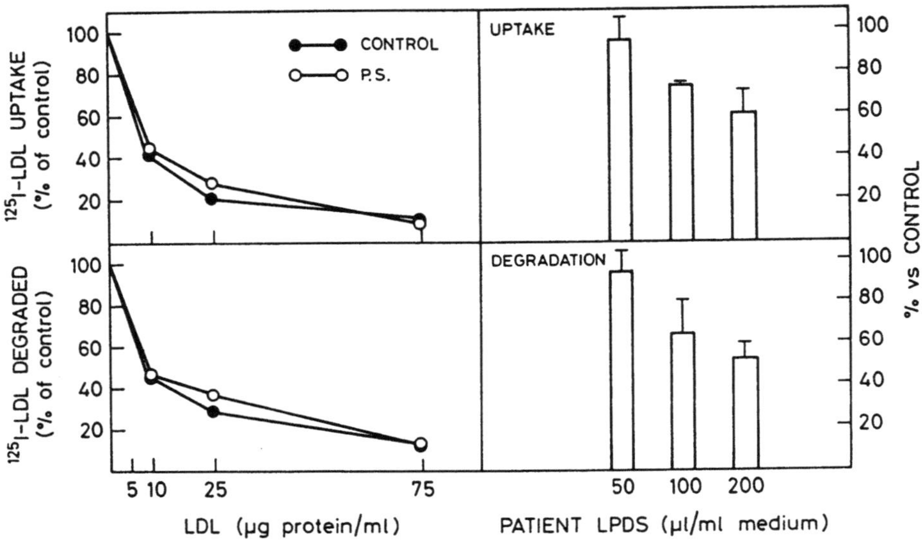

Fig. 6. Effect of LDL and LPDS from patient P.S. and a control on the uptake and degradation of normal ^{125}I-LDL by cultured skin fibroblasts.

Left panels. Confluent cells were preincubated for 24 h in a medium containing 10% control LPDS and then incubated for 5 h with the indicated concentrations of unlabeled LDL in the presence of normal ^{125}I-LDL (7.5 μg/ml). Data are the average of triplicates that did not differ by more than 10%.

Right panels. Confluent cells were preincubated at 37°C with the indicated amounts of control and P.S. LPDS. After 24 h, ^{125}I-LDL (7.5 μg/ml) were added to the medium and the incubation continued for a further 5 h. The protein content of control and P.S. LPDS was 16 mg/ml. Specific uptake and degradation were determined as described in Materials and Methods. Data are the average of triplicates that did not differ by more than 10%.

DISCUSSION

A number of clinical and epidemiological studies have documented that elevated cholesterol levels are associated with an increased morbidity and mortality from CHD (Stamler,

1979; Lowering Blood Cholesterol to Prevent Heart Disease. Consensus Conference, 1985). Clinical trials with lipid lowering drugs have reported fewer CHD events in intervention groups vs controls (Lipid Rasearch Clinics Program, 1984; Frick et al., 1987; Brown et al., 1990; Kane et al., 1990).

Recently HMG-CoA reductase inhibitors, a class of powerful cholesterol-lowering drugs, became available (Grundy, 1988). The inhibition of cholesterol synthesis by vastatins triggers a compensatory increase in LDL receptor expression, that accounts for the lowering of LDL levels (Bilheimer et al., 1983; Illingworth and Sexton, 1984; Havel et al., 1987; Grundy, 1988; Grundy et al., 1990). The control of LDL levels largely depends upon the LDL-receptor interaction, as best documented in familial hypercholesterolemia. Hypercholesterolemias resulting from binding defects of apo B-100, the physiological ligand for the receptor (Innerarity et al., 1987; Corsini et al., 1991), or from circulating factors (i.e. antibodies) that could interfere with the LDL-receptor interaction (Corsini et al., 1986), have also been described. Drugs such as vastatins, that primarily act by increasing the LDL receptor expression, may be of limited use in treating such patients and may, at least in part, account for the 4-5% "poor responders" reported in large clinical trials with lovastatin (Stossel, 1988; The Lovastatin Study Group IV, 1990). In this study we report the characterization of 11 patients with a poor response (< 15%) to simvastatin.

Patients defined as hyporesponders (only to 9-15% decrease of LDL cholesterol) to treatment with simvastatin were referred to us by different Italian Lipid Clinics.

Two possible causes were investigated:

a) the presence of binding-defective LDL,
b) the presence of serum factors that could interfere with the LDL-receptor interaction.

LDL obtained from three (N.N., R.F., Z.I.) out of 11 patients were less effective (1.5-3 fold) than LDL from normolipidemic subjects in competing "in vitro" with normal ^{125}I-LDL for uptake and degradation to human skin fibroblasts.

These patients are members of families with primary hyperlipoproteinemia in which the affected members possess LDL with binding abnormalities. DNA analysis failed to show the presence of the 3500 mutation, thus ruling out the diagnosis of FDB. While chemical composition and size (by FPLC) of LDL from patients N.N. and Z.I. were normal, LDL from R.F. showed an enrichment with triglyceride (relative to protein and cholesterol) and an abnormal size by FPLC (data not shown). These abnormalities are present in LDL from hypertriglyceridemic patients and are believed to be responsible for the poor binding of these lipoproteins to the LDL receptor (Kleinman et al., 1985). This binding abnormality may account for the poor response of patient R.F. to simvastatin.

The reasons underlying the poor response of patient N.N. and Z.I. are unknown. Since apo B size, as judged by SDS gel electrophoresis, was normal, we cannot exclude the possibility that mutations other than the 3500 affecting apo B binding to LDL receptor might have occurred. We are currently exploring this possibility. Whatever the mechanism could be in these subjects binding defective LDL (with or without the 3500 mutation) segregate with hypercholesterolemia and a poor response to simvastatin. With Probucol (1.0 g/die), a hypocholesterolemic drug that does not act via an induction of LDL receptor expression (Trezzi et al., 1984; Naruszewicz et al., 1984), effectively lowered total, LDL and HDL cholesterol in patient N.N. (28%, 31%, and 28% respectively).

Hypercholesterolemia may also derive from the presence of factors impeding the interaction between LDL and their receptors (Corsini et., 1986). Autoantibodies to lipoprotein components as well as to the LDL receptor (Lewis et al., 1975; Riesen and Noseda 1975; Beaumont and Beaumont, 1977; Corsini et al., 1986; Feingold et al., 1989) have been reported, and these conditions are related either to hyperlipidemia with increased concentrations of cholesterol or to the presence of xanthomas. The lipoprotein-deficient sera of two non responders to simvastatin (L.A. and P.S.) contain a factor(s) that inhibits the receptor-mediated uptake and degradation of LDL by human skin fibroblasts. Their LDL are normal in chemical composition and in their ability to interact with LDL receptors in an "in vitro" cell binding assay. The experimental protocol we used does not allow to differentiate between the presence of factors interacting with LDL or with their receptor. Preliminary data indicate that autoantibodies to apo B cannot account for the effect of the patients serum on LDL binding.

The causal relationship between serum factor(s) and hypercholesterolemia in these patients, although likely, is not demonstrated. Alternative hypotheses cannot be ruled out; for example LPDS obtained from these patients (P.S. and L.A.) may have an intrinsically reduced ability to up regulate the LDL receptor expression, that may be responsible for the reduced receptor interaction. Further studies are required to identify the factor(s) responsible for the poor interaction of LDL with their receptor.

Other patients were poor responders to simvastatin treatment. However, neither binding-defective LDL nor serum factor(s) interfering with LDL-receptor interaction were detected. Other causes may underlie the lack of response to HMG-CoA reductase inhibitors in these patients. An escape phenomenon has been described with pravastatin in FH patients (Yamamoto et al., 1988). This rebound has been attributed to an induction of HMG-CoA reductase (Brown et al., 1978; Endo et al., 1979; Reihner et al., 1990). The effect of vastatins on cell cholesterol metabolism appears to be inversely related to the ability of cells to maintain sterol synthesis by increasing HMG-CoA reductase mass. The potential for an induction of HMG-CoA reductase activity, balancing the inhibitory effect of lovastatin, may be responsible for the resistance to the cholesterol-lowering effect of lovastatin (Stossel, 1988; The Lovastatin Study Group IV, 1990). It has been reported in fact that up to 5% of patients treated with the maximal therapeutic dose of lovastatin (80 mg/day) achieve a less than 20% reduction in LDL cholesterol (Stossel, 1988; The Lovastatin Study Group IV, 1990).

In summary, we have investigated hypercholesterolemic patients experiencing only slight reductions of total and LDL cholesterol after simvastatin treatment. In 5 out of 11 of these patients binding-defective LDL and serum factor(s) interfering with the LDL-receptor interaction were identified these abnormalities could account for the lack of response.

ACKNOWLEDGMENTS

The Authors wish to thank Miss Maddalena Marazzini for typing the manuscript. This work was supported, in part, by grants from Progetto Finalizzato Invecchiamento, publication n° 923144, and Biotecnologie e Biostrumentazione (CNR), and from MURST. We also thank Merck Sharp & Dohme (Roma, Italy) for continuous support during this study.

REFERENCES

Beaumont JL, Beaumont V (1977) Autoimmune hyperlipidemia. Atherosclerosis 26: 405-418.

Bilheimer DW, Eisenberg S, Levy RI (1972) The metabolism of very low density lipoprotein proteins. I. Preliminary in vitro and in vivo observations. Biochim Biophys Acta. 260: 212-221.

Bilheimer DW, Grundy SM, Brown MS, Goldstein JL (1983) Mevinolin and colestipol stimulate receptor-mediated clearance of low density lipoprotein from plasma in familial hypercholesterolemia heterozygotes. Proc

Natl Acad Sci USA 80: 4124-4128.
Brown MS, Faust JR, Goldstein JL, Kaneko I, Endo A (1978) Induction of 3-hydroxy-3-methylglutaryl coenzyme A reductase activity in human fibroblasts incubated with compactin (ML-236B), a competitive inhibitor of the reductase. J Biol Chem 253: 1121-1128.
Brown MS, Goldstein JL (1981) Lowering plasma cholesterol by raising LDL receptors. N Engl J Med 305: 515-517.
Brown MS, Goldstein JL (1986) A receptor-mediated pathway for cholesterol homeostasis. Science 232: 34-47.
Brown MS, Goldstein JL (1987) The hyperlipoproteinemias and other disorders of lipid metabolism. In: Braunwald E, Isselbacher KJ, Petersdorf RG, Wilson JD, Martin JB, Fauci AS (eds) Harrison's Principles of Internal Medicine. McGraw-Hill, New York, 11th Edition, 1650-1661.
Brown G, Albers JJ, Fisher LD, et al (1990) Regression of coronary artery disease as a result of intensive lipid-lowering therapy in men with high levels of apolipoprotein B. N Engl J Med 323: 1289-1298.
Corsini A, Roma P, Sommariva D, Fumagalli R, Catapano AL (1986) Autoantibodies to the low density lipoprotein receptor in a subject affected by severe hypercholesterolemia. J Clin Invest 78: 940-946.
Corsini A, McCarthy BJ, Granata A, et al (1991) Familial defective apo B-100, characterization of an Italian family. Eur J Clin Invest 21: 389-397.
Endo A, Tsujita Y, Kuroda M, Tanzawa K (1979) Effects of ML-236B on cholesterol metabolism in mice and rats: lack of hypocholesterolemic activity in normal animals. Biochim Biophys Acta 575: 266-276.
Feingold KR, Castro GR, Ishikawa Y, Fielding PE, Fielding CR (1989) Cutaneous xanthoma in association with paraproteinemia in the absence of hyperlipidemia. J Clin Invest 83: 796-802.
Fisher RA, Yates F (1953) In: Oliver and Boyd (eds) Statistical Tables for Biological, Agricultural and Medical Research, 4th Edition, 60.
Frick MH, Elo O, Haapa K, et al (1987) Helsinki Heart Study: primary-prevention trial with gemfibrozil in middle-aged men with dyslipidemia N Engl J Med 317: 1237-1245.
Grundy SM (1988) HMG-CoA reductase inhibitors for treatment of hypercholesterolemia. N Engl J Med 319: 24-33.
Grundy SM, Vega GL (1990) Causes of high blood cholesterol. Circulation 81: 412-427.
Grundy SM, Vega GL, Garg A (1990) Use of 3-hydroxy-3-methylglutaryl coenzyme A reductase inhibitors in various forms of dyslipidemia. Am J Cardiol 66: 31B-38B.
Goldstein JL, Basu SK, Brown MS (1983) Receptor-mediated endocytosis of the low density lipoproteins in cultured cells. Methods Enzymol 98: 241-260.
Goldstein JL, Brown MS (1989) Familial hypercholesterolemia. In: Scriver CR, Beaudet AL, Sly WS, Valle D (eds) The Metabolic Basis of Inherited Disease. McGraw-Hill, New

York, 6th Edition, 1: 1215-1250.
Hagemenas FC, Pappu AS, Illingworth DR (1990) The effects of simvastatin on plasma lipoproteins and cholesterol homeostasis in patients with heterozygous familial hypercholesterolaemia. Eur J Clin Invest 20: 150-157.
Havel RJ, Eder HA, Bragdon JH (1955) The distribution and chemical composition of ultracentrifugally separated lipoproteins in human serum. J Clin Invest 34: 1345-1354.
Havel RJ, Hunninghake DB, Illingworth DR, et al (1987) A multicenter study of lovastatin (mevinolin) in the therapy of heterozygous familial hypercholesterolemia. Ann Intern Med 107: 609-615.
Illingworth D, Sexton G (1984) Hypocholesterolemic effects of mevinolin in patients with heterozygous familial hypercholesterolemia. J Clin Invest 74: 1972-1978.
Innerarity TL, Mahley RW (1978) Enhanced binding by cultured human fibroblasts of apo E-containing lipoproteins as compared with low density lipoproteins. Biochemistry 17: 1440-1447.
Innerarity TL, Weisgraber KH, Arnold KS, et al (1987) Familial defective apolipoprotein B-100: low density lipoproteins with abnormal receptor binding. Proc Natl Acad Sci USA 84: 1650-1661.
Innerarity TL, Mahley RW, Weisgraber KH, et al (1990) Familial defective apolipoprotein B-100: a mutation of apolipoprotein B that causes hypercholesterolemia. J Lipid Res 31: 1337-1349.
Kane JP, Malloy MJ, Ports TA, Phillips NR, Diehl JC, Havel RJ (1990) Regression of coronary atherosclerosis during treatment of familial hypercholesterolemia with combined drug regimens. JAMA 264: 3007-3012.
Kleinman Y, Eisenberg S, Oschry Y, Gavish D, Stein O, Stein Y (1985) Defective metabolism of hypertriglyceridemic low density lipoprotein in cultured human skin fibroblasts. J Clin Invest 75: 1796-1803.
Laemmli UK (1970) Cleavage of structural proteins during the assembly of the head of bacteriophage T4. Nature 227: 680.
Lewis LA, De Wolfe VG, Butkus A, Page IH (1975) Autoimmune hyperlipidemia in a patient. Am J Med 59: 208-218.
Lipid Research Clinics Program (1984) Reduction in incidence of coronary heart disease. JAMA 251: 31-364.
Lowering Blood Cholesterol to Prevent Heart Disease. Consensus Conference (1985) JAMA 253: 2080-2086.
Naruszewicz M, Carew TE, Pittman RC, Witztum JL, Steinberg D (1984) A novel mechanism by which probucol lowers low density lipoprotein levels demonstrated in the LDL receptor-deficient rabbit. J Lipid Res 25: 1206-1213.
Raveh D, Israeli A, Arnon R, Eisenberg S (1990) Effects of lovastatin therapy on LDL receptor activity in circulating monocytes and on structure and composition of plasma lipoproteins. Atherosclerosis 82: 19-26.
Reihner E, Rudling M, Stahlberg D, et al (1990) Influence of pravastatin, a specific inhibitor of HMG-CoA reductase, on hepatic metabolism of cholesterol. N

Engl J Med 323: 224-228.
Riesen W, Noseda G (1975) Antibodies against lipoproteins in man, occurence and biological significance. Klin Wochenschr 53: 353-361.
Saiki RK, Gelfand DH, Stoffel S, et al (1988) Primer-directed enzymatic amplification of DNA with a thermostable DNA polymerase. Science 239: 487-491.
Soria LF, Ludwig EH, Clarke HRG, Vega GL, Grundy SM, McCarthy BJ (1989) Association between a specific apolipoprotein B mutation and familial defective apolipoprotein B-100. Proc Natl Acad Sci USA 86: 587-591.
Stamler J (1979) Population studies. In: Levy R, Rifkind B, Dennis B, Ernst N (eds) Nutrition, Lipids, and Coronary Heart Disease. Raven Press, New York, 25-88
Stossel TP (1988) A multicenter comparison of lovastatin and cholestyramine therapy for severe primary hypercholesterolemia. JAMA 260: 359-366.
The Lovastatin Study Group IV (1990) A multicenter comparison of lovastatin and probucol for treatment of severe primary hypercholesterolemia. Am J Cardiol 66: 22B-30B.
Trezzi E, Roma P, Bernini F, Fumagalli R, Catapano AL (1984) Effects of probucol on the in vivo plasma clearance of human low density lipoprotein receptors in vitro. Atherosclerosis 52: 309-316.
Vega GL, Grundy SM (1986) In vivo evidence for reduced binding of low density lipoproteins to receptors as a cause of primary moderate hypercholesterolemia. J Clin Invest 78: 1410-1414.
Yamamoto A, Yokoyama S, Yamamura T (1988) Escape phenomenon occurs by lowering cholesterol with a hydroxymethylglutaryl coenzyme A (HMG-CoA) reductase inhibitor in patients with familial hypercholesterolemia. Atherosclerosis 71: 257-260.

MOLECULAR GENETICS OF THE FAMILIAL CHYLOMICRONEMIA SYNDROME

Silvia Santamarina-Fojo
Molecular Disease Branch
National Heart, Lung and Blood Institute
National Institutes of Health
9000 Rockville Pike, Bldg. 10/7N117
Bethesda, MD 20892

Familial chylomicronemia, a rare genetic syndrome inherited as an autosomal recessive trait, is characterized by severe fasting hypertriglyceridemia and massive accumulations of chylomicrons in plasma (Brunzell 1989, Santamarina-Fojo 1991). Affected individuals often present early in childhood with recurrent episodes of abdominal pain and/or pancreatitis frequently precipitated by the ingestion of a fatty meal. Hepatosplenomegaly and mild elevations in liver transaminases may result from lipid accumulation in these organs. Phagocytosis of chylomicrons by macrophages in the skin results in the formation of eruptive xanthomas, and on funduscopic examination, lipemia retinalis may be evident in patients with triglyceride levels greater than 1000 mg/dl. The major morbidity associated with this disorder is recurrent episodes of pancreatitis, which in some patients have resulted in pancreatic insufficiency. The two major molecular defects in the familial chylomicronemia syndrome are a deficiency of lipoprotein lipase (LPL), or of apolipoprotein (apo) C-II. Lipoprotein lipase, a glycoprotein located on the luminal surface of capillary vessels, catalyzes the hydrolysis of triglycerides present in circulating chylomicrons and very low density lipoproteins (VLDL) to mono- and di-glycerides and to free fatty acids. ApoC-II, a 79 amino acid protein present in chylomicrons, VLDL, and HDL, serves as a cofactor for LPL and is essential for normal triglyceride hydrolysis.

Various kindreds with familial chylomicronemia due to apoC-II deficiency have been described (Breckenridge 1978, Santamarina-Fojo 1991). The underlying molecular defects that lead to a deficiency of apoC-II have been well characterized in several of these families. Figure 1 summarizes the mutations identified in the apoC-II gene of patients presenting with the familial chylomicronemia syndrome.

NATO ASI Series, Vol. H 73
Human Apolipoprotein Mutants III
Edited by C. R. Sirtori, G. Franceschini, B.H. Brewer Jr.

A.

FAMILY	MUTATION/RESIDUE	RESTRICTION ENZYMES	PLASMA ApoC-II
Hamburg	G → C / intron 2 donor splice defect	Dde I / Hph I	tr
Nijmegen	G deletion / frameshift / val_{18} → Stop	Hph I	–
Padova	C → A / tyr_{37} → Stop	Rsa I	tr
Bari	C → G / tyr_{37} → Stop	Rsa I	–
$Paris_1$	A → G / met_{-22} → val	–	–
$Paris_2$	C → T / arg_{-19} → Stop	Nla III	–
Toronto	T deletion / frameshift / leu_{75} → Stop	NR	+
St. Michael	NR / frameshift with extension to 96 aa	NR	+
Venezuela	C deletion / –82bp T → G / –28bp C deletion / gln_2 → Stop AT deletion / polyadenylation	NR	–
Japan	C deletion / gln_2 → Stop T → C / ser_{21} → pro C → T / gln_{70} → Stop A → T / glu_{79} → val C → T / 3' untranslated	NR	–

NR-Not reported
tr-trace

B.

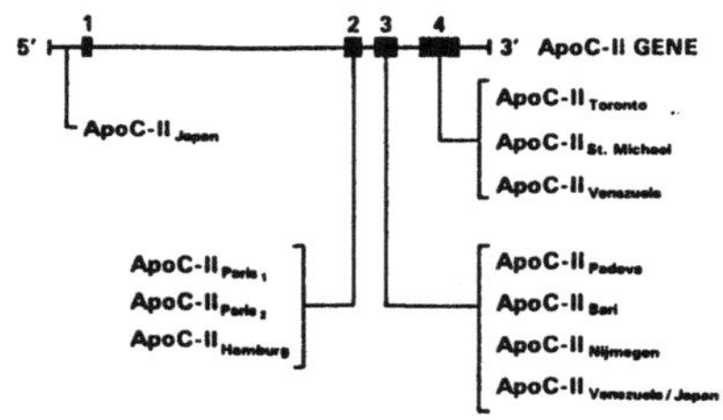

Figure 1. **Panel A summarizes the molecular defects identified in the apoC-II gene. The location of these mutations in the apoC-II gene is indicated in panel B. NR, not reported; tr, trace.**

Most of the defects identified to date are single point mutations that result in reduced expression of the apolipoprotein. Thus, the majority of individuals affected with this disorder have markedly reduced or totally absent plasma apoC-II as determined by immunoblotting and/or immunoassay. Included in this group are patients with molecular defects that lead to the introduction of a premature stop codon (apoCII$_{Nijmegen}$, apoC-II$_{Padova}$, apoC-II$_{Bari}$ and apoC-II$_{Paris2}$) (Fojo 1988a, Fojo 1989a, Crecchio 1990, Parrott 1992), the substitution of the initiation methionine codon (apoC-II$_{Paris1}$) (Fojo 1989b) and a donor splice site mutation (apoC-II$_{Hamburg}$) (Fojo 1988b). In apoC-II$_{Toronto}$ and apoC-II$_{St. Michael}$, frameshift mutations result in significant plasma expression of mutant, non-functional apolipoproteins (Connelly 1987a, Connelly 1987b). As expected, all these mutations result in the inability to synthesize and/or express a functional apoC-II. The defects in the Japanese and Venezuelan kindreds are unique in that four to five different mutations were identified in the apoC-II gene of the homozygous probands, resulting in a deficiency of apoC-II (Xiong 1991). With the exception of these two kindreds, there is no evidence for a founder gene effect in apoC-II deficiency.

A.

MUTANT LPL	MUTATION/RESIDUE	RESTRICTION ENZYMES	PLASMA LPL
LPL_{61}	T → A / tyr → Stop	NR	NR
LPL_{86}	NR / trp → arg	NR	+
LPL_{102}	pentanucleotide insertion at position 102 in exon 5/frameshift with extension to 146 a.a.	NR	+
LPL_{106}	C → T / gln → Stop	NR	NR
LPL_{136}	NR / his → arg	NR	+
LPL_{142}	G → A / gly → glu	NR	+
LPL_{156}	A → G / asp → gly G → A / asp → asn	Taq I Taq I	NR NR
LPL_{157}	C → G / pro → arg	Pvu II	+
LPL_{176}	G → A / ala → thr	SfaN I	+
LPL_{188}	G → A / gly → glu	Sau96 I Ava II	+
LPL_{194}	T → C / ile → thr	–	+
LPL_{204}	C → G / asp → glu	Hinc II	+
LPL_{207}	C → T / pro → leu	–	+
LPL_{216}	T → A / cys → ser	NR	NR
LPL_{224}	G deletion/frameshift/ val_{224} → Stop	NR	NR
LPL_{243}	G → A / arg → his	Hae I Hha I Eco47 III	+
LPL_{244}	T → A / ser → thr	NR	NR
LPL_{250}	NR / asp → asn	NR	NR
LPL_{262}	C → A / tyr → Stop T → C / tyr → his	NR Asu I	NR –
LPL_{382}	G → A / trp → Stop	Mbo II	+
LPL_{447}	C → G / ser → Stop	NR	+
Intron 2 acceptor splice site defect	C → A / –	NR	NR
Intron 2 donor splice site defect	G → A / –	–	–
2 Kb insertion	partial duplication of exon 6	many	–
3 Kb deletion	deletion of exon 3	many	NR
6 Kb deletion	deletion of exons 3-5	many	–

NR-Not reported

B.

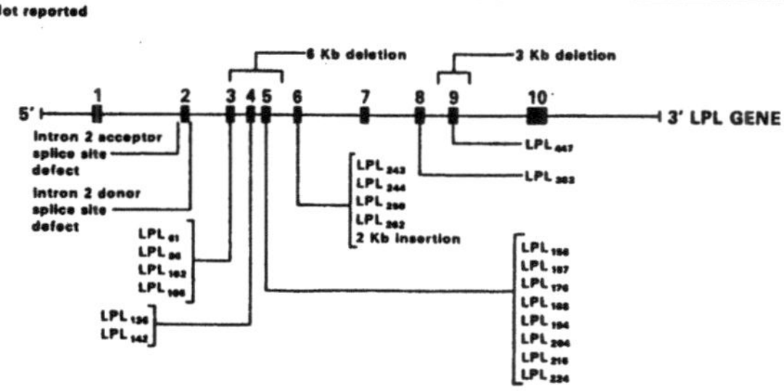

Figure 2. Panel A summarizes the molecular defects identified in the LPL gene. The location of these mutations in the LPL gene is indicated in panel B. NR, not reported.

The majority of patients presenting with the familial chylomicronemia syndrome have LPL deficiency, a relatively common disorder with a frequency of 1:5000-1:10000 in some populations (Gagne 1989). A large number of mutations in the LPL gene that lead to the expression of familial chylomicronemia syndrome have been identified and are summarized in Figure 2. Three major gene rearrangements involving a 2 kb, 3 kb, and 6 kb insertion (Langlois 1989, Benlian 1991, Devlin 1990), as well as several splice mutations (Hata 1990 and Gotoda 1991) and mutations resulting in the introduction of a premature stop codon (61, 102, 106, 224, 262, 382 and 447) have been described (Gotoda 1991, Henderson 1990, Emi 1990, Deeb 1991a, Funke 1990 and Hata 1990). These defects result in markedly reduced expression or absolute deficiency of post-heparin plasma LPL. In contrast, many missense mutations resulting in the synthesis and expression of non-functional LPL variants have been reported. These include LPL mutants 86, 136, 142, 156, 157, 176, 188, 194, 204, 207, 216, 243, 244, and 250 (Gotoda 1991, Deeb 1991, Ishimuro-Oka 1991, Ameis 1990, Faustinella 1991, Ma 1991, Bruin 1990, Beg 1990, Paulweber 1991, Emi 1990, Monsalve 1990, Dichek 1991, Henderson 1991, Ma 1991 and Lohse 1991). The exact nature and location of the defects in the LPL gene are summarized in Figure 2. Many of these mutations lead to the substitution of hydrophobic by more polar residues within the central homology region of LPL. Mutations altering residues 188 (Monsalve 1990) and 207 (Ma 1991) as well as the 2 kb insertion (Langlois 1989, Devlin 1990) account for a high percentage of LPL defects in certain populations, suggesting the presence of a founder gene.

Defects in LPL and apoC-II have been clearly implicated as the underlying pathophysiologic mechanism leading to chylomicronemia syndrome (Santamarina-Fojo 1991). The role that LPL and apoC-II may play in the development of other dyslipoproteinemias, however, remains unclear. Recent identification and characterization of the underlying molecular defects that lead to a deficiency of LPL and apoC-II in patients with chylomicronemia have provided new insights into the potential contribution of these two proteins in the development of other dyslipidemias. In heterozygotes for LPL gene defects, for example, carrier status for the LPL^{188} mutation was significantly associated with a form of familial hypertriglyceridemia characterized by increased plasma triglyceride, VLDL, and apoB and decreased LDL and HDL concentrations (Paulweber 1991, Wilson 1990). The LPL^{207} allele was detected in 11 out of 180 (6.1%) hyperlipidemic patients and 0 out of 170 (0%) of normolipidemic control patients in French Canadians (Ma 1991). Both the LPL^{188} and LPL^{207} mutants have been identified as a frequent cause of LPL deficiency in different populations (Monsalve 1990, Ma 1991). In addition, Brunzell and co-workers (Babirak 1989) have suggested that the heterozygous form of LPL deficiency as identified by post-heparin LPL plasma levels might predispose an individual to the development of familial combined hyperlipidemia.

Similar findings have been described in individuals heterozygous for two different variants of apoC-II. ApoC-II_{Gln55}, a functional apoC-II mutant, is present in 12% of Americans of African ancestry (Menzel 1986) and was found at least in heterozygous Nigerian blacks to be associated with lower total and LDL cholesterol levels as well increased fasting triglyceride levels (Sepehrnia 1989) but not with the development of chylomicronemia. In this study, the authors reported an interaction between the apoE and apoC-II loci modulating total cholesterol and LDL cholesterol levels, suggesting a potential role of the apoC-II locus in lipid disorders. Interestingly, a similar digenic interaction between the non-functional variant apoC-$II_{Toronto}$ and apoE-4 modulating plasma triglyceride levels has been proposed to account for the association between hypertriglyceridemia and the carrier status for both defects by Hegele et al. (Hegele 1991).

Although analysis of these kindreds has provided increasing evidence for a potential association between the heterozygous state for LPL and/or apoC-II deficiency and dyslipidemias other than familial chylomicronemia, the role of LPL and apoC-II gene mutations in the development of various genetic lipid disorders, and perhaps even premature cardiovascular disease remains to be established.

REFERENCES

Ameis D, Kobayashi J, Davis RC, Ben-Zeev O, Malloy MJ, Kane JP, Lee G, Wong H, Greten H, Havel RJ, Schotz MC (1990) Familial hyperchylomicronemia associated with a missense mutation in the lipoprotein lipase gene. *J Clin Invest* 87:1165--1170

Babirak SP, Iverius PH, Fujimoto WY, Brunzell JD (1989) Detection and characterization of the heterozygote state for lipoprotein lipase deficiency. *Arteriosclerosis* 9:326--334

Beg OU, Meng MS, Skarlatos SI, Previato L, Brunzell JD, Brewer HB, Jr, Fojo SS (1990) Lipoprotein lipase Bethesda: a single amino acid substitution (Ala-176-->Thr) leads to abnormal heparin binding and loss of enzymic activity. *Proc Natl Acad Sci USA* 87:3474--3478

Benlian P, Loux N, De Gennes JL (1991) A homozygous deletion of exon 9 in the lipoprotein lipase gene causes type I hyperlipoproteinemia. *Arteriosclerosis and Thrombosis* 11:1465a

Breckenridge WC, Little JA, Steiner G, Chow A, Poapst M (1978) Hypertriglyceridemia associated with deficiency of apolipoprotein C-II. *N Engl J Med* 298:1265--1273

Bruin T, Kastelein JJP, Monsalve MU (1990) A novel missense mutation in the gene for lipoprotein lipase (LPL) underlying severe hypercholesterolemia in a kindred of Dutch descent. In: *Control of Blood Cholesterol* 55th Meeting of the European Atherosclerosis Society pp. 91A

Brunzell JD (1989) Familial lipoprotein lipase deficiency and other causes of the chylomicronemia syndrome. In: Scriver CR, Beaudet AL, Sly WS, Valle D, eds. The Metabolic Basis of Inherited Disease. 6th ed. New York, NY:McGraw-Hill International Book Co. 1165-1180

Connelly PW, Maguire GF, Little JA (1987) Apolipoprotein $CII_{St.\ Michael}$. Familial apolipoprotein CII deficiency associated with premature vascular disease. *J Clin Invest* 80:1597-1606

Connelly PW, Maguire GF, Hofmann T, Little JA (1987) Structure of apolipoprotein $C\text{-}II_{Toronto}$, a nonfunctional human apolipoprotein. *Proc Natl Acad Sci USA* 84-270-273

Crecchio C, Capurso A, Pepe G (1990) Identification of the mutation responsible for a case of plasmatic apolipoprotein CII deficiency (Apo CII_{Bari}). *Biochem Biophys Res Commun* 168:1118--1127

Deeb SS, Reina M, Peterson J, Takata K, Kajiyama G, Brunzell JD (1991) Gene mutations in patients with lipoprotein lipase deficiency. *Arteriosclerosis and Thrombosis* 11:1418a

Devlin RH, Deeb S, Brunzell J, Hayden MR (1990) Partial gene duplication involving exon-Alu interchange results in lipoprotein lipase deficiency. *Am J Hum Genet* 46:112--119

Dichek H, Fojo SS, Beg OU, Skarlatos SI, Brunzell JC, Cutler GB, Jr., Brewer HB, Jr. Identification of two separate allelic mutations in the LPL gene of an LPL deficient patient. *J Biol Chem* 266:473--477

Emi M, Wilson DE, Iverius PH, Wu L, Hata A, Hegele R, Williams RR, Lalouel JM (1990) Missense mutation (Gly-->Glu188) of human lipoprotein lipase imparting functional deficiency. *J Biol Chem* 265:5910--5916

Emi M, Hata A, Robertson M, Iverius PH, Hegele R, Lalouel JM (1990) Lipoprotein lipase deficiency resulting from a nonsense mutation in exon 3 of the lipoprotein lipase gene. *Am J Hum Genet* 47:107--111

Fojo SS, De Gennes JL, Chapman J, Parrott C, Lohse P, Kwan SS, Truffert J, Brewer HB, Jr (1989b) An initiation codon mutation in the apoC-II gene (apoC-II_{Paris}) of a patient with a deficiency of apolipoprotein C-II. *J Biol Chem* 264:20839--20842

Fojo SS, Stalenhoef AF, Marr K, Gregg RE, Ross RS, Brewer HB,Jr (1988a) A deletion mutation in the ApoC-II gene (ApoC-$II_{Nijmegen}$) of a patient with a deficiency of apolipoprotein C-II. *J Biol Chem* 263:17913--17916

Fojo SS, Lohse P, Parrott C, Baggio G, Gabelli C, Thomas F, Hoffman J, Brewer HB, Jr (1989a) A nonsense mutation in the apolipoprotein C-II_{Padova} gene in a patient with apolipoprotein C-II deficiency. *J Clin Invest* 84:1215--1219

Fojo SS, Beisiegel U, Beil U, Higuchi K, Bojanovski M, Gregg RE, Greten H, Brewer HB, Jr (1988b) Donor splice site mutation in the apolipoprotein (Apo) C-II gene (Apo C-$II_{Hamburg}$) of a patient with Apo C-II deficiency. *J Clin Invest* 82:1489--1494

Funke H, Wiebusch H, Paulweber B, Assmann G (1990) Identification of the molecular defect in a patient with type I hyperlipoproteinemia. *Arteriosclerosis* 10:830a.

Faustinella F, Chang A, Van Biervlet JP (1991) Catalytic triad residue mutation (Asp156-->Gly) causing familial lipoprotein lipase deficiency. Co-inheritance with a nonsense mutation (Ser447-->Ter) in a Turkish family. *J Biol Chem* 266:14418--14424

Gagne C, Brum LD, Julien P, Moorjani S, Lupien PJ (1989) Primary lipoprotein lipase activity deficiency: clinical investigation of a French Canadian population. *Can Med Assoc J* 140:405--411

Gotoda T, Yamada N, Kawamura M, Kozaki K, Mori N, Ishibashi S, Shimano H, Takaku F, Yazaki Y, Furuichi Y, Murase T (1991) Heterogeneous mutations in the human lipoprotein lipase gene in patients with familial lipoprotein lipase deficiency. *J Clin Invest* 88:1856--1864

Hata A, Robertson M, Emi M, Lalouel JM (1990) Direct detection and automated sequencing of individual alleles after electrophoretic strand separation: identification of a common nonsense mutation in exon 9 of the human lipoprotein lipase gene. *Nucleic Acids Res* 18:407--411

Hata A, Mitsuru E, Luc G, Basdevant A, Gambert P, Iverius P-H, Lalouel J-M (1990) Compound heterozygote for lipoprotein lipase deficiency Ser->Thr244 and transition in 3 splice site intron 2 (AG->AA) in the lipoprotein lipase gene. *Am J Hum Genet* 47:721--726

Hegele RA, Breckenridge WC, Cox DW, Maguire GF, Little JA, Connelly PW (1991) Interaction between variant apolipoproteins C-II and E that affects plasma lipoprotein concentrations. *Arteriosclerosis and Thrombosis* 11:1303--1309

Henderson HE, Ma Y, Hassan MF, Monsaive MV, Marais AD, Winkler F, Gubernator K, Peterson J, Brunzell JD, Hayden MR (1991) Amino acid substitution (Ile194-->Thr) in exon 5 of the lipoprotein lipase gene causes lipoprotein lipase deficiency in three unrelated probands. *J Clin Invest* 87:2005--2011

Henderson HE, Devlin R, Peterson J, Brunzell JD, Hayden MR (1990) Frameshift mutation in exon 3 of the lipoprotein lipase gene causes a premature stop codon and lipoprotein lipase deficiency. *Mol Biol Med* 7:511--517

Ishimura-Oka K, Semenkovich CF, Faustinella F (1991) Identification of compound heterozygotes for lipoprotein lipase deficiency in three unrelated families. Arteriosclerosis and Thrombosis 11:1415a

Langlois S, Deeb S, Brunzell JD, Kastelein JJ, Hayden MR (1989) A major insertion accounts for a significant proportion of mutations underlying human lipoprotein lipase deficiency. *Proc Natl Acad Sci USA* 86:948--952

Lohse P, Beg OU, Brunzell JD, Santamarina-Fojo S, Brewer HB, Jr (1991) Familial chylomicronemia: identification of a unique patient homozygote for two separate mutations in the LPL gene. *Arteriosclerosis and Thrombosis* 1:1415a

Ma Y, Bruin T, Tuzgol S, Wilson BI, Roederer G, Liu M-S, Davignon J, Kastelein JJP, Brunzell JD, Hayden MR (1992) Two naturally occurring mutation at the first and second bases of codon aspartic acid 156 in the proposed catalytic triad of human lipoprotein lipase. *J Biol Chem* 267:1918--1923

Ma Y, Henderson HE, Ven Murthy MR, Roederer G, Monsalve MV, Clarke LA, Normand T, Julien P, Gagne C, Lambert M, Davignon J, Lupien PJ, Brunzell J, Hayden MR (1991) A mutation in the human lipoprotein lipase gene as the most common cause of familial chylomicronemia in French Canadians. *N Engl J Med* 324:1761--1766

Menzel HJ, Kane JP, Malloy MJ, Havel RJ (1986) A variant primary structure of apolipoprotein C-II in individuals of African descent. *J Clin Invest* 77:595--601

Monsalve MV, Henderson HE, Roederer G, Julien P, Deeb S, Kastelein JJ, Peritz L, Devlin R, Bruin T, Murthy MRV, Gagne C, Davignon J, Lupien PJ, Brunzell JD, Hayden MR (1990) A missense mutation at codon 188 of the human lipoprotein lipase gene is a frequent cause of lipoprotein lipase deficiency in persons of different ancestries. *J Clin Invest* 86:728--734

Parrott CL, Alsayed N, Rebourcet R, Santamarina-Fojo S (1991) ApoC-II_{Paris2}: a premature termination mutation in the signal peptide of apoC-II resulting in the familial chylomicronemia syndrome. *J Lipid Res* 33:361-367

Paulweber B, Wiebusch H, Miesenboeck G, Funke H, Assmann G, Hoelzl B, Sippl MJ, Friedl W, Patsch JR, Sandhofer F (1991) Molecular basis of lipoprotein lipase deficiency in two Austrian families with type I hyperlipoproteinemia. *Atherosclerosis* 86:239--250

Santamarina-Fojo S, Brewer HB, Jr (1991) The familial hyperchylomicronemia syndrome. New insights into underlying genetic defects. *JAMA* 265:904--908

Sepehrnia B, Kamboh MI, Adams-Campbell LL, Bunker CH, Nwankwo M, Majumder PP, Ferrell RE (1989) Genetic studies of human apolipoproteins. XI. The effect of the apolipoprotein C-II polymorphism on lipoprotein levels in Nigerian blacks. *J Lipid Res* 30:1349--1355

Wilson DE, Emi M, Iverius P-H, Hata A, Wu LL, Hillas E, Williams RR, Lalouel J-M (1990) Phenotypic expression of heterozygous lipoprotein lipase deficiency in the extended pedigree of a proband homozygous for a missense mutation. *J Clin Invest* 86:735--750

Xiong WJ, Li W-H, Posner I, Yamamura T, Yamamoto A, Gotto AM,Jr, Chang L (1991) No severe bottleneck during human evolution: evidence from two apolipoprotein C-II deficiency alleles. *Am J Hum Genet* 48:383--389

Rare mutations in the APOE gene associated with dominant mode of inheritance of Familial Dysbetalipoproteinemia (FD).

Louis M. Havekes, Monique Mulder, Peter de Knijff, Arn van den Maagdenberg and Rune R. Frants.

Gaubius Laboratory IVVO-TNO, P.O. Box 430, 2300 AK Leiden and Department of Human Genetics, P.O. Box 9503, 2300 RA Leiden, The Netherlands.

Introduction

Apolipoprotein E (apoE) is a major constituent of chylomicron- and VLDL-remnants. It plays a major role in the clearance of these lipoproteins as it serves as a ligand for recognition by hepatic lipoprotein receptors. With isoelectric focusing (IEF) apoE can be separated into three commonly occuring isoforms e.g. E2, E3 and E4.

Patients with familial dysbetalipoproteinemia (FD) show an increased plasma level of atherogenic chylomicron- and VLDL-remnants. The increased level of these lipoproteins leads to an enhanced risk of premature atherosclerosis and is supposed to be due to both an increased synthesis and a decreased clearance.

FD with E2E2 phenotype

Most FD patients show homozygosity for the apoE2 variant. The major underlying metabolic defect in these patients is an impaired clearance of the remnant particles, due to a defective binding of apoE2 to the hepatic lipoprotein receptors.

NATO ASI Series, Vol. H 73
Human Apolipoprotein Mutants III
Edited by C. R. Sirtori, G. Franceschini, B.H. Brewer Jr.

In the general population the frequency of E2E2 homozygosity is about 1%. However, only about 4% of all E2E2 homozygotes will eventually develop FD, mostly after the third decade in men and after the menopause in women. This implies that FD is a recessively inherited multifactorial disease. In addition to E2E2 homozygosity, additional genetical and/or environmental factors are required for the expression of the disease. It is commonly assumed that these additional factors are somehow related to the synthesis of chylomicrons and VLDL like nutrition, diabetes, obesity and hormonal status (1). We addressed the question whether this multifactorial character was (partly) due to additional heterogeneity in the E*2 allele.

To investigate whether additional heterogeneity in the APOE gene is responsible for this reduced penetrance of FD, we evaluated the exact mutation in both E*2 alleles of 60 E2E2 homozygous FD patients. By using a PCR-based RFLP analysis (with restriction enzyme HhaI) (2) we were able to show that all FD patients are homozygous for the common E*2(arg158→cys) mutation. This indicates that factors not related to the E*2 allele determine the expression of FD in E2E2 homozygotes.

We measured the binding efficiency to the LDL receptor of HepG2 cells of the lipoprotein fraction with density d<1.019 g/ml isolated from E2E2 homozygous FD patients before and after treatment with Gemfibrozil. Treatment with Gemfibrozil normalizes VLDL and IDL cholesterol levels in these patients. However, this normalization is not accompanied with an improved binding of the d<1.019 lipoprotein fraction to the LDL receptor. The absence of an effect of treatment with Gemfibrozil on the binding efficiency is also reflected by the absence of an effect of Gemfibrozil treatment on the plasma level of apoE. We hypothesize that normal VLDL and IDL cholesterol levels in E2E2 homozygotes reflect a reduced production rate of VLDL particles rather than an improved binding efficiency of the d<1.019 lipoprotein fraction to the LDL receptor.

FD with E3E2 phenotypes

Besides to 60 FD patients with E2E2 homozygosity we found 6 FD probands with the E3E2 phenotype. DNA sequence analyses revealed that all these probands carry the E2(lys146→gln) mutation as described before (3). In addition, family analyses showed that almost all E*2(lys146→gln) allele carriers express FD, although with a variable degree of severity. Thus, in contrast to the common E*2(arg158→cys) allele, the E*2(lys146→gln) allele is associated with a dominantly inherited form of FD with high penetrance (4). The exact mechanism of the dominant behaviour of the E2(lys146→gln) variant in expression of FD is still subject to speculation. However, we obtained some evidence that the $d < 1.019$ lipoprotein fraction from E*2(lys146→gln) allele carriers is defective in lipolysis by bovine milk lipoprotein lipase. As VLDL lipoproteins become more efficient in binding to the LDL receptor after treatment with lipoprotein lipase, this relative defect as substrate for lipoprotein lipase could explain the relative defect of E2(lys146→gln) containing $d < 1.019$ lipoprotein fraction in its receptor-mediated clearance.

Although this variant has originally been described by Rall et al (3) in two sibs, we suggest this variant to be a Dutch variant because of the number of probands found in the Dutch population. This hypothesis is further sustained by the finding of Dr. Marais in Cape Town, South Africa. He found a severe FD patient with heterozygosity for the E*2(lys146→gln) allele and originating from the Dutch speaking part of the Belgium population which is genetically closely related to the Dutch population. Genealogical studies are currently being performed to evaluate whether these E*2(lys146→gln) allele carrying FD patients share common ancestry.

FD with E3E3 phenotypes

Six FD patients have been found with the E3E3 phenotype. These patients exhibit heterozygosity for the E*3-Leiden allele which is characterized by a partial gene duplication of codons 120-126 (5,6). For five probands family analyses have been carried out. Genealogical studies revealed that these five probands share common ancestry in the second half of the 17th century (7).

We collected blood from 128 family members, including 42 E*3-Leiden allele carriers. All E*3-Leiden allele carriers show characteristics of FD like increased levels of cholesterol in the d<1.019 lipoprotein fraction, increased ratios of cholesterol over triglyceride in this fraction and strongly increased levels of plasma apoE. Although all E*3-Leiden allele carriers express FD there is a variable degree of severity. Multiple linear regression analysis revealed that this variability in expression of FD can be explained by age or body mass index. Gender did not show any significant additional influence on the expression of FD nor did it influence the age of onset of FD. This is in strong contrast with FD with E2E2 homozygosity.

Subdividing the group of E*3-Leiden allele carriers according to the second, common APOE allele showed that the E*2 allele significantly enhanced the expression of FD, whereas the E*4 allele had the opposite effect. However, irrespective of the effect of the common second APOE allele on the severeness of FD, heterozygosity for the APOE*3-Leiden allele was absolutely associated with the expression of this disease. We conclude that in APOE*3-Leiden allele carriers FD is a dominantly inherited monofactorial disease, i.e., the presence of normally functioning apoE molecules in the plasma does not prevent the age- or body mass index-related expression of this disease.

By using quantitative western blotting after isoelectric focusing of samples with and without cysteamine treatment, we were able to measure the protein ratio E3-Leiden / common E variant. In complete plasma the mean ratio E3-Leiden / common E was about 3:1. In the HDL density fraction this ratio was about 1:1, whereas the E3-Leiden isoform largely predominates in the VLDL-IDL density fraction (mean ratio is 7:1). This large predominance of E3-Leiden in the VLDL-IDL fraction renders these lipoproteins into "apparent homozygous" E3-Leiden particles and could therefore be the mechanism underlying the dominant behaviour of E3-Leiden in the expression of FD.

Screening for new apoE variants

Using a combination of isoelectric focusing (8) and genotyping with the allele-specific restriction enzymes like HhaI (2) enables screening for the presence of new mutations in the APOE gene. By comparison the results of both methods, typing discrepancies will indicate the presence of new mutations in the APOE gene. However, for screening purposes this combination of techniques is not 100% informative and too laborious. For screening 120 hyperlipoproteinemic and normolipidemic individuals for the presence of new apoE variants, we introduced the technique of denaturing gradient gel electrophoresis (DGGE,9). With this technique we were able to screen for the presence of mutations in pre-determined parts of exon 4 of the APOE gene. Subsequently, we used the direct sequencing method to determine the exact nature of these four mutations: E2(val236→glu), E1(leu252→glu), E4(cys112→arg; arg274→his) and E4(ser296→arg). Preliminary family analyses revealed no cosegregation of FD or any other type of hyperlipoproteinemia with any of these new variants. At present we can not rule out the possibility that mutations in the APOE gene can cause phenotypes different from FD. The method of DGGE offers an efficient method for screening large numbers of individuals with respect to the occurance of new mutations in predetermined parts of the APOE gene.

Conclusion

FD is a genetically heterogeneous disease. In case of E2E2 homozygosity, FD is recessively inherited, whereas additional factors not related to the APOE gene are required for the expression (low penetrance or multifactorial). In case of heterozygosity for rare apoE variants like E2(lys146→gln) and E3-Leiden, the disease is dominantly inherited with a high or almost absolute rate of penetrance. By using DGGE techniques we found four new mutants in the APOE gene that were not clearly associated with the expression of any form of hyperlipoproteinemia.

References

1 Mahley RW, and Rall SC. 1989. Type III hyperlipoproteinemia (dysbetalipoproteinemia): the role of the apolipoprotein E in normal and abnormal lipoprotein metabolism. In The Metabolic Basis of Inherited Disease 6th edition. Scriver CR, Beaudet AL, Sly WS, and Valle D, editors. McGraw-Hill Book Co., Inc., New York. 1195-1213.

2 Hixson SA, Vernier DT. 1990. Restriction isotyping of human apolipoprotein E by gene amplification and cleavage with HhaI. J. Lipid Res. 31: 545-548.

3 Rall SC, Weisgraber KH, Innerarity TL, Bersot TP, Mahley RW, and Blum CB. 1983. Identification of a new structural variant of human apolipoprotein E, E2(Lys146→Gln), in a type III hyperlipoproteinemic subject with the E3/2 phenotype. J. Clin. Invest. 72: 1288-1297.

4 Smit M, de Knijff P, van der Kooij-Meijs, Groenendijk C, van den Maagdenberg AMJM, Gevers Leuven JA, Stalenhoef AFH, Stuyt PMJ, Frants RR, and Havekes LM. 1990. Genetic heterogeneity in familial dysbetalipoproteinemia: the E2(lys146→gln) variant results in a dominant mode of inheritance. J. Lipid Res. 31: 45-53.

5 Wardell MR, Weisgraber KH, Havekes LM, and Rall SC. 1989. Apolipoprotein E3-Leiden contains a seven-amino acid insertion that is a tandem repeat of residues 121-127. J. Bioil. Chem. 264: 21205-21210.

6 Van den Maagdenberg AMJM, de Knijff P, Stalenhoef AFH, Gevers Leuven JA, Havekes LM, and Frants RR. 1989. Apolipoprotein E*3-Leiden allele results from a partial gene duplication in exon 4. Biochem. Biophys. Res. Commun. 165: 851-857.

7 De Knijff P, van den Maagdenberg AMJM, Stalenhoef AFH, Gevers Leuven JA, Demacker PNM, Kuyt LP, Frants RR, and Havekes LM. 1991. Familial dysbetalipoproteinemia associated with apolipoprotein E3-Leiden in an extended multigeneration pedigree. J. Clin. Invest. 88: 643.

8 Havekes LM, de Knijff P, Beisiegel U, Havinga J, Smit M, and Klasen E. 1987. A rapid micro-method for apolipoprotein E phenotyping directly in serum. J. Lipid Res. 28: 445-463.

9 Sheffield VC, Cox DR, Lerman LS, and Myers RM. 1989. Attachment of a 40-base-pair G+C-rich sequence (GC-clamp) to genomic DNA fragments by the polymerase chain reaction results in improved detection of singl-base changes. Proc. Natl. Acad. Sci. USA 86: 232-236.

IDENTIFICATION OF COMMON GENETIC POLYMORPHISMS THAT DETERMINE PLASMA LEVELS OF APOAI AND HDL-C

Steve Humphries, Vilmundur Gudnason, Hiroko Paul-Hayase, N. Saha#, Maryvonne Rosseneu*.

Centre for the Genetics of Cardiovascular Disorders, Rayne Institute, University College Hospital, London WC1E 6JJ. * Clinical Chemistry, A-Z St Jan Belgium. # Department of physiology, Faculty of Medicine, National University of Singapore, Kent Ridge, Singapore.

GENETIC DETERMINANTS OF PLASMA APOAI AND HDL-C

A large number of epidemiological studies have demonstrated that low serum levels of high density lipoprotein cholesterol (HDL-C) and apolipoprotein AI (apo AI) are risk factors for the development of coronary artery disease (CAD). Various environmental factors including exogenous gonadal steroids, alcohol intake, stress, infection, the amount of exercise and some drug therapy can significantly affect plasma HDL-C levels. However, there is also good evidence for the involvement of genetic variation in determining these traits, and the heritability of HDL-C and apo AI has been estimated in various studies to be between 0.43 and 0.66 (Hamsten et al 1986). Recently, evidence for a major gene determining individual differences in quantitative levels of apo AI was demonstrated in a population-based sample by applying biometrical techniques (Moll et al 1989).

The apo AI gene is located in a cluster with the genes for apo CIII and apo AIV on chromosome 11 (Karathanasis et al 1986). The genes have been cloned, and many Restriction Fragment Length Polymorphisms (RFLPs) of the genes reported. These RFLPs have been used as markers to study the genetic determinants of lipid metabolism and CAD risk and a number of studies have demonstrated association between these RFLPs and plasma levels of lipids, lipoproteins and apoproteins (reviewed in Humphries et al 1988 and Table 1). Although all three apolipoproteins coded for by this gene cluster are known to be involved either directly or indirectly in HDL metabolism, there is, as yet, no specific evidence for a mechanism by which variation in the gene cluster may be affecting HDL-cholesterol concentration.

Recently we and others have identified a G to A substitution occuring 75bp upstream from the start of transcription of the apoAI gene. The "A" allele was identified in an individual

NATO ASI Series, Vol. H 73
Human Apolipoprotein Mutants III
Edited by C. R. Sirtori, G. Franceschini, B.H. Brewer Jr.

selected from a kindred where several members had significantly elevated levels of apoAI and HDL-C (Jeenah et al 1989). The G to A sequence change destroys an MspI cleavage site in this region of the gene, and genotyping can be carried out using the Polymerase Chain Reaction (PCR) technique Figure 1. We have previously shown that in a sample of 94 healthy men from the UK the "A" allele was associated with elevated apoAI and HDL-C

TABLE 1. COMPARISON OF PUBLISHED ASSOCIATION STUDIES BETWEEN APOPROTEIN LEVELS AND APO AI-CIII-AIV RFLPS IN HEALTHY INDIVIDUALS

Polymorphisms	Traits	Rare allele Assoc'n	Sample (Age in years) (1)	Reference
MspI	Apo AI	↑ $p<0.006$	136F (45-59)	Pagani et al 1990
G/A	Apo AI	↑ NS	108M (27-51)	Pagani et al 1990
	Apo AI	↑ $p<0.025$	96M (45-59)	Jeenah et al 1991
-75bp	Apo AI	↑ $p<0.025$	162M (7-23)	Paul-H et al 1992
PstI	Apo AI	↑ $p<0.050$	110M (45-59)	Kessling et al 1988a
	Apo AI	None	118M (> 55)	Paulweber et al 1988
	Apo AI	↑ NS	103M (> 60)	Wile et al 1989
	Apo AI	↓ NS	145MF(53-40)	Ordovas et al 1991
	Apo AI	↓ $p<0.056$	162M (7-23)	Paul-H et al 1992
SstI	Apo AI	↓ NS	145MF(53-40)	Ordovas et al 1988
	Apo AI	None	154MF(> 60)	Shoulders et al 1988
	Apo AI	↑ NS	58F (> 20)	Anderson et al 1989
	Apo AI	↓ NS	162M (7-23)	Paul-H et al 1992
	Apo CIII	↑ $p<0.040$	154MF(> 60)	Shoulders et al 1991
	Apo CIII	↑ NS	162M (7-23)	Paul-H et al 1992
	Apo AIV	↑ NS	79MF(> 60)	Shoulders et al 1991
	Apo AIV	↓ NS	162M (7-23)	Paul-H et al 1992
PvuIIA-CIII	Apo AI	↓ $p<0.025$	209MF(8-10)	Xu et al 1990
(2)	Apo AI	↓ NS	162M (7-23)	Paul-H et al 1992
	Apo CIII	↓ $p<0.023$	162M (7-23)	Paul-H et al 1992
PvuIIB-AIV	Apo AI	↓ NS	92M (45-59)	Kessling et al 1988b
	Apo AI	↑ NS	145MF(53-40)	Ordovas et al 1991
	Apo AI	↓ $p<0.010$	206MF(8-10)	Xu et al 1990
	Apo AI	↓ $p<0.035$	162M (7-23)	Paul-H et al 1992
	Apo CIII	↓ $p<0.085$	162M (7-23)	Paul-H et al 1992
XbaI-AIV	Apo AI	None	145MF(53-40)	Ordovas et al 1991
	Apo AI	None	162M (7-23)	Paul-H et al 1992
	Apo AIV	None	162M (7-23)	Paul-H et al 1992

(1) M = male, F = female. NS Not Significant
(2) PvuII-CIII and PvuII-AIV show strong linkage disequilibrium.
↑ Associated with high levels , ↓ Associated with low levels of the trait

(Table 2) explaining respectively 6% and 4.0% of the sample variance (Jeenah et al 1989)). To confirm this association we next examined a sample of young men from Brugge.

Children have had less exposure to some of the environmental factors which contribute to differences in plasma lipid levels in adults such as smoking, alcohol consumption and CAD, though lipid levels may be more influenced by developmental factors such as hormone levels than adults. 162 individuals aged 8-18 years were selected and fasting lipids, lipoproteins,

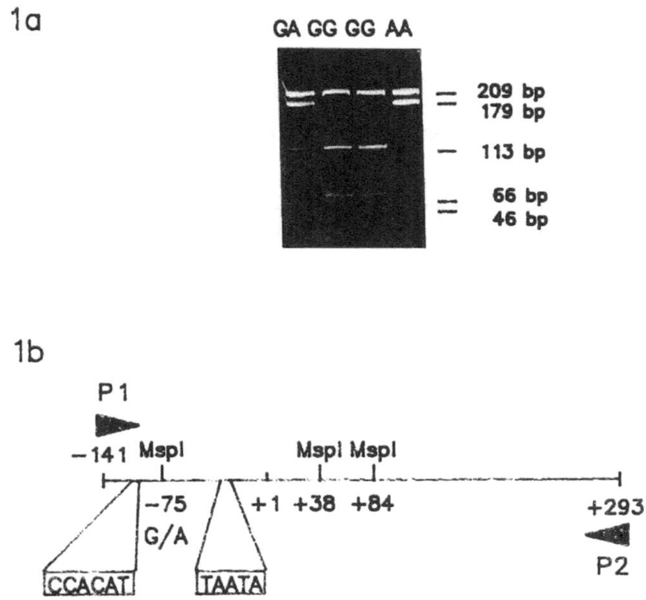

Figure 1. G/A genotyping of the apo AI gene promoter. a) Ethidium bromide stain of a 1% gel of DNA from individuals of different genotype after PCR amplification and MspI digestion. b) Map of the apoAI promoter. Position of the constant and variable cutting sites for MspI are shown. The positions of the start of transcription of the apo AI gene is indicated as +1 and the location of the CAAT and TATA base elements is shown. P1 and P2 indicate the positions of the oligonucleotides used for the PCR amplification (from 9).

apolipoproteins and testosterone levels measured. Lipid data was adjusted for age, BMI and testosterone values, explaining 3.6% and 21.7% of the sample variance in apo AI and HDL-C respectively. As shown in Table 1, there was a significant association observed, with those individuals with one or more "A" allele having higher levels of apo AI and HDL-C. In this sample the G/A substitution explained 3.6% of the sample variance of apo AI (Paul-Hayase et al 1992). This confirms the observations in the Bristol sample and that reported in individuals in Italy (Pagani et al 1990).

Recently we have studied two other samples of healthy men of from Singapore of Chinese origin (Paul-Hayase et al in preparation) and from Iceland (Sigurdsson et al 1992). The frequency of the "A" allele is significantly higher in the Chinese than in the other samples (0.26 versus 0.11), but in both samples the "A" allele is associated with significantly higher levels of apoAI and HDL-C (Table 2). Taken together these studies demonstrate unequivically that this polymorphism is associated with a small, but significant effect on apoAI and HDL-C levels, explaining roughly 5% of the sample variance, after adjusting for the effects of the age and BMI.

TABLE 2. MEAN (+SE)OF APO AI AND HDL-C LEVELS IN MALES WITH DIFFERENT G/A GENOTYPES

Sample (ref)	Genotype	No	Apo AI (mg/dl)	HDL-(mmol/l)
Bristol	GG	76	128±4	1.18±0.05
Jeenah et al	GA + AA	18	151±13**	1.43±0.13*
Brugge	GG	99	120±2.	1.32±0.03
Paul-H et al	GA + AA	42	125±3**	1.14±0.06
Chinese	GG	62	134±5	1.12±.04
Paul-H et al	GG + AA	69	152±5**	1.24±.04**
Iceland	GG	70	144±4	1.17±.03
Siggurdson et al	GA + AA	28	157±6**	1.31±.06***

* $p < 0.05$, ** $p < 0.025$ *** $p < 0.005$

It is known that the promoter region between nucleotides -256 bp and -41 bp upstream from the transcription start site of the apo AI gene is necessary and sufficient for maximal levels of expression in HepG2 but not in Caco-2 cells (Sastry et al 1988)). The A substitution at -75 bp is located between the CACAT sequence and the TAAATA box of the transcriptional start site of apo AI gene, and creates a 6 bp perfect repeat CAGGGC-CA*GGGC. Because of its position in the promoter region of the apo AI gene, it is likely that the G/A sequence change itself may have a direct effect on the transcription of apo AI gene, and thereby the secretion of apo AI either from the liver or the intestine. A recent study using DNase I footprint analysis on the proximal part of the apo AI promoter has identified four protected areas (Panagiota et al 1991), with one of the protein binding regions being from -128bp to -

77bp. This is in close proximity to the G/A sequence change at -75bp and our hypothesis is that the G/A substitution alters the affinity for a transacting nuclear protein which affects the transcription rate of the apo AI gene and thus alters the rate of synthesis of apo AI from the liver and intestines. Experiments are in progress to test the effect of the substitution on promoter function and on the binding of nuclear proteins to the two sequences. Preliminary data using CAT assays suggest that the "A" allele (-256 to +180bp fragment) has a roughly 20% greater promoter strength in HepG2 cells compared to the "G" allele (Paul et al unpublished observations). Although these data are compatible with the hypothesis that the A substitution itself is causing an increase in apo AI synthesis by a direct positive effect on gene transcription, it is also possible that the A substitution is in linkage disequilibrium with a second functionally important mutation elsewhere in the gene cluster, or even in some other unidentified linked gene. However, if correct the mechanism we suggest would be one of the first reported examples of a frequent (10%) sequence variant in the promoter region of an apoprotein gene with an effect on gene transcription.

GENE-ENVIRONMENT INTERACTION AND LEVELS OF HDL-C AND APOAI

Levels of HDL-C and apo AI are known to be influenced by a number of factors, including diet, exercise, sex hormones, alcohol intake, age and BMI. Cigarette smoking is well recognised as a risk factor of CAD, and has been shown to have a lowering effect on both HDL-C and apo AI (Haffner et al, 1985). This effect is reversible, and within two weeks from cessation of smoking the levels of HDL-C can return to normal (Tuomilehto et al 1986). To examine the possibility of interaction between G/A genotype and environmental factors we have extended the analysis of non-smoking men from Iceland to include smokers and women. In the sample examined 30% of men and 35% of women were current smokers, and in both men and women, lower levels of HDL-C and apo AI were observed in those who smoked compared to those who did not smoke (Sigurdsson et al, 1992). The association between genotype and HDL-C and apoAI was analysed separately in the four groups, and results are presented in Table 3. The G/A substitution was only significantly associated with differences in levels of HDL-C and apo AI in non-smoking men. Those who had the A allele had 11% higher levels of HDL-C and 9.7% higher levels of apo AI, and after adjustment for age and BMI, the percentage of sample variance (R^2 x 100) accounted for by genotype was 13.9% and 8% respectively. There was a 15% difference in apo AI levels and 20% difference in HDL-C levels between smokers and nonsmokers in those with one or more A allele.

Table 2. ADJUSTED MEAN (±SEM) LEVELS OF HDL-C AND APO AI IN MEN AND WOMEN WITH DIFFERENT G/A GENOTYPE, IN SMOKERS AND NON SMOKERS.

(A)	MEN	Non Smokers	Smokers	Prob
All	n	104	45	
	HDL-C (mmol/l)	1.21 (0.03)	1.06 (0.03)	p=0.002
	Apo AI (mg/dl)	147.2 (2.2)	138.9 (2.5)	p=0.023
GG	n	76	34	
	HDL-C (mmol/l)	1.17 (0.03)	1.07 (0.04)	p=0.032
	Apo AI (mg/dl)	143.5 (2.4)	140.4 (2.8)	N.S.
GA + AA	n	28	11	
	HDL-C (mmol/l)	1.31 (0.07)*	1.04 (0.07)	p=0.023
	Apo AI (mg/dl)	157.4 (4.4)**	134.4 (5.4)	p=0.005
(B)	**WOMEN**	**Non Smokers**	**Smokers**	**Prob**
All	n	107	59	
	HDL-C (mmol/l)	1.48 (0.03)	1.37 (0.04)	p=0.028
	Apo AI (mg/dl)	165.7 (2.4)	155.9 (2.8)	p=0.011
GG	n	87	44	
	HDL-C (mmol/l)	1.49 (0.04)	1.35 (0.04)	p=0.024
	Apo AI (mg/dl)	166.4 (2.7)	156.1 (3.4)	p=0.024
GA + AA	n	20	15	
	HDL-C (mmol/l)	1.43 (0.09)	1.40 (0.07)	N.S.
	Apo AI (mg/dl)	162.9 (5.2)	155.5 (4.5)	N.S.

* p=0.029, difference in mean levels of HDL-C between genotypes in non smokers by ANOVA

** p=0.004, difference in mean levels of Apo AI between genotypes in non smokers by ANOVA

In those with only the G allele the difference between smokers and nonsmokers only reached statistical significance for HDL-C though the trend could be seen for the apo AI as well. To test for interaction between genotype, gender and smoking, a two-way analysis of variance was performed. For apo AI, significant interaction ($p<0.02$) was found between genotype and gender in non-smokers, and between genotype and smoking in males ($p<0.02$). For HDL-C a similar trend was seen for interaction between gender, smoking and genotype but these differences did not reach statistical significance. There was no significant difference in levels of other lipid parameters between genotypes of the apo AI gene in either men or women (data not shown).

Our current hypothesis is that the A substitution increases the binding affinity of this sequence for a positive transcription factor resulting in a direct effect on apo AI gene transcription in the liver or intestine. The observed difference in response to the G/A substitution between men and women is likely to be through the effect of sex hormones, since levels of HDL-C and apo AI are positively associated with endogenous levels of testosterone in adult men (Khaw et al 1991). One possibility is thus that the effect which testosterone has on the synthesis of apo AI in the liver and intestine, is mediated by altering either the level or affinity of the nuclear transcription factor proteins which are involved in binding to the DNA sequences around -75bp of the apoAI promoter.

Several studies have shown evidence for interaction between genotype and smoking in the determination of apo AI and HDL-C levels (Kondo et al 1989, Kaprio et al 1989). Our data in the men provides evidence for significant interaction between the G/A genotype and smoking in the determination of these traits and similar observations have been made in our laboratory in a group of healthy Chinese men (Paul-Hayase et al unpublished observations). The effect of smoking on lowering levels of apoAI and HDL-C is well documented, but the mechanism of this effect is unknown. Apo AI is a negative acute phase protein (Bienvenu et al 1988) and the action of growth factors such as transforming growth factor beta on gene expression in the liver may explain this effect, possibly acting through post-transcriptional mechanisms affecting mRNA stability (Morrone et al 1989). The data suggests that, by some mechanism, smoking overrides the positive effect of the A allele on levels of apo AI and HDL-C, with the result that men carrying the A allele would be predicted to change from having amongst the highest levels of HDL-C and apo AI to amongst the lowest levels if they are smokers.

Based on the reported protective effect associated with elevated apoAI and HDL-C (Castelli et al 1986) the G/A polymorphism would therefore be a marker for one of the genetic factors contributing to CAD risk in men. The effect of the genotype-smoking interaction on relative risk of CAD can be estimated by extrapolation from data which show that 1% change in apo AI levels is associated with a 2% change in a relative risk of CAD (Castelli et al 1986). Based on this data, those men with an A allele would have a risk of CAD roughly 80% of that of individuals who lack the A allele, but if these same individuals smoke cigarettes, their relative risk will increase by 1.3 fold because of the large decrease (15%) in levels of apo AI they experience, and also increase by 1.8 fold because of the direct affect associated with smoking, for a total increase in relative risk of 2.3 fold. Similar estimates were obtained using HDL-C as a predictor of relative risk. Understanding of the mechanism of the interaction between this gene polymorphism and environmental factors, will be important in determining the etiology of low plasma HDL-C and apo AI levels and, in particular, in identifying those smoking individuals who would lower their risk of CAD substantially by cessation of smoking.

ACKNOWLEDGEMENTS

Support for this work was provided by the British Heart Foundation (grant RG16), the Icelandic Heart Association and Icelandic Council of Science, and the Charing Cross Sunley Research Trust. We thank Ms Gina Deeley for help in preparing the manuscript.

REFERENCES

Anderson RA, Burns TL, Lee J, Swenson D, Bristow JL (1989) Restriction fragment length polymorphisms associated with abnormal lipid levels in an adolescent population. Atherosclerosis 77:227-237

Bienvenu J, Deshaires P, Bernon H, Armanet P, Peristeris P, Lepape A, Perdrix JP (1988) [Serum amyloid protein A (SAA) and HDL. Clinical implications and surgical resuscitation.] Ann Biol Clin (Paris) 46:343-6

Castelli WP, Garrison RJ, Wilson PWF, Abbot RD, Kalousdian S, Kannel WB (1986) Incidence of coronary heart disease and lipoprotein cholesterol levels: the Framingham study. JAMA 256:2835-8

Haffner SM, Applebaum-Bowden D, Wahl PW, Hoover JJ, Warnick GR, Albers JJ, Hazzard WR (1985) Epidemiological correlates of high density lipoprotein subfractions, apolipoprotein AI, AII and D and lecithin cholesterol acyltransferase. Effects of smoking, alcohol and adiposity. Arteriosclerosis 5:169-77

Hamsten A, Iselius L, Dahlen G, Fraire U (1986) Genetic and cultural inheritance of serum lipids low and high density lipoprotein cholesterol and serum apolipoproteins AI, AII and B. Atherosclerosis 60:199-208

Humphries SE (1988) DNA polymorphisms of the apolipoprotein genes -their use in the investigation of the genetic component of hyperlipidaemia and atherosclerosis. Atherosclerosis 72:89-108

Jeenah M, Kessling A, Miller N and Humphries SE (1989) G to A substitution in the promoter region of the Apolipoprotein AI and high density lipoprotein cholesterol concentrations. Mol Biol Med 7:233-241

Kaprio J, Ferrell RE, Kottke BA, Sing CF (1989) Smoking and reverse cholesterol transport: evidence for gene-environment interaction. Clinic Genet 36:266-8

Karathanasis SK (1986) Apolipoprotein multigene family: tandem organisation of apolipoprotein AIV, AI and CIII genes. Proc Natl Acad Sci USA 82:6374-6378

Kessling AM, Rajput-Williams J, Bainton D, Scott J, Miller N, Baker I, Humphries SE (1988) DNA polymorphisms of the apolipoprotein AII and AI-CIII-AIV genes: a study in men selected for differences in high density lipoprotein cholesterol concentration. Am J Hum Genet 42:458-67

Kessling AM, Temple A, Taylor A, Hutson J, Hidalgo A, Hidalgo A, Humphries S.E. (1988b) A PvuII polymorphism in the 5' flanking region of the apo AIV gene: detection of genetic variation determining apo AI and HDL-cholesterol concentration. Human Genetics 78: 237-239.

Khaw K-T, Barrett-Conner E (1991) Endogenous sex hormones, high density lipoprotein cholesterol, and other lipoprotein fractions in men. Arteriosclerosis and Thrombosis 11:489-94

Kondo I, Berg K, Drayna D, Lawn R (1989) DNA polymorphism at the locus for human cholesteryl ester transfer protein (CETP) is associated with high density lipoprotein cholesterol and apolipoprotein levels. Clinic Genet 35:49-56

Moll PP, Michels VV, Weidman WH and Kottke BA (1989) Genetic determination of plasma apolipoprotein and AI in a population-based sample. Am J Human Genet 44:124-139

Morrone G, Cortese R Sorrentino V (1989) Post-transcriptional control of negative acute phase genes by transforming growth factor beta. EMBO-J 8:3767-71

Ordovas JM, Civeira F, Genest J Jr, Craig S, Robbins AH, Meade T, Pocovi M, Frossard PM, Masharani U, Wilson PWF, Salem DN, Ward RH, Schaefer EJ (1991) Restriction fragment length polymorphisms of the apolipoprotein AI,CIII, AIV gene locus. Relationships with lipids, apolipoproteins and premature coronary artery disease. Atherosclerosis 87:75-86

Pagani F, Sidoli A, Giudici GA, Barenghi L, Vergani C, Baralle, FE (1990) Human apolipoprotein AI promoter polymorphism: association with hyperlipidaemia. J Lipid Res 31:1371-1377

Panagiota P, Ogami K, Ramji DP, Nicosia A, Monaci P, Cladaras C, Zannis VI (1991) Promoter elements and factors involved in hepatic transcription of the human apoA-I gene positive and negative regulators bind to overlapping sites. J Biological Chemistry 266:5790-5797

Paul-Hayase H, Rosseneu M and Humphries SE (1992) Polymorphisms in the apolipoprotein (apo) AI-CIII-AIV gene cluster: detection of genetic variation determining plasma apo AI, apo CIII and apo AIV concentrations. Hum Genet 88:439-446

Paul-Hayase H, Saha N, Jeenah M, Humphries SE (1992) The G/A substitution in the ApoAI promoter and elevated apoAI levels: Association in men and not in women and test of promoter function (in preparation)

Reichl D, Miller NE (1989) Pathophysiology of reverse cholesterol transport: insights from inherited disorders of lipoprotein metabolism. Arteriosclerosis 9:785-97

Sastry KN, Seedorf U and Karanthansis DK (1988) Different cis-acting DNA elements control expression of the human apolipoprotein AI gene in different cell types. Mol and Cell Biol 8: 605-614

Shoulders CC, Harry PJ, Lagrost L, White SE, Shah NF, North JD, Gilligan M, Gambert P, Ball MJ (1991) Variation at the apo Ai-CIII-AIV gene complex is associated with elevated plasma levels of apo CIII. Atherosclerosis 87:239-247

Sigurdsson G, Baldursdottir A, Sigvaldason H, Agnarsson U, Thorgeirsson G, Sigfusson N (1992) The predictive value of apolipoproteins in a prospective survey for coronary heart disease in men (submitted)

Sigurdsson G jr, Gudnason V, Sigurdsson G, Humphries S (1992) Interaction between a polymorphism of the apolipoprotein AI promoter ad smoking determines plasma levels of high lipoprotein and apolipoprotein AI. Arteriosclerosis (in press)

Tuomilehto J, Tanskanen A, Salonen JT, Nissinen A, Koskela K (1986) Effect of smoking and stopping smoking on serum high-high density lipoprotein cholesterol levels in a representative population sample. Prev Med 15:35-45

Wile DB, Barbir M, Gallagher J, Myant NB, Ritchie CD, Thompson GR, Humphries SE (1989) Apolipoprotein AI gene polymorphisms: frequency in patients with coronary artery disease and healthy controls and association with serum apo AI and HDL-cholesterol concentration. Atherosclerosis 78:9-18

Xu C-F, Nanjee MN, Savill J, Talmud P, Angelico F, Del Ben M, Antonini R, Mazzarella B, Miller N, Humphies SE (1990) Variation at the apolipoprotein (apo) AI-CIII-AIV gene cluster and apo B gene locus is associated with lipoprotein and apolipoprotein levels in Italian children. Am J Hum Genet 47:429-439

Structure-Function Relationships of Apolipoprotein A-I Variants

Arnold von Eckardstein[1], Ana Jonas[2], Graciela Castro[3], Jean-Charles Fruchart[3], Harald Funke[1], and Gerd Assmann[1,4]

1: Institut für Klinische Chemie und Laboratoriumsmedizin, Zentrallaboratorium
Westfälische Wilhelms-Universität Münster
Albert-Schweitzer-Str. 33
D-4400 MÜNSTER
F.R.G.

Introduction

Apolipoprotein A-I (apo A-I) is synthesized by hepatic and intestinal cells as preproapo A-I encompassing 267 amino acids of known sequence. Intracellular cleavage of 18 aminoterminal amino acids results in proapolipoprotein A-I that is the secreted isoform of apo A-I abundant in chylomicrons, intestinal and hepatic HDL precursors. A yet unidentified metalloenzyme removes the aminoterminal hexapeptid. The resulting mature apo A-I is the major protein component of HDL, the cofactor of the cholesterol esterifying enzyme lecithin:cholesterol-acyltransferase (LCAT), and the putative ligand of a specific HDL receptor which helps to promote cholesterol efflux from many cells. Because of its *in vitro* functions apo A-I is supposed to be a key protein in reverse cholesterol transport, by which excess cholesterol is transported from peripheral cells to the liver for final deposition into the biliary tract (reviewed in Tall et al. 1990, Johnson et al. 1991). This model is supported by the epidemiological finding of an inverse correlation between HDL-cholesterol serum concentration and the risk for myocardial infarction (Gordon and Rifkind 1989), by the finding of an increased prevalence of low HDL-cholesterol in kindreds with a high incidence of myocardial infarction (Breslow 1988), and by the observation of prevented atherosclerosis in transgenic mice overexpressing the human apo A-I gene (Rubin et al. 1991).

2: Department of Biochemistry, University of Illinois at Urbana-Champaign, U.S.A.
3: Département d'Etudes des Lipides et des Lipoprotéines, Institut Pasteur, Lille, France
4: Institut für Arterioskleroseforschung an der Universität , Münster , F.R.G.

NATO ASI Series, Vol. H 73
Human Apolipoprotein Mutants III
Edited by C. R. Sirtori, G. Franceschini, B.H. Brewer Jr.

The conformation and the structural requirements in apo A-I necessary to fulfill its functions are as yet little understood. From computer modellings of the apo A-I primary structure, a secondary structure consisting of a series of homologous and amphipathic *alpha*-helices has been postulated. Except for two helices encompassing 11 amino acids (residues 88_98 and residues 209_218) the alpha-helices are composed of 22 amino acids and terminated by beta-turn structures at positions 88, 99, 121, 143, 165, 187, 209 and 220. It has also been suggested that these amphipathic helices mediate the protein's biological functions in LCAT-activation and lipid binding (reviewed in Li et al. 1988, Segrest et al. 1992). One possibility for the identification of functional domains in apo A-I is to analyse structural variants of apo A-I with known amino acid changes for their biological activities *in vitro*. We have recently elucidated the structural changes in eight apo A-I variants that were identified in 16 families by electrophoretic screening of 32,000 newborns (von Eckardstein et al. 1989, 1990): 3:Pro->Arg (1x), 4:Pro->Arg (1x), 89:Asp->Glu (1x), 107:Lys->0 (4x), 107:Lys->Met (2x), 139:Glu->Gly (2x), 147:Glu->Val (1x), 165:Pro->Arg (4x) and 198:Glu->Lys (1x). In subsequent studies we analysed these variants for their impact on lipid metabolism in the carriers as well as for their properties in secondary structure, lipid binding, LCAT activation, cholesterol efflux promotion from cells, and binding to cellular apo A-I binding sites.

Lipid metabolism in apo A-I variant carriers

In extensive family studies, only one of the eight apo A-I variants was found to quantitatively affect lipoprotein concentrations in the heterozygous carriers, namely apo A-I(165:Pro->Arg) (von Eckardstein et al. 1989). This variant was identified in four unrelated families (heterozygote frequency in Germany: 1:8,000). Compared to unaffected family members, the mean HDL-cholesterol and apo A-I plasma concentrations were decreased by 40 % in the heterozygous carriers. Frequently, the plasma levels of both parameters were below the 5th percentiles of sex matched controls in the PROCAM population (Assmann & Schlulte 1986). The decreased concentration of apo A-I could be explained by the decreased concentration of the variant allele product relative to the normal allele product in the heterozygous carriers. Despite low HDL-cholesterol associated with this apo A-I variant, family histories in the four kindreds affected with apo A-I(165:Pro->Arg) did not reveal any increased frequency of myocardial infarction.

Interestingly, the point mutation underlying apo A-I(165:Pro->Arg) is cis-located with a rare Pst1 restriction site in the apo A-I/C-III/A-IV gene cluster. A previous population study observed that this rare P2 allele, which is characterized by a 3.3. kb fragment instead of a usual 2.2. kb fragment, was associated with coronary heart disease and low HDL-cholesterol concentrations. (Ordovas et al. 1986). In the German population, we could not reestablish these associations (Assmann et al. 1989). Sequence analysis of the apo A-I gene in two carriers of the P2 allele did not help to identify the apo A-I(165:Pro->Arg) mutation. It thus can be ruled out that inthe German population the P2 allele predicts the presence of the apo A-I(165:Pro->Arg) mutation and thereby low HDL-cholesterol concentrations (Funke, von Eckardstein, Assmann, unreported observations).

Conformational Properties

The impact of the amino acid changes in the apo A-I variants was tested by intrinsic tryptophan fluorescence and by circular dichroism studies of reconstituted HDL (rHDL) (Jonas et al. 1991, 1992; see also article by Jonas et al. in this volume). rHDL were obtained by the cholate dialysis method associating palmitoyloleyl phospahtidylcholine (POPC) or dipalmitoyl phosphatidylcholine (DPPC) with either normal or variant apo A-I isoforms that were isolated from the plasma of the heterozygous carriers. Only one variant - apo A-I(107:Lys->0) - differed from normal apo A-I. Furthermore, this variant was slightly less stable towards increasing concentrations of guanidinium hydrochloride and towards increasing temperature than normal apo A-I. These findings were not unexpected since the elimination of a single amino acid from an amphipathic alpha-helix should change the orientation of subsequent residues by 100° and thus should change the conformation of the protein (Rall et al. 1984).

Also the substitution of the proline residue 165 by an arginine should influence the conformation of apo A-I since in secondary structure models this amino acid is central for the formation of a beta-turn between two adjacent alpha-helices and since this beta turn is expected to be eliminated by the replacement of proline through arginine. However, rHDL containing apo A-I(165:Pro->Arg) did not differ from normal apo A-I by their tryptophan fluorescence properties (Jonas et al. 1991). Whether or not apo A-I(165:Pro->Arg) is conformationally different in other particles remains to be established. Some support for an altered secondary structure of apo A-I(165:Pro->Arg) comes from our studies on methionine sulfoxide formation in apo A-I variants: In both apo A-I(165:Pro->Arg) and apo A-I(107:Lys->0) we found significant changes in the relative oxidation of methionine residues 112 and 148 which in normal apo A-I and other apo A-I variants are oxidized at a constant ratio (von Eckardstein et al. 1991).

Lipid binding Properties

The association properties of the apo A-I variants with phosphatidylcholines were analysed by nondenaturing gradient gel electrophoresis of rHDL before and after incubation with LDL (Jonas et al. 1991, 1992; see also article by Jonas et al. in this volume). Normal and all but one variant apo A-I isoforms formed particles with POPC which had a mean diameter of 9.6 nm and which after incubation with LDL were rearranged to particles with mean diameters of 7.7 and 10.4 nm. Apo A-I(107:Lys->0) formed particles of 9.4 nm mean diameter which only little changed in the presence of LDL.

LCAT-cofactor activity

rHDL containing normal or variant apo A-I isoforms were compared towards their substrate properties for LCAT. rHDL with apo A-I(107:Lys->0), apo A-I(165:Pro->Arg), and apo A-I(198:Glu->Lys) showed reaction kinetics that differed from those containing the respective normal allele product from the respective heterozygous variant carrier (Jonas et al. 1991, 1992; see also article by Jonas et al. in this volume). rHDL containing these variants were less suitable substrates compared to the rHDL containing the respective normal allele products from the heterozygous variant carriers. However, the kinetic constants varied highly between all apo A-I preparations and, therefore, the observed differences were not significant if compared to the average of all normal apo A-I preparations. Also another study found apo A-I(107:Lys->0) decreased in its LCAT cofactor activity (Rall et al. 1984) so that at least this variant appears to be a defective LCAT cofactor. However, the metabolic relevance of this impairment is questionable as the in vivo LCAT activity in heterozygous apo A-I(107:Lys->0) variant carriers was found normal.

Cholesterol efflux promotion by rHDL containing apo A-I variants

It is generally accepted now that unesterified cholesterol diffuses from the cell membrane to HDL through unspecific lipid-lipid interactions and that HDL apolipoproteins are not necessary for this step (reviewed in Johnson et al. 1991). However, in adipocytes, macrophages and fibroblasts, the mobilization of intracellular cholesterol pools and their translocation to the cell membrane is much facilitated after the binding of HDL to the cell surface (reviewed in Johnson et al. 1991). The nature of the cellular HDL binding site is still a matter of debate since ligand blotting studies identified putative HDL receptors of various molecular masses ranging from 70 to 120 kDa (reviewed in Johnson et al. 1991). Also the ligands of the putative receptor are controversially discussed, especially the question whether only apo A-I or also other apolipoproteins display intrinsic activities, i.e. help to translocate cholesterol from intracellular pools to the plasma membrane. The controversial results reported so far may have originated from the heterogeneity of mechanisms through which different cell types regulate cholesterol trafficking between different subcellular compartments. Studies aimed at the characterization of structural domains in a defined ligand, thus as in apo A-I, should therefore use cell types in which the release of cholesterol from intracellular pools is not diversely regulated. Such a modell has been introduced with murine Ob1771 adipocytes which store exogenous cholesterol in a non-lysosomal compartment and which release cholesterol upon binding of HDL that contain apo A-I and/or apo A-IV but not apo A-II or apo Cs (Barbaras et al. 1987, Steinmetz et al. 1991, Theret et al. 1991). This specifity implies a high structural constraint in the ligands of the adipocyte HDL binding sites, which in turn is an ideal prerequisite for studies aimed at the characterization of functional domains in apo A-I. We therefore used Ob 1771 adipocytes which were preloaded with radiolabeleled cholesterol via LDL, to compare the cholesterol efflux promoting activities of rHDL containing normal or variant apo A-I (von Eckardstein et al. 1992). rHDL were obtained by the cholate dialysis method associating dimyristoyl phospahtidylcholine (DMPC) with either normal or variant apo A-I isoforms that were isolated from the plasma of the heterozygous carriers. All but one apo A-I variants promoted cholesterol efflux as effectively as did normal apo A-I. rHDL containing apo A-I(165:Pro->Arg) were 30 % less effective cholesterol efflux promotors than rHDL with normal apo A-I (von Eckardstein et al. 1992).

Binding to cells

The binding of rHDL containing apo A-I variants to apo A-I binding sites on Ob 1771 adipocytes was normal in all cases including apo A-I((165:Pro->Arg) (von Eckardstein et al. 1992). Therefore, the decreased cholesterol efflux promotion by apo A-I(165:Pro->Arg) must be caused by post binding events, e.g. by differences in the signal transduction activity.

Conclusions

In two of the apo A-I variants analysed by us the structural changes led to conformational and/or functional changes: apo A-I(107:Lys->0) and apo A-I(165:Pro->Arg). However, although in both cases the amino acid changes predict to result in conformational changes in the protein's alpha-helical structure (Rall et al. 1984, von Eckardstein et al. 1989), the structural, functional, and metabolic consequences are different (table I): In discoidal rHDL, the expected conformational changes were only found for apo A-I(107:Lys->0). Despite its conformational and functional changes *in vitro*, apo A-I(107:Lys->0) appears to be without any metabolic consequences *in vivo*: The heterozygous carriers are not affected by any lipid disorders (Utermann et al. 1982). By contrast, the amino acid change in apo A-I(165:Pro->Arg) leads to low HDL-cholesterol concentrations in the heterozygous carriers (von Eckardstein et al. 1989). As cholesterol efflux promotion is impaired for apo A-I(165:Pro->Arg) but not for apo A-I(107:Lys->0),

Table I: Comparison of Functional Properties in Apo A-I($Lys_{107}\rightarrow 0$) and Apo A-I($Pro_{165}\rightarrow Arg$)

		$Lys_{107}\rightarrow 0$	$Pro_{165}\rightarrow Arg$
lipid values in the carriers		normal	40% lower HDL-C. and apo A-I
conformation:	α-helical content	decreased	?
	Trp fluorescence	altered	normal
	stability in GndHCl	lower	normal
lipid binding in discoidal rHDL		altered	normal
LCAT activation		K_m increased , V_{max} normal	K_m increased , V_{max} normal
cholesterol efflux promotion		normal	decreased
binding to cells		normal	normal

this disability is possibly central in the pathogenesis of low HDL cholesterol in apo A-I(165:Pro->Arg) carriers. If so, this would imply that cholesterol efflux promotion from cells is a central function of apo A-I in HDL metabolism.

Acknowledgements

The projects were supported by grants from the Deutsche Forschungsgemeinschaft to A.v.E. (Ec 116,1-1) and from the NIH to A.J. (HL-16059 and HL-29939).

References

Assmann, G, Schulte, H (1986) PROCAM-Trial. Prospective cardiovascular Münster trial. Panscientia Verlag, Hedingen/Zürich, Swiss.

Assmann, G, Schulte, H, Funke, H, von Eckardstein, A, Schmitz, G, Robenek, H (1989) High density lipoproteins and atherosclerosis. in Crepaldi, G, Gotto, AM, Manzato, E, Baggio, G (eds.) Atherosclerosis VIII. Excerpta Medica, Amsterdam, pp. 341-351.

Barbaras, R, Puchois, P, Fruchart, JC, and Ailhaud, G (1987) Cholesterol efflux from cultured adipose cells is mediated by LpA-I particles but not by LpA-I:A.II particles. *Biochem. Biophys. Res. Commun.* **142**:63-70.

Breslow, JL (1988). Apolipoprotein genetic variation and human disease. *Physiol. Rev.* **68**:85-132.

Gordon, D, Rifkind, BM (1989). Current concepts: High density lipoproteins - the clinical implications of recent studies. *N. Engl. J. Med.* **321**:1311-1315.

Johnson, WJ, Mahlberg, FH, Rothblat, GH, Phillips, MC (1991) Cholesterol transport between cells and high density lipoproteins. *Biochim. Biophys. Acta.* **1085**:273-298.

Jonas, A, von Eckardstein, A, Kezdy, KE, Steinmetz, A, and Assmann, G (1991). Structural and functional properties of reconstituted high density lipoprotein discs prepared with six apolipoprotein A-I variants. *J. Lipid Res.* **32**:97-106

Jonas, A, von Eckardstein, A, Churgay, L, Mantulin, WW Assmann, G (1992) Structural and functional properties of natural and chemical variants of apolipoprotein A-I. *J. Lipid Res.* 1992, *submitted*

Li, WH, Tanimura, M, Luo, CC, Datta, S, and Chan, L (1988) The apolipoprotein multigene family: Biosynthesis, structure, structure-function relationships and evolution. *J. Lipid Res.* **29**:245-271.

Ordovas, JM, Schaefer, EJ, Salem, D, Ward, RH, Glueck, CJ, Vergani, C, Wilson, PJ, Karathanasis, SK (1986) Apolipoprotein A-I gene polymorphism with premature coronary artery disease and familial hypoalphalipoproteinemia. *N. Engl. J. Med.* **314**:671-677.

Rall, SC, Weisgraber, KH, Mahley, RW, Ogawa, Y, Fielding, CJ, Utermann, G, Haas, J, Steinmetz, A, Menzel, HJ, and Assmann, G (1984) Abnormal lecithin:cholesterol-acyltransferase activation by a human apolipoprotein A-I variant in which a single lysine residue is deleted. *J. Biol. Chem.* **259**:10063-10070.

Rubin, EM, Krauss, RM; Spangler, EA, Verstuyft, JG, Clift, SM (1991) Inhibition of early atherogenesis in transgenic mice by human apolipoprotein A-I. *Nature* **353**:265-267.

Segrest, JP, Jones, MK, DeLoof, H, Brouilette, CG, Venkatachalapathi, YV, Anatharamaiah, GM (1992) The amphipathic helix in the excahngeable apolipoproteins: a review of esecondary strucutre and function. *J. Lipid Res.* **33**:141-166.

Steinmetz, A, Barbaras, R, Ghalim, N, Clavey, V, Fruchart, J-C, Ailhaud, G (1990). Human apolipoprotein A-IV binds to apolipoprotein A-I/A-II receptor sites and promotes cholesterol efflux from adipose cells. *J. Biol. Chem.* **265**:7859-7863.

Tall, AR (1990) Plasma high density lipoproteins. Metabolism and relationship to atherogenesis. *J. Clin. Invest.* **86**:379-384.

Theret, N, Delbart, C, Aquie, G, Fruchart, JC, Vassaux, G, and Ailhaud, G (1990) Cholesterol efflux from adipose cells is coupled to diacylglycerol production and protein kinase C activation. *Biochem. Biophys. Res. Commun.* **137**:1361-1367.

Utermann, G, Steinmetz, A, Paetzold, R, Wilk, J, Feussner, G, Kaffarnik, H, Müller-Eckhardt, C, Seidel, D, Vogelberg, KA, and Zimmer, F (1982). Apolipoprotein A-I(Marburg): Studies on two kindreds with a mutant of apolipoprotein A-I. *Hum. Genet.* **61**:329-337.

von Eckardstein, A, Funke, H, Henke, A, Altland, K, Benninghoven, A, and Assmann, G (1989). Apolipoprotein A-I variants: Naturally occurring substitutions of proline residues affect plasma concentrations of apolipoprotein A-I. *J. Clin. Invest.* **84**:1722-1730.

von Eckardstein, A, Funke, H, Walter, M, Altland, K, Benninghoven, A, and Assmann, G (1990). Structural analysis of human apolipoprotein A-I variants: amino acid substitutions are nonrandomly distributed throughout the apolipoprotein A-I primary structure. *J. Biol. Chem.* **265**:8610-8617.

von Eckardstein, A, Walter, M, Holz, H, Benninghoven A, Assmann, G (1991): Site specific methionine sulfoxide formation is the strucutral basis of chromatographic heterogeneity of apolipoproteins A-I, C-II, and C-III. *J. Lipid Res.* **32**:1465-1475.

von Eckardstein, A, Castro, G, Duverger, N, Theret, N, Duchateau, P, Fruchart, JC, Ailhaud, G, Wybranska, I, Assmann, G (1992) Interaction of Reconstituted High Density Lipoprotein Discs Containing Human Apolipoprotein A-I (Apo A-I) Variants with Murine Adipocytes: Evidence for Reduced Cholesterol Efflux Promotion by Apo A-I(165:Pro->Arg). *J. Biol. Chem., submitted.*

SEX-RELATED DIFFERENCES IN THE PHENOTYPIC EXPRESSION OF THE APOLIPOPROTEIN A-I_{Milano} MUTANT

Guido Franceschini, Laura Calabresi, Gemma Gianfranceschi, Cesare R. Sirtori
Center E. Grossi Paoletti, Institute of Pharmacological Sciences, University of Milano, Italy

Apolipoprotein A-I_{Milano} (apo A-I_M) is a molecular variant of apo A-I characterized by a cysteine for arginine substitution at the position 173 in the primary sequence of apo A-I (Franceschini et al, 1980; Weisgraber et al, 1980, 1983). The presence of a single free cysteine in the mutant apolipoprotein leads to the formation of disulfide-linked homodimers (A-I_M/A-I_M) and heterodimers with apo A-II (A-I_M/A-II). The presence of these high molecular weight forms of the variant, slowly catabolized *in vivo* (Gregg et al, 1989), confers an increased metabolic stability to the carriers' high density lipoproteins (HDL), both *in vitro* (Franceschini et al, 1990a) or *in vivo* (Calabresi et al, 1992b). In an *in vitro* system, the interconversion of A-I_M HDL is impaired vs the case of control HDL, possibly due to a decreased capacity to accept/exchange lipids with triglyceride-rich lipoproteins (TGRLP); this same effect is observed *in vivo*, during the postprandial phase, when a pronounced postprandial rise of plasma and TGRLP triglyceride levels does not result in a parallel increase in the transfer of TGRLP triglycerides to HDL.

Thirty-six carriers of the A-I_M variant, aged between 1 and 79 years and all heterozygotes for the mutant apolipoprotein, have been identified, all descending from a single couple who lived in a small village in Northern Italy in the second half of XVIII century (Gualandri et al, 1985). In contrast to other recognized molecular variants of apo A-I (von Eckardstein et al, 1990), with the possible exception of the $Pro_{165} \rightarrow Arg$ mutant (von Eckardstein et al, 1989), the presence of the apo A-I_M variant has a significant effect on lipoprotein structure/function. The A-I_M carriers show multiple abnormalities in the lipid/lipoprotein pattern, mainly affecting the high density lipoprotein system (Franceschini et al, 1980, 1985; Gualandri et al, 1985): the plasma levels of HDL are constantly reduced compared to control subjects. Heterogeneous abnormal HDL_3, generally small in size and with a low flotation rate, are present in carriers' plasma, while plasma HDL_2 levels are markedly reduced (Franceschini et al,

NATO ASI Series, Vol. H 73
Human Apolipoprotein Mutants III
Edited by C. R. Sirtori, G. Franceschini, B.H. Brewer Jr.

1982, 1987). In addition to the abnormalities in the HDL fraction, the A-I_M carriers show a higher prevalence of hypertriglyceridemia compared to close, non-affected relatives (Franceschini et al, 1985; Gualandri et al, 1985). The intriguing aspect in this syndrome is that, in spite of an unfavorable lipid/lipoprotein profile (Manninen et al, 1992) the apo A-I_M carriers have a very low incidence of coronary artery disease (Gualandri et al, 1985).

In spite of a well defined, typical lipid/lipoprotein pattern in the A-I_M carriers, a significant variability has been detected among the individual carriers (Franceschini et al, 1981, 1985). In particular, the HDL-cholesterol (HDL-C) levels are distributed over a relatively wide range (6-35 mg/dl) and are closely related to other lipoprotein parameters, such as the plasma triglyceride levels and the HDL particle size distribution (Franceschini et al, 1987). To gain further insight into the mechanisms responsible for this heterogeneity in the phenotypic expression of the mutant apolipoprotein, we examined the sex-related differences in the lipid/lipoprotein profile and possible correlations between the major lipid/lipoprotein parameters, with particular attention to the ratio between the mutant and the normal apolipoprotein.

Sex-related differences in the lipid/lipoprotein pattern in the A-I_M carriers

In this study we examined the plasma lipid/lipoprotein levels in 28 adult A-I_M carriers, 12 females and 16 males. For most of the examined parameters a mean value, calculated from at least 3 measurements in the last three years, was considered for each individual carrier. The difference between females and males was tested by one-way ANOVA; a probability value of less than 0.05 (two-tailed) was considered as significant.

The anthropometric and lipid/lipoprotein levels in the examined subjects are reported in Table 1. The two groups did not differ for age and BMI. The only significant differences between females and males were in the HDL levels. Very similar mean values were found for plasma total and low density lipoprotein (LDL) cholesterol, and apo B levels. LDL particles of normal size were found in both females and males. Plasma triglyceride levels were higher compared to nonaffected subjects from the same families (Franceschini et al, 1985; Gualandri et al, 1985), but very similar between female and male carriers. Cholesterol esterification, evaluated as cholesterol esterification rate and plasma LCAT concentration (Table 1), was similarly impaired in both sexes compared to controls (Franceschini et al, 1990b). Plasma CETP levels, measured in a small subset of carriers, were also similar in the two groups.

Table 1. Anthropometric and lipid/lipoprotein parameters in female (n=12) and male (n=16) A-I_M carriers (mean ± SEM).

	Females	Males
Age (y)	43.8 ± 4.9	44.8 ± 4.8
BMI (kg/m²)	24.2 ± 0.8	24.1 ± 0.8
Total cholesterol (mg/dl)	189.9 ± 10.0	180.7 ± 11.8
Free cholesterol (%)	33.1 ± 1.0	33.3 ± 0.7
Triglycerides (mg/dl)	188.8 ± 21.3	175.6 ± 20.2
LDL-cholesterol (mg/dl)	130.9 ± 7.6	130.1 ± 8.8
LDL size (nm)	25.5 ± 0.2	25.5 ± 0.1
Apolipoprotein B (mg/dl)	108.3 ± 8.8	102.4 ± 7.3
HDL cholesterol (mg/dl)	21.3 ± 2.1	15.5 ± 1.6*
HDL_2-cholesterol (mg/dl)	3.6 ± 0.5	2.3 ± 0.3*
HDL_3-cholesterol (mg/dl)	17.7 ± 1.7	13.2 ± 1.3*
Apolipoprotein A-I (mg/dl)	94.3 ± 6.8	70.5 ± 6.7*
Apolipoprotein A-II (mg/dl)	24.8 ± 1.8	21.2 ± 1.3
Apolipoprotein A-I_M/A-I_N	1.03 ± 0.07	2.00 ± 0.22*
HDL particle score (AU)	2.48 ± 0.14	2.82 ± 0.11*
CER (nmol/ml/h)	23.7 ± 4.1	19.6 ± 1.5
LCAT (μg(ml)	3.9 ± 0.7 (n=7)	3.0 ± 0.1 (n=10)
CETP (μg/ml)	1.0 ± 0.1 (n=3)	0.9 ± 0.1 (n=3)

* significantly different from females

By contrast, HDL-C was 27% lower in males, due to a reduction of both HDL_2-C and HDL_3-C, by 36% and 25%, respectively. Plasma apo A-I levels, evaluated by immunoturbidimetry, which estimates the concentration of total apo A-I (normal A-I plus the various molecular forms of the mutant) (Weisgraber et al, 1980), were also significantly lower in males (-25%), while the reduction of apo A-II (-15%) did not reach significance ($p=0.12$); due to these changes, the A-I/A-II ratio was significantly lower in males (3.36±0.27 vs 3.85±0.16), possibly reflecting a more pronounced reduction of plasma Lp A-I than Lp A-I:A-II (Bekaert et al, 1992).

A dramatic difference between female and male carriers was detected in the mutant/normal apolipoprotein ratio, analyzed by isoelectric focusing under reducing conditions (Franceschini et al, 1981); in females the contribution of the two apolipoproteins to total plasma A-I was nearly identical, while males had a two-fold higher concentration of the mutant vs normal apo A-I (Table 1). This highly significant difference ($p=0.0014$) is probably responsible for the pronounced dissimilarity in the HDL particle size distribution between females and males. We previously showed that the abnormal HDL particle distribution in the carriers can be differentiated into three HDL patterns (Franceschini et al, 1987): pattern I is the most abnormal, characterized by a predominance of small HDL_{3b} particles; in pattern III, defined by a relatively normal HDL distribution, HDL_{3a} are the major component; the intermediate pattern II is characterized by a nearly equivalent contribution by HDL_{3a} and HDL_{3b}. Pattern I was significantly more frequent among males, while most of female carriers had a relatively normal HDL particle distribution (Pattern III, Table 2).

Table 2. Individuals with different HDL patterns among female and male $A\text{-}I_M$ carriers.

	Females	Males
Pattern I	3 (25.0%)	10 (62.5%)
Pattern II	2 (16.7%)	3 (18.7%)
Pattern III	7 (58.3%)	3 (18.7%)

HDL patterns have been described in Franceschini et al, 1987.

When the entire HDL particle distribution was evaluated by assigning each subject an HDL particle score (Calabresi et al, 1992a), so that a higher value reflects a distribution shifted toward particles of smaller size, a significantly lower mean value was found in males vs females (Table 1).

Since the post-menopausal age is associated with significant changes in lipid/lipoprotein metabolism (Seed, 1991), pre- (n=8) and post-menopausal (n=4) female carriers were analyzed separately. The only significant differences between the two groups were in plasma total and LDL cholesterol levels, both higher in post-menopausal (total: 224.2±8.2 mg/dl; LDL: 153.3±9.6 mg/dl) vs pre-menopausal women (172.7±8.9 mg/dl and 119.7±6.1 mg/dl, respectively). Differently from what observed in non carrier females (Masarei et al, 1980), post-menopausal carriers showed higher plasma HDL levels vs pre-menopausal women (Table 3), although only the difference in HDL_2-C reached borderline statistical significance ($p=0.057$). The $A\text{-}I_M$/A-I apolipoprotein ratio was almost identical in the two groups (pre: 1.01±0.07; post: 1.08±0.11).

Table 3. Plasma HDL levels (mg/dl) in pre-menopausal (n=8) and post-menopausal (n=4) $A\text{-}I_M$ carriers

	Pre-menopausal	Post-menopausal
HDL-cholesterol	19.4 ± 1.6	25.0 ± 3.6
HDL_2-cholesterol	3.1 ± 0.3	4.9 ± 0.8
HDL_3-cholesterol	17.4 ± 1.7	20.8 ± 2.4
Apolipoprotein A-I	87.2 ± 6.1	108.2 ± 9.8
Apolipoprotein A-II	23.2 ± 1.5	27.7 ± 2.7

In summary, the male carriers have a higher mutant/normal apo A-I ratio, lower apo A-I and HDL-C levels, and smaller HDL particles, compared to females, both pre- or post-menopausal. To investigate the role of the $A\text{-}I_M$/A-I ratio in determining the alterations in the HDL system typically found in the $A\text{-}I_M$ carriers, a series of correlations was calculated.

Relationship between the A-I_M/A-I ratio and lipid/lipoprotein parameters

The results of simple correlation analyses performed on the data from the whole series of 28 subjects are reported in Table 4. The apo A-I_M/A-I ratio was strongly correlated with several HDL parameters, including HDL-C, HDL_2-C, HDL_3-C and apo A-I levels (Figure 1); the correlation with plasma apo A-II levels was weaker, but still statistically significant.

Table 4. Correlations between the mutant/normal apo A-I ratio and lipid/lipoprotein parameters in 28 A-I_M carriers.

	r	p
HDL-cholesterol	- 0.60	0.0007
HDL_2-cholesterol	- 0.57	0.0017
HDL_3-cholesterol	- 0.68	< 0.0001
Apolipoprotein A-I	- 0.72	< 0.0001
Apolipoprotein A-II	- 0.41	0.031
HDL particle score	0.76	< 0.0001
HDL pattern	- 0.64	0.0003

The mutant to normal apo A-I ratio was also related to the type of HDL particles present in the individual's plasma, as demonstrated by the negative correlation with the HDL pattern (I to III) and by the positive correlation with the HDL particle score (the higher the score, the smaller the particles). The correlations with other lipid/lipoprotein, or enzymatic parameters were non significant, except for a weak, positive correlation with plasma triglyceride levels ($r=0.44$, $p<0.05$)

Conclusions

The present study revealed a characteristic sex-related difference in the concentration of various HDL parameters in the A-I_M carriers. Females had significantly higher

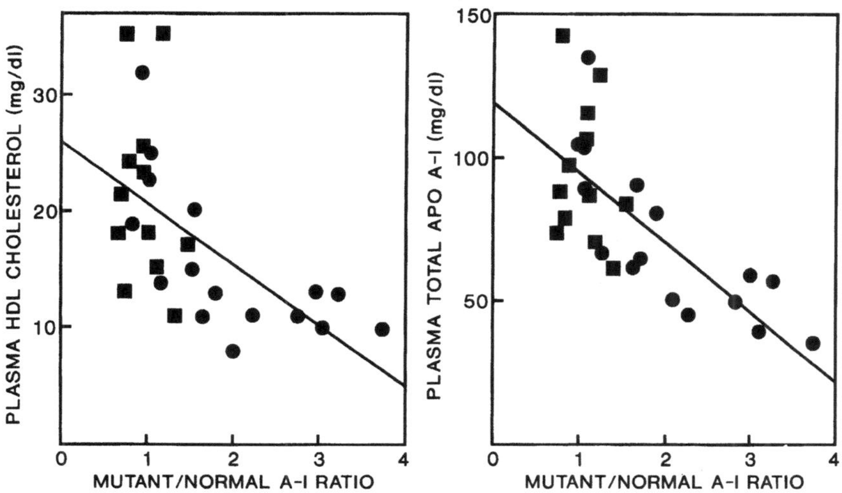

Figure 1. Relationship between the mutant/normal apo A-I ratio and plasma HDL-C and total A-I levels in 28 A-I_M carriers, 12 females (■) and 16 males (●).

HDL-C and apo A-I levels, and larger HDL particles than males. These sex-related differences are consistent with those found in normal individuals (Albers et al, 1976), but seem to be amplified in the A-I_M carriers, possibly due to the common, well defined genetic background of these subjects, all belonging to the same large kindred (Gualandri et al, 1985). The mechanism responsible for the higher HDL levels usually seen in women is generally thought to be related to the inhibitory action of estrogens on hepatic lipase (HL)(Tikkanen et al, 1982), leading to a reduced turnover of HDL; the dissociation of apo A-I from HDL is prevented (Clay et al, 1991), further contributing to the elevated plasma apo A-I levels (Goldberg et al, 1990). By contrast, the increased HL activity in males results in lower HDL levels vs females. The hypothesis of a regulatory role of sex hormones on plasma HDL levels is confirmed by epidemiological data showing significant reductions of HDL in post-menopausal women (Seed, 1991). This seems not to be the case in the A-I_M condition; post-menopausal females had the highest plasma HDL-C and apo A-I levels, while displaying the expected increase in plasma LDL-C vs pre-menopausal women.

The main finding of the present study was the disclosure of the strong effect of the mutant/normal apo A-I ratio on the HDl system in the individual carriers. A high variability in this ratio was already revealed in the original A-I_M family (Franceschini et al, 1981) and now confirmed by the investigation of a larger number of A-I_M carriers; it is of interest to note that this variability was significantly higher among male (CV=44%) vs female (CV=23%) carriers (Fig. 1). The mutant/normal apo A-I ratio was strongly correlated with several HDL parameters, including HDL-C and apo A-I levels, HDL subfraction distribution and particle size. In male carriers, a relative abundance of the mutant apolipoprotein, which is catabolized faster than normal apo A-I (Gregg et al, 1989), would lead to low plasma levels of total A-I at steady state; the formation of apo A-I_M dimers should increase, and abnormal HDL, enriched in A-I_M dimers, should accumulate in plasma. By contrast, in female carriers, the predominance of normal apo A-I, with a slower catabolism vs the mutant, would result in higher, closer to normal apo A-I levels and HDL particle distribution. The minimal, non-significant sex-related difference in plasma apo A-II levels, together with the weaker correlation between the mutant/normal apo A-I ratio and apo A-II, vs A-I, levels indicate that the metabolism of apo A-II is less affected by the presence of the apo A-I mutant. This is consistent with the recently described minor sex-related difference in Lp-A-I:A-II levels compared to a marked decrease of Lp-A-I in male vs female carriers (Bekaert et al, 1992). All together these findings indicate that the individual capacity to regulate the proportion of mutant vs normal apo A-I plays a major role in determining the individual HDL pattern, and that sex, by unknown mechanisms and in addition to other factors, particularly in males, significantly contributes to this regulation.

A second major finding of this study was that, in spite of the 25-30% lower plasma HDL-C and apo A-I levels in male vs female carriers, plasma triglyceride concentrations were similarly elevated in both sexes. Moreover, and differently from normal subjects, no correlation could be found between fasting triglyceride and HDL-C levels, either in the whole series, or by separate analysis of female and male data. A completely different situation was found in the postprandial state, where the A-I_M carriers showed a three-fold higher postprandial lipemia, which was inversely correlated, similarly to controls, with fasting plasma HDL-C and apo A-I levels (Calabresi et al, 1992). These findings argue for a different co-regulation of plasma HDL and triglyceride levels in the fasting and postprandial state, in the A-I_M carriers. In the fasting state plasma HDL levels are mainly regulated by changes in the protein composition of HDL, and not by the lipid content, as occurs in normal subjects (Brinton et al, 1991). In the postprandial state, the presence of a reduced plasma pool of HDL, already enriched in triglycerides, leads to an impaired transfer/exchange of triglyceride between TGLRP and HDL, with a progressive accumulation of triglyceride enriched particles in plasma (Calabresi et al, 1992). A contrasting situation occurs in normal subjects, particularly in those with coronary disease (Groot et al, 1991), where an increased postprandial lipemia, due to a defective lipolysis, results in low plasma HDL levels, possibly as a result of increased triglyceride transfer to HDL_2 with subsequent hydrolysis by hepatic lipase (Patsch et al, 1984). In the $A\text{-}I_M$ syndrome, a primary defect in HDL structure leads to low plasma HDL levels and reduced triglyceride transfer to HDL; this condition is associated with elevated fasting and postprandial triglyceride level, but apparently good vascular conditions.

Part of the reported research work was supported by the Consiglio Nazionale delle Ricerche of Italy [PF Ingegneria Genetica and Invecchiamento (cod 923237)]

References

Albers JJ, Wahl PW, Cabana VG, Hazzard WR, Hoover JJ (1976) Quantification of apolipoprotein A-I of human plasma high density lipoprotein. Metabolism 25:633-644

Bekaert ED, Alaupovic P, Knight-Gibson CS, Franceschini G, Sirtori CR (1992) Apolipoprotein $A\text{-}I_{Milano}$. The sex-related differences in the concentration and composition of apoA and apoB-containing lipoprotein particles. J Lipid Res, in press

Brinton EA, Eisenberg S, Breslow JL (1991) Increased apo A-I and apo A-II fractional catabolic rate in patients with low high density lipoprotein-cholesterol levels with or without hypertriglyceridemia. J Clin Invest 87:536-544

Calabresi L, Banfi C, Sirtori CR, Franceschini G (1992a) Apolipoprotein A-II modulates HDL remodeling in plasma. Biochim Biophys Acta 1124:195-198

Calabresi L, Cassinotti M, Gianfranceschi G, Safa O, Sirtori CR, Franceschini G (1992b) Increased postprandial lipemia in the apolipoprotein $A\text{-}I_{Milano}$ carriers. Arterioscl Thromb, in press

Clay MA, Newnham HH, Barter PJ (1991) Hepatic lipase promotes a loss of apolipoprotein A-I from triglyceride-enriched human high density lipoproteins during incubation in vitro. Arterioscler Thromb 11:415-422

Franceschini G, Sirtori CR, Capurso A, Weisgraber K, Bersot T, Mahley RW (1980) AI_{Milano} apoprotein. Decreased high density lipoprotein cholesterol levels with significant lipoprotein modifications in an Italian family. J Clin Invest 66:892-900

Franceschini G, Sirtori M, Gianfranceschi G, Sirtori CR (1981) Relation between the HDL apoproteins and AI isoproteins in subjects with the AI_{Milano} abnormality. Metabolism 30:502-509

Franceschini G, Frosi TG, Manzoni C, Gianfranceschi G, Sirtori CR (1982) High density lipoprotein-3 heterogeneity in subjects with the apo AI_{Milano} variant. J Biol Chem 257:9926-9930

Franceschini G, Sirtori CR, Bosisio E, Gualandri V, Orsini GB, Mogavero AM, Capurso A (1985) Relationship of the phenotypic expression of the AI_{Milano} apoprotein with plasma lipid and lipoprotein patterns. Atherosclerosis 58:159-174

Franceschini G, Calabresi L, Tosi C, Sirtori CR, Fragiacomo C, Noseda G, Gong EL, Blanche PJ, Nichols AV (1987) Apolipoprotein AI_{Milano}. Correlation between high density lipoprotein subclass distribution and triglyceridemia. Arteriosclerosis 7:426-435

Franceschini G, Calabresi L, Tosi C, Gianfranceschi G, Sirtori CR, Nichols AV (1990a) Apolipoprotein AI_{Milano}. Disulfide linked dimers increase high density lipoprotein stability and hinder particle interconversion in carrier plasma. J Biol Chem 265:12224-12231

Franceschini G, Baio M, Calabresi L, Sirtori CR, Cheung MC (1990b) Apolipoprotein AI-Milano. Partial lecithin:cholesterol acyltransferase deficiency due to low levels of a functional enzyme. Biochim Biophys Acta 1043:1-6

Goldberg IJ, Blaner WS, Vanni TM, Moukides M, Ramakrishnan R (1990) Role of lipoprotein lipase in the regulation of high density lipoprotein apolipoprotein metabolism. Studies in normal and lipoprotein lipase-inhibited monkeys. J Clin Invest 86:463-473

Gregg RE, Roma P, Bojanovski D, Schaefer JR, Zech LA, Kindt MR, Meng MS, Ronan R, Franceschini G, Sirtori CR, Brewer HB (1989) The kinetics of apolipoprotein A-I metabolism in humans with altered levels of high density lipoproteins. In: Crepaldi G, Gotto AM, Manzato E, Baggio G, eds. Atherosclerosis VIII. Amsterdam: Excerpta Medica 285-289

Groot PHE, van Stiphout WAHJ, Krauss XH, Jansen H, van Tol A, van Ramshorst E, Chin-On S, Hofman A, Cressewell SR, Havekes L (1991) Postprandial lipoprotein metabolism in normolipidemic men with and without coronary artery disease. Arterioscler Thromb 11:653-662

Gualandri V, Franceschini G, Sirtori CR, Gianfranceschi G, Orsini GB, Cerrone A, Menotti A (1985) AI_{Milano} apoprotein. Identification of the complete kindred and evidence of a dominant genetic transmission. Am J Human Genet 37:1083-1097

Manninen V, Tenkanen L, Koskinen P, Huttunen JK, Manttari M, Heinonen OP, Frick MH (1992) Joint effects of serum triglycerides and LDL cholesterol and HDL cholesterol concentrations on coronary heart disease risk in the Helsinki Heart Study. Circulation 85:37-45

Masarei JRL, Armstrong BK, Skinner MW (1980) HDL cholesterol and sex hormone status. Lancet 1:208

Patsch JR, Prasad S, Gotto AMJr, Bengtsson-Olivecrona G (1984) Postprandial lipemia. A key for the conversion of high-density $lipoprotein_2$ into high-density $lipoprotein_3$ by hepatic lipase. J Clin Invest 74:2017-2023

Seed M (1991) Sex hormones, lipoproteins, and cardiovascular risk. Atherosclerosis 90:1-7

von Eckardstein A, Funke H, Henke A, Atland K, Benninghoven A, Assmann G (1989) Naturally occurrong substitutions of proline residues affect plasma concentration of apolipoprotein A-I. J Clin Invest 84:1722-1730

von Eckardstein A, Funke H, Walter M, Atland K, Benninghoven A, Assmann G (1990) Structural analysis of human apolipoprotein A-I variants. Amino acid substitutions are nonrandomly distributed throughout the apolipoprotein A-I primary sequence. J Biol Chem 265:8610-8617

Weisgraber KH, Bersot TP, Franceschini G, Sirtori CR, Mahley RW (1980) AI_{Milano} apoprotein. Isolation and characterization of a cysteine containing variant of the A-I apoprotein AI from human high density lipoproteins. J Clin Invest 66:901-907

Weisgraber KH, Rall SC, Bersot TP, Mahley RW, Franceschini G, Sirtori CR (1980) Apolipoprotein AI_{Milano}. Detection of normal AI in affected subjects and evidence for a cysteine for arginine substitution in the variant AI. J Biol Chem 258:2508-2513

HUMAN APOLIPOPROTEIN A-IV

H. Bryan Brewer, Jr.
Molecular Disease Branch
National Heart, Lung and Blood Institute
National Institutes of Health
9000 Rockville Pike, Bldg. 10/7N117
Bethesda, MD 20892

Peter Lohse
Dept. of Clinical Chemsistry
Klinikum Grobhadern
Marchioninistr. 15
D-8000 Munich 70
Germany

Human apolipoprotein A-IV is a 376 amino acid glycoprotein with a molecular weight of 46,000 Daltons, synthesized predominantly in the small intestine (Weisgraber 1978; Beisiegel 1979; Green 1979; Utermann 1979; Green 1980; Weinberg 1983; Karathanasis 1986; Elshourbagy 1986; Gordon 1984). ApoA-IV is found in plasma and lymph mainly unassociated with lipoproteins (Fidge, 1980; Bisgaier 1985; Weinberg 1985; Ohta 1985; Ghiselli 1986) with only small quantities of apoA-IV detected in chylomicrons, very low density and high density lipoproteins. The affinity of apoA-IV to these lipoprotein particles appears to be weak, since apoA-IV is readily displaced by other apolipoproteins (Weisgraber 1978; Rifici 1985) and may easily be dissociated by analytical procedures (e.g. ultracentrifugation; (Green 1980; Bisgaier 1985; Lagrost 1989)). Using high performance gel filtration, however, Lagrost et al. (Lagrost 1989) reported approximately 70% of apoA-IV was present in the larger-sized high density lipoprotein subfraction.

In humans, several isoforms of apoA-IV have been described in both Caucasian (Utermann 1982; Menzel 1982a; Menzel 1982b; Schamaun 1985; Zannis 1985; Kamboh 1987; Bisgaier 1987; Menzel 1988; de Knijff 1988; Eichner 1989) and Black populations (Sepehrnia 1988; Menzel 1990a). In order to to gain insight into the genetic basis for apoA-IV polymorphism and the frequency of the apoA-IV isoforms we phenotyped 761 unrelated blood donors of mixed ethnic origin by isoelectric focusing (pH range 5-6) of delipidated plasma samples, followed by immunoblotting with a polyclonal rabbit anti-human apoA-IV antibody. Four major plasma isoproteins of apoA-IV were identified by these techniques. ApoA-IV-1, the more common isoprotein, was present in 98% of either homozygous or heterozygote individuals tested. ApoA-IV-2, containing one additional basic charge

NATO ASI Series, Vol. H 73
Human Apolipoprotein Mutants III
Edited by C. R. Sirtori, G. Franceschini, B.H. Brewer Jr.

was detected in approximately 15% of individuals while the most basic isoprotein, apoA-IV-3 and the most acidic isoform, apoA-IV-0, were each identified in 0.66% of subjects evaluated.

DNA sequence analysis of PCR-amplified DNA of the apoA-IV alleles revealed several point mutations and an insertion as the underlying structural alterations at the molecular level resulting in apoA-IV isoprotein heterogeneity. The results of these studies are summarized in Table I.

TABLE 1. The apolipoprotein (apo)A-IV isoproteins and their underlying nucleotide and amino acid substitutions.

ApoA-IV allele	Mutation	Amino acid	RFLP	N
1	-	-	-	12
1	ACT→TCT	Thr^{347}→Ser	*HinfI*	8
2	CAG→CAT	Gln^{360}→His	-	13
0	4-Amino-acid insertion	Between residues 361 and 362	-	4
0	AAG→GAG	Lys^{167}→Glu	*AvaI*	1
	CAG→CAT	Gln^{360}→His	-	
3	GAG→AAG	Glu^{165}→Lys	*SacI*	3
3	GAG→AAG	Glu^{230}→Lys	*StyI*	2

N, number of alleles analyzed; RFLP, restriction-fragment length polymorphism.

The apoA-IV-1 allele is considered to be the parent allele because of its high frequency in all populations studied to date. In addition, sequence analysis demonstrated the presence of a second major allelic variant of the apoA-IV-1 isoprotein. A common nucleotide substitution converts threonine-347 (ACT) to serine (TCT), thereby eliminating a restriction site for the enzyme Hinf I. The resulting isoprotein is indistinguishable from the more frequent "normal" apoA-IV-1 isoprotein on isoelectric focusing gels (Lohse 1990a).

In the apoA-IV-2 allele, we identified a G-to-T substitution that changes residue 360 of the mature apoA-IV, glutamine (CAG), to histidine (CAT), thereby adding one positive charge unit to the apoA-IV-1 isoprotein (pI 4.97) and creating the more basic apoA-IV-2 isoprotein (pI 5.02).

Four out of the five apoA-IV-0 alleles sequenced had a 12-nucleotide in-frame insertion that was identical in its nucleotide and amino acid sequence with the second of four highly conserved repeats in the carboxyl-terminal region of apoA-IV (Lohse 1990b). This additional repeat codes for glutamic acid (GAA) followed by three glutamines (CAG). The four amino acid insertion is located between residues 361 and 362, producing a four amino acid longer and one charge unit more acidic apoA-IV-0 isoprotein (pI 4.92). One of the apoA-IV-0 alleles sequenced identified two point mutations converting lysine-167 (AAG) to glutamic acid (GAG), and glutamine-360 (CAG) to histidine (CAT). The double mutant is probably the result of a second mutation within the apoA-IV-2 allele.

The apoA-IV-3 isoprotein (pI 5.08) was shown to result from the replacement of a negatively charged glutamic acid (GAG) by a positively charged lysine (AAG) at two different locations. One point mutation affected amino acid 165 of the mature apoA-IV (Lohse 1990a) while the other was present in two individuals and involved residue 230 (Lohse 1990b). Both base substitutions created an RFLP due to the loss of a cleavage site for the enzyme Sac I in the apoA-IV-3(Glu^{165}->Lys) allele and the presence of an additional cutting site for Sty I in the apoA-IV-3 (Glu^{230}->Lys) allele (Table I).

The effect of the different isoproteins on plasma cholesterol, triglycerides, HDL cholesterol, and apolipoprotein concentrations has been studied in several populations (Menzel 1988; de Knijff 1988; Eichner 1989; Menzel 1990b). In three of the studies (Menzel 1988; de Knijff 1988; Menzel 1990b), the apoA-IV-2/1 phenotype was found to be associated with increased HDL cholesterol and decreased plasma triglyceride levels. The apoA-IV-2 isoprotein significantly correlated with higher levels of HDL cholesterol in non-fasting samples from Tyrol (Menzel 1988) and fasting samples from Iceland (Menzel 1990b), whereas the triglyceride concentrations were observed to be significantly reduced in fasting samples from white USA females (Eichner 1989) and Icelanders (Menzel 1990b). Analysis of lipid and apolipoprotein concentrations in five non-fasting subjects with the apoA-IV-1/0 phenotype revealed no statistically significant differences when compared with apoA-IV-1/1 homozygotes (de Knijff 1988). Similarly, the plasma concentrations for 10 non-fasting subjects with the apoA-IV-2/2 phenotype were also analyzed. Mean values of cholesterol, triglycerides, apoA-I, apoA-IV, apoB and apoE from these individuals were not statistically significantly different from those of apoA-IV-1/1 homozygotes (de Knijff 1988). The same study permitted the assessment of the correlation of apoA-IV-3 isoprotein with lipid and apolipoprotein levels in non-fasting subjects with the apoA-IV-3/1 phenotype because of the high frequency of this allele in the Dutch population. The results clearly demonstrated the heterozygosity for the apoA-IV-3 allele is associated with apoA-IV plasma concentrations of approximately one-third of normal, while the other lipid values in these subjects were within the normal range. This suggests that the presence of the apoA-IV-3 isoprotein has an effect on the metabolism of apoA-IV-1.

The precise metabolic function of apoA-IV remains unclear. The protein has been proposed to play a role in the biogenesis and catabolism of intestinally derived triglyceride-rich lipoproteins (Gordon 1982), and in cholesterol metabolism as an activator of lecithin-cholesterol acyltransferase (Steinmetz 1985). Further *in vitro* evidence suggested that apoA-IV modulates the activation of lipoprotein lipase by the cofactor apoC-II (Goldberg 1990), potentially by rapidly dissociating from triglyceride-rich particles as the C-apolipoproteins are transferred from HDL (Weinberg 1985). The protein's most important role may be in reverse cholesterol transport as a structural component of the HDL particle and as one possible ligand for the putative HDL receptor (Stein 1986; Savion 1987; Savion 1988; Weinberg 1990; Steinmetz 1990).

Kinetic studies on apoA-IV have been performed in control human subjects and in patients with hypertriglyceridemia. Radiolabeled apoA-IV in control subjects was rapidely catabolized from plasma with a fractional catabolic rate (FCR) of approximately 2.4 1/d which is faster thatn apoA-I with an FCR of about .24. ApoA-IV was very readily dissociated from the plasma lipoproteins and a significant fraction was present in the lipoprotein free fraction. ApoA-IV was catabolized slower in hypertriglyceridemic patients (FCR 1.2 1/d) that in controls. The delay in FCR of apoA-IV in patients with hypertriglyceridemia was apparently due to a delay in the dissociation of apoA-IV from the triglyceride rich lipoproteins to the lipoprotein free fraction (Verges 1991).

Investigation of the genetic basis for human plasma apoA-IV polymorphism has let to an enhanced understanding of the underlying molecular basis for the differnt plasma isoproteins. Knowledge of the structure of the various isoproteins will be useful in the analysis of the functional properties of apoA-IV both *in vitro* and *in vivo*. These studies and additional kinetic studies will provide new information on the specific function of this unusual apolipoprotein in lipoprotein metabolism.

REFERENCE

Beisiegel U, Utermann G (1979) An apolipoprotein homolog of rat apolipoprotein A-IV in human plasma. Isolation and partial characterization. *Eur J Biochem* 93:601-608

Bisgaier CL, Sachdev OP, Megna L, Glickman RM (1985) Distribution of apolipoprotein A-IV in human plasma. *J Lipid Res* 26:11-25

Bisgaier CL, Lee ES, Glickman RM (1987) A method to screen apolipoprotein polymorphisms in whole plasma: description of apolipoprotein A-IV variants in dyslipidemias and a reassessment of apolipoprotein A-I in Tangier disease. *Biochim Biophys Acta* 918:242-249

de Knijff P, Rosseneu M, Beisiegel U, de Keersgieter W, Frants RR, Havekes LM (1988) Apolipoprotein A-IV polymorphism and its effect on plasma lipid and apolipoprotein concentrations. *J Lipid Res* 29:1621-1627

Eichner JE, Kuller LH, Ferrell RE, Kamboh MI (1989) Phenotypic effects of apolipoprotein structural variation on lipid profiles: II. Apolipoprotein A-IV and quantitative lipid measures in the healthy women study. *Genet Epidemiol* 6:493-499

Elshourbagy NA, Walker DW, Boguski MS, Gordon JI, Taylor JM (1986) The nucleotide and derived amino acid sequence of human apolipoprotein A-IV mRNA and the close linkage of its gene to the genes of apolipoproteins A-I and C-III. *J Biol Chem* 261:1998-2002

Fidge NH (1980) The redistribution and metabolism of iodinated apolipoprotein A-IV in rats. *Biochim Biophys Acta* 619:129-141

Ghiselli G, Krishnan S, Beigel Y, Gotto AM,Jr (1986) Plasma metabolism of apolipoprotein A-IV in humans. *J Lipid Res* 27:813-827

Goldberg IJ, Scheraldi CA, Yacoub LK, Saxena U, Bisgaier CL (1990) Lipoprotein. ApoC-II activation of lipoprotein lipase. Modulation by apolipoprotein A-IV. *J Biol Chem* 265:4266-4272

Gordon JI, Smith DP, Alpers DH, Strauss AW (1982) Cloning of a complementary deoxyribonucleic acid encoding a portion of rat intestinal preapolipoprotein AIV messenger ribonucleic acid. *Biochemistry* 21:5424-5431

Gordon JI, Bisgaier CL, Sims HF, Sachdev OP, Glickman RM, Strauss AW (1984) Biosynthesis of human preapolipoprotein A-IV. *J Biol Chem* 259:468-474

Green PH, Glickman RM, Saudek CD, Blum CB, Tall AR (1979) Human intestinal lipoproteins. Studies in chyluric subjects. *J Clin Invest* 64:233-242

Green PH, Glickman RM, Riley JW, Quinet E (1980) Human apolipoprotein A-IV. Intestinal origin and distribution in plasma. *J Clin Invest* 65:911-919

Kamboh MI, Ferrell RE (1987) Genetic studies of human apolipoproteins. I. Polymorphism of apolipoprotein A-IV. *Am J Hum Genet* 41:119-127

Karathanasis SK, Yunis I, Zannis VI (1986) Structure, evolution, and tissue-specific synthesis of human apolipoprotein AIV. *Biochemistry* 25:3962-3970

Lagrost L, Gambert P, Boquillon M, Lallemant C (1989) Evidence for high density lipoproteins as the major apolipoprotein A-IV-containing fraction in normal human serum. *J Lipid Res* 30:1525-1534

Lohse P, Kindt MR, Rader DJ, Brewer HB,Jr (1990a) Molecular basis for human plasma apolipoprotein A-IV polymorphism. *Circulation* 425 (suppl III):1689A

Lohse P, Kindt MR, Rader DJ, Brewer HB,Jr (1990b) Human plasma apolipoproteins A-IV-0 and A-IV-3. Molecular basis for two rare variants of apolipoprotein A-IV-1. *J Biol Chem* 265:12734-12739

Menzel HJ, Kladetzky RC, Assmann G (1982a) One-step screening method for the polymorphism of apolipoproteins A-I, A-II, and A-IV. *J Lipid Res* 23:915-922

Menzel HJ, Kovary PM, Assmann G (1982b) Apolipoprotein A-IV polymorphism in man. *Hum Genet* 62:349-352

Menzel HJ, Boerwinkle E, Schrangl-Will S, Utermann G (1988) Human apolipoprotein A-IV polymorphism: frequency and effect on lipid and lipoprotein levels. *Hum Genet* 79:368-372

Menzel HJ, Dwyer K, Sandholzer C, Schrangl-Will S, Dieplinger H, Hoye E, Lackner C, Utermann G (1990a) New electrophoretic methods for the detection of apolipoprotein variants. In: Lenfant C, Albertini A, Paoletti R Catapano AL eds.*Biotechnology of Dyslipoproteinemias: Applications in Diagnosis and Control. Atherosclerosis Reviews*. ed. 20 New York: Raven Press 171-178

Menzel HJ, Sigurdsson G, Boerwinkle E, Schrangl-Will S, Dieplinger H, Utermann G (1990b) Frequency and effect of human apolipoprotein A-IV polymorphism on lipid and lipoprotein levels in an Icelandic population. *Hum Genet* 84:344-346

Ohta T, Fidge NH, Nestel PJ (1985) Studies on the in vivo and in vitro distribution of apolipoprotein A-IV in human plasma and lymph. *J Clin Invest* 76:1252-1260

Rifici VA, Eder HA, Swaney JB (1985) Isolation and lipid-binding properties of rat apolipoprotein A-IV. *Biochim Biophys Acta* 834:205-214

Savion N, Gamliel A, Tauber JP, Gospodarowicz D (1987) Free apolipoproteins A-I and A-IV present in human plasma displace high-density lipoprotein on cultured bovine aortic endothelial cells. *Eur J Biochem* 164:435-443

Savion N, Gamliel A (1988) Binding of apolipoprotein A-I and apolipoprotein A-IV to cultured bovine aortic endothelial cells. *Arteriosclerosis* 8:178-186

Schamaun O, Olaisen B, Teisberg P, Gedde-Dahl T,Jr Ehnholm C (1985) Genetic studies of apolipoprotein A-IV by two-dimensional electrophoresis. In: Peeters H ed.*Protides of the Biological Fluids*. ed. 33 New York: Pergamon 471-474

Sepehrnia B, Kamboh MI, Adams-Campbell LL, Nwankwo M, Ferrell RE (1988) Genetic studies of human apolipoproteins. VII. Population distribution of polymorphisms of apolipoproteins A-I, A-II, A-IV, C-II, E, and H in Nigeria. *Am J Hum Genet* 43:847-853

Stein O, Stein Y, Lefevre M, Roheim PS (1986) The role of apolipoprotein A-IV in reverse cholesterol transport studied with cultured cells and liposomes derived from an ether analog of phosphatidylcholine. *Biochim Biophys Acta* 878:7-13

Steinmetz A, Utermann G (1985) Activation of lecithin: cholesterol acyltransferase by human apolipoprotein A-IV. *J Biol Chem* 260:2258-2264

Steinmetz A, Barbaras R, Ghalim N, Clavey V, Fruchart JC, Ailhaud G (1990) Human apolipoprotein A-IV binds to apolipoprotein A-I/A-II receptor sites and promotes cholesterol efflux from adipose cells. *J Biol Chem* 265:7859-7863

Utermann G, Beisiegel U (1979) Apolipoprotein A-IV: a protein occurring in human mesenteric lymph chylomicrons and free in plasma. Isolation and quantification. *Eur J Biochem* 99:333-343

Utermann G, Feussner G, Franceschini G, Haas J, Steinmetz A (1982) Genetic variants of group A apolipoproteins. Rapid methods for screening and characterization without ultracentrifugation. *J Biol Chem* 257:501-507

Verges B, Rader DJ, Zech L, Kindt M, Gambert P, Brewer HB,Jr (1991) In vivo metabolism of human apolipoprotein A-IV in type V hyperlipoproteinemia. *Arteriosclerosis and Thrombosis* 11:1946a

Weinberg RB, Scanu AM (1983) Isolation and characterization of human apolipoprotein A-IV from lipoprotein-depleted serum. *J Lipid Res* 24:52-59

Weinberg RB, Spector MS (1985) Human apolipoprotein A-IV: displacement from the surface of triglyceride-rich particles by HDL2-associated C-apoproteins. *J Lipid Res* 26:26-37

Weinberg RB, Patton CS (1990) Binding of human apolipoprotein A-IV to human hepatocellular plasma membranes. *Biochim Biophys Acta* 1044:255-261

Weisgraber KH, Bersot TP, Mahley RW (1978) Isolation and characterization of an apoprotein from the d less than 1.006 lipoproteins of human and canine lymph homologous with the rat A-IV apoprotein. *Biochem Biophys Res Commun* 85:287-292

Zannis VI, Breslow JL (1985) Genetic mutations affecting human lipoprotein metabolism. *Adv Hum Genet* 14:125-215, 383-6

Subject Index

NATO ASI Series H

Vol. 38: The Molecular Biology of Autoimmune Disease.
Edited by A.G. Demaine, J-P. Banga, and A.M. McGregor. 404 pages. 1990.

Vol. 39: Chemosensory Information Processing.
Edited by D. Schild. 403 pages. 1990.

Vol. 40: Dynamics and Biogenesis of Membranes.
Edited by J. A. F. Op den Kamp. 367 pages. 1990.

Vol. 41: Recognition and Response in Plant-Virus Interactions.
Edited by R. S. S. Fraser. 467 pages. 1990.

Vol. 42: Biomechanics of Active Movement and Deformation of Cells.
Edited by N. Akkas. 524 pages. 1990.

Vol. 43: Cellular and Molecular Biology of Myelination.
Edited by G. Jeserich, H. H. Althaus, and T. V. Waehneldt. 565 pages. 1990.

Vol. 44: Activation and Desensitization of Transducing Pathways.
Edited by T. M. Konijn, M. D. Houslay, and P. J. M. Van Haastert. 336 pages. 1990.

Vol. 45: Mechanism of Fertilization: Plants to Humans.
Edited by B. Dale. 710 pages. 1990.

Vol .46: Parallels in Cell to Cell Junctions in Plants and Animals.
Edited by A. W Robards, W. J . Lucas, J . D. Pitts, H . J . Jongsma, and D. C. Spray. 296 pages. 1990.

Vol. 47: Signal Perception and Transduction in Higher Plants.
Edited by R. Ranjeva and A. M. Boudet. 357 pages. 1990.

Vol. 48: Calcium Transport and Intracellular Calcium Homeostasis.
Edited by D. Pansu and F. Bronner. 456 pages. 1990.

Vol. 49: Post-Transcriptional Control of Gene Expression.
Edited by J. E. G. McCarthy and M. F. Tuite. 671 pages. 1990.

Vol. 50: Phytochrome Properties and Biological Action.
Edited by B. Thomas and C. B. Johnson. 337 pages. 1991.

Vol. 51: Cell to Cell Signals in Plants and Animals.
Edited by V. Neuhoff and J. Friend. 404 pages. 1991.

Vol. 52: Biological Signal Transduction.
Edited by E. M . Ross and K . W. A. Wirtz. 560 pages. 1991.

Vol. 53: Fungal Cell Wall and Immune Response.
Edited by J. P. Latge and D. Boucias. 472 pages. 1991.

Vol. 54: The Early Effects of Radiation on DNA.
Edited by E. M. Fielden and P. O'Neill. 448 pages. 1991.

Vol. 55: The Translational Apparatus of Photosynthetic Organelles.
Edited by R. Mache, E. Stutz, and A. R. Subramanian. 260 pages. 1991.

NATO ASI Series H

Vol. 56: **Cellular Regulation by Protein Phosphorylation.**
Edited by L. M. G. Heilmeyer, Jr. 520 pages. 1991.

Vol. 57: **Molecular Techniques in Taxonomy.**
Edited by G . M . Hewitt, A. W. B. Johnston, and J. P. W. Young . 420 pages. 1991.

Vol. 58: **Neurocytochemical Methods.**
Edited by A. Calas and D. Eugene. 352 pages. 1991.

Vol. 59: **Molecular Evolution of the Major Histocompatibility Complex.**
Edited by J. Klein and D. Klein. 522 pages. 1991.

Vol. 60: **Intracellular Regulation of Ion Channels.**
Edited by M. Morad and Z. Agus. 261 pages. 1992.

Vol. 61: **Prader-Willi Syndrome and Other Chromosome 15q Deletion Disorders.**
Edited by S. B. Cassidy. 277 pages. 1992.

Vol. 62: **Endocytosis. From Cell Biology to Health, Disease and Therapie.**
Edited by P. J. Courtoy. 547 pages. 1992.

Vol. 63: **Dynamics of Membrane Assembly.**
Edited by J. A. F. Op den Kamp. 402 pages. 1992.

Vol. 64: **Mechanics of Swelling. From Clays to Living Cells and Tissues.**
Edited by T. K. Karalis. 802 pages. 1992.

Vol. 65: **Bacteriocins, Microcins and Lantibiotics.**
Edited by R. James, C. Lazdunski, and F. Pattus. 530 pages. 1992.

Vol. 66: **Theoretical and Experimental Insights into Immunology.**
Edited by A. S. Perelson and G. Weisbuch. 497 pages. 1992.

Vol. 67: **Flow Cytometry. New Developments.**
Edited by A. Jacquemin-Sablon. 1993.

Vol. 68: **Biomarkers. Research and Application in the Assessment of Environmental Health.** Edited by D. B. Peakall and L. R. Shugart. 138 pages. 1993.

Vol. 69: **Molecular Biology and its Application to Medical Mycology.**
Edited by B. Maresca, G. S. Kobayashi, and H. Yamaguchi. 271 pages. 1993.

Vol. 70: **Phospholipids and Signal Transmission.**
Edited by R. Massarelli, L. A. Horrocks, J. N. Kanfer, and K. Löffelholz. 448 pages. 1993.

Vol. 71: **Protein Synthesis and Targeting in Yeast.**
Edited by A. J. P. Brown, M. F. Tuite, and J. E. G. McCarthy. 425 pages. 1993.

Vol. 72: **Chromosome Segregation and Aneuploidy.**
Edited by B. K. Vig. 425 pages. 1993.

NATO ASI Series H

Vol. 73: **Human Apolipoprotein Mutants III. In Diagnosis and Treatment.**
Edited by C. R. Sirtori, G. Franceschini, B. H. Brewer Jr. 302 pages. 1993.

Vol. 74: **Molecular Mechanisms of Membrane Traffic.**
Edited by D. J. Morré, K. E. Howell, and J. J. M. Bergeron. 429 pages. 1993.

Vol. 75: **Cancer Therapy. Differentiation, Immunomodulation and Angiogenesis.**
Edited by N. D'Alessandro, E. Mihich, L. Rausa, H. Tapiero, and T. R. Tritton. 299 pages. 1993.

Vol. 76: **Tyrosine Phosphorylation/Dephosphorylation and Downstream Signalling.**
Edited by L. M. G. Heilmeyer Jr. 388 pages. 1993.

FSC
www.fsc.org
MIX
Papier aus verantwortungsvollen Quellen
Paper from responsible sources
FSC® C105338

If you have any concerns about our products,
you can contact us on
ProductSafety@springernature.com

In case Publisher is established outside the EU,
the EU authorized representative is:
Springer Nature Customer Service Center GmbH
Europaplatz 3, 69115 Heidelberg, Germany

Printed by Libri Plureos GmbH
in Hamburg, Germany